Biology of Extracellular Matrix

For further volumes:
http://www.springer.com/series/8422

Preface

The extracellular matrix (ECM) is a simple descriptive term that belies the complex nature, variety, and versatility of the components and their interactions in this tissue compartment. Far from being a simple "space-filler" or "ground substance," the ECM is fundamentally responsible for the structural integrity of all multicellular organisms. Selective expression and interactions between ECM components determine the architecture and physical properties of the matrix. Furthermore, both through direct receptor-mediated interactions with cells and by binding and regulating availability of potent growth factor molecules, the matrix influences cell adhesion, migration, and differentiation.

Proteins of the ECM tend to be modular and multifunctional, allowing dynamic bridging between ECM components and facilitating physical organization of tissues. Since the first appearance of these ECM proteins, many of them relatively early in metazoan evolution, there has been a great expansion in their variety and versatility including, in higher chordates and vertebrates, the introduction of "new" proteins such as fibronectin, elastin, and the microfibril-associated glycoproteins (MAGPs). This volume gathers together the current state of knowledge of the functions and interactions of these ECM proteins and their evolutionary histories and offers insights into mechanisms underlying the remarkable explosion of diversity and functionality through metazoan evolution.

In Chap. 1, Adams provides a general introduction to metazoan evolution, the appearance of the ECM as a simple "glue" between cells, and its subsequent expansion and functional diversification. She briefly reviews the function of several of the members of the ECM family of proteins and points out the importance of modularity, domain acquisition, and oligomerization for function. She also addresses the question of possible pre-metazoan origins of at least some matrix proteins.

Further details of possible mechanisms for generating diversity are addressed by Wada in Chap. 2. In particular, he focuses on evidence for exon-based domain shuffling as a mechanism for assembly of "new" matrix components, giving examples from cartilage and cell adhesion proteins. He also points out the importance of gene and genome duplication events followed by divergence of sequence and function and the possible emergence of novel motifs from noncoding genomic sequences.

In Chap. 3, Exposito and Lethias address the early metazoan origins of collagen as a simple tandem repeat sequence of Gly–Pro–Pro. They provide a model, involving sequence reduplication, exon shuffling, and rounds of genomic duplication, for the expansion of this ancestral sequence to generate the large family of collagens and collagen-like proteins now present in all invertebrates and vertebrates.

Chapter 4 provides information on the evolutionary history of elastin, an ECM protein with the unusual properties of extensibility and elastic recoil that emerged rather late in evolution, coinciding with the development of pulsatile, high-pressure, closed circuit circulatory systems in vertebrates. In this chapter, Keeley discusses sequence motifs and styles that appear to be required for such elastomeric properties, identify both domains shared by all elastins and domains that are highly variable between species, and propose a model for expansion and diversification of elastin sequences among species through multiple and differential replication of exon pairs at a specific site in the elastin gene.

In Chap. 5, Jensen et al. introduce the fibrillin family of proteins, which have ancient roots in metazoan evolution. Fibrillins are the principal component of microfibrils, distinctively beaded filaments present not only in association with elastin in elastic tissues but also independently providing structural integrity in other tissues. Indeed it has been proposed that, before the appearance of elastin, these microfibrillar filaments were responsible for extensibility and recoil in vascular and other tissues. The authors discuss the similarities in domain structure and arrangements between fibrillins and members of the latent TGF-β binding protein (LTBP) family and review current knowledge on the organization and assembly of microfibrils and their role in regulating cell activity through binding of growth factors.

In Chap. 6, Segada describes the fibulin and MAGP families of microfibrillar-associated proteins, the former present at the base of metazoan evolution and the latter only appearing in chordates. He introduces the important concept of coevolution of a "suite of genes" within a functional module and the additional evolutionary considerations that may be a consequence of such functional linkages. He then traces the appearance of additional members of the fibulin and MAGP families through exon shuffling and gene duplication events, adding further diversity and subtlety to their range of functions, including important interactions with elastin in the formation of elastic fibers.

Chapter 7, written by Baratta et al., describes the functions and evolutionary origins of the SPARC and tenascin families of proteins. Unlike the previously considered proteins, which either form the fundamental building blocks of polymeric ECM assemblies or are integrated into these assemblies, SPARCs and tenascins function as soluble, diffusible components of the matrix. Both of these proteins have important roles in regulating signaling pathways, particularly those involved in angiogenesis and fibrosis.

In Chap. 8, Roberts and Frazier introduce the thrombospondin family of proteins, which also have origins early in metazoan evolution. They discuss the development of modularity and oligomerization and the remarkable ability of these bridging proteins to interact with a wide variety of cellular receptors as well as with other matrix components.

In Chap. 9, Johnson et al. describe the pre-metazoan origin of integrin-like cell adhesion receptors and their critical role in the increasing complexity of multicellular organisms throughout metazoan evolution. They review the earliest appearance of integrin-like domains, the diversification of their binding to both extracellular and intracellular partners, and their coevolution with other cellular and matrix components.

Several common themes emerge. These include the fundamental role of modularity in matrix proteins, both for their multiplicity and selectivity of interactions as well as for the facilitation of further diversification through processes such as exon duplication, exon shuffling, and gene and genome duplication. Overlying all of these is the recognition that the ECM consists of an integrated network of interactions, imposing additional coevolutionary constraints. While further sequence data and analyses will no doubt clarify the complexity and answer at least some of the many remaining questions, these contributions will provide the reader with an comprehensive overview of current knowledge of the origins of the remarkable functional diversity of matrix proteins.

Toronto ON, Canada Fred W. Keeley
St. Louis, MO, USA Robert P. Mecham

Contents

Chapter 1
Extracellular Matrix Evolution: An Overview

Josephine C. Adams

Abstract The extracellular matrix (ECM) is a central mediator of multicellularity in the metazoa. ECM bridges between cells, provides mechanical support to tissues and is a source of signalling cues that are important for cell survival, proliferation, differentiation and migration. A structured ECM is apparent in most metazoa, from sponges to humans, with particular complexity in the ECM of vertebrates, where there are many tissue-specific matrix assemblies and innovations of fibril systems. Understanding how the ECM and its associated "adhesome" evolved as a system has important implications for understanding its normal functions throughout life and for applications such as tissue engineering and design of synthetic matrices. This chapter overviews the evolution of central components of basement membranes and connective ECM in light of recent advances from comparative genomics, with emphasis on the numerous components that appear to have originated on the metazoan stem lineage.

Abbreviations

BLAST	Basic local alignment search tool
ECM	Extracellular matrix
EGF	Epidermal growth factor
EST	Expressed sequence tag
FN	Fibronectin
TSP	Thrombospondin
TSR	Thrombospondin type 1 domains

J.C. Adams (✉)
School of Biochemistry, University of Bristol, Medical Sciences Building, Bristol BS8 1TD, UK
e-mail: jo.adams@bristol.ac.uk

F.W. Keeley and R.P. Mecham (eds.), *Evolution of Extracellular Matrix*,
Biology of Extracellular Matrix, DOI 10.1007/978-3-642-36002-2_1,

1.1 Introduction: Extracellular Matrix as a Mediator of Multicellularity in the Metazoa

Multicellularity has evolved in many lineages of life on earth, being selected for by benefits of larger organism size, possibilities for separation of germ and somatic cells and the specialised differentiation of somatic cell lineages enabling the formation of distinct tissues (Rokas 2008). The most ancient multicellular fossils are derived from bacterial cells that resemble modern filamentous cyanobacteria and multicellular eukaryotes are apparent in the fossil record from around 800 million years ago (Tomitani et al. 2006; Knoll et al. 2006; Marshall and Valentine 2010). The mechanisms by which multicellularity is achieved are diverse and need not be complex. For example, two related volvocine algae, *Chlamydomonas reinhardtii* (unicellular) and *Volvox carteri* (multicellular with two cell types), are remarkably similar in gene content, yet multicellularity in *Volvox* is supported by the enlargement of gene families for secreted pherophorins and gametolysin/*Volvox* matrix metalloproteases that participate in formation of an extracellular matrix (ECM) between cells. In this instance, multicellularity has been achieved simply by expansion of the ability to assemble "glue" between the cells (Prochnik et al. 2010).

In the metazoa, cells that lack a cell wall are conspicuous for particularly elaborate forms of multicellularity with large numbers of distinct somatic cell types (Rokas 2008). Cell interactions in metazoa are achieved by two major mechanisms. One is direct cell–cell adhesion, mediated by cell–cell adhesion receptors of which the cadherins are a major family; the other is secretion of large extracellular glycoproteins to form an insoluble bridging network (the ECM) anchored to cell surfaces by cell-ECM adhesion receptors, of which the integrins are a major family. Both integrins and cadherins are represented in unicellular eukaryotes: cadherins emerged before the divergence of choanoflagellates and metazoa (Hulpiau and van Roy 2009; Nichols et al. 2012), and integrins are present in an early diverging opisthokont and a unikont protozoan (Sebé-Pedrós et al. 2010). The functions of these adhesion receptors in unicellular eukaryotes are intriguing questions. Cadherins of *Monosiga brevicollis* localise with F-actin and are postulated to function in capture of bacteria during feeding (Abedin and King 2008). Several key structural and signalling components of focal adhesions are co-conserved with the protist integrins, suggesting that intracellular signalling activities are important to integrin function in these unicellular eukaryotes, whatever the nature of the integrin extracellular binding partners might be (Sebé-Pedrós et al. 2010) (Fig. 1.1). This might explain why the functions of metazoan integrins are so closely linked to basic processes such as cell cycle progression and cell survival (Wickström et al. 2011).

As will be discussed in this chapter, the large, secreted, multidomain ECM components that are viewed as "classic" integrin ligands appear to be specific to the metazoa (Fig. 1.1). Therefore, the evolution of ECM was a crucial innovation on the metazoan stem lineage that must have been accompanied by repurposing or adaptation of the activities of integrins and cadherins to support multicellularity.

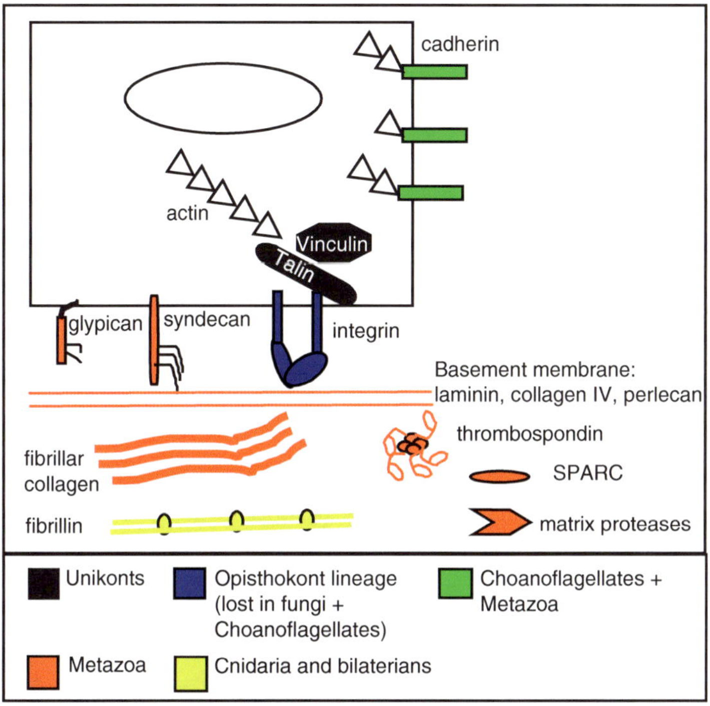

Fig. 1.1 The patchwork origin of the metazoan adhesome. Schematic depiction of exemplary features of the adhesome found in an early diverging metazoan. Molecules are colour-coded according their evolutionary origin. Data are drawn from recent comparative genomics as discussed in this chapter and the references therein. Not to scale

This chapter will consider in overview the origins of important domains of ECM components and of the ECM molecules that characterise metazoan interstitial ECM and basement membranes.

1.2 Pre-metazoan Origins of Some Important ECM Protein Domains

The single sentence definition of metazoan ECM proteins is that they are large, secreted, multidomain and often oligomeric proteins with capacity to form extracellular fibrils and supramolecular networks. The repeated nature of many of their constituent domains (see examples in Fig. 1.2) contributes greatly to the lengths of ECM polypeptides and the ability of single molecules to engage in multivalent interactions. Oligomerisation is achieved in the collagen family by the collagen triple helix, and in other ECM components, including laminins, thrombospondins (TSPs), tenascins and matrilins, by heptad repeat coiled-coil domains. Both of these

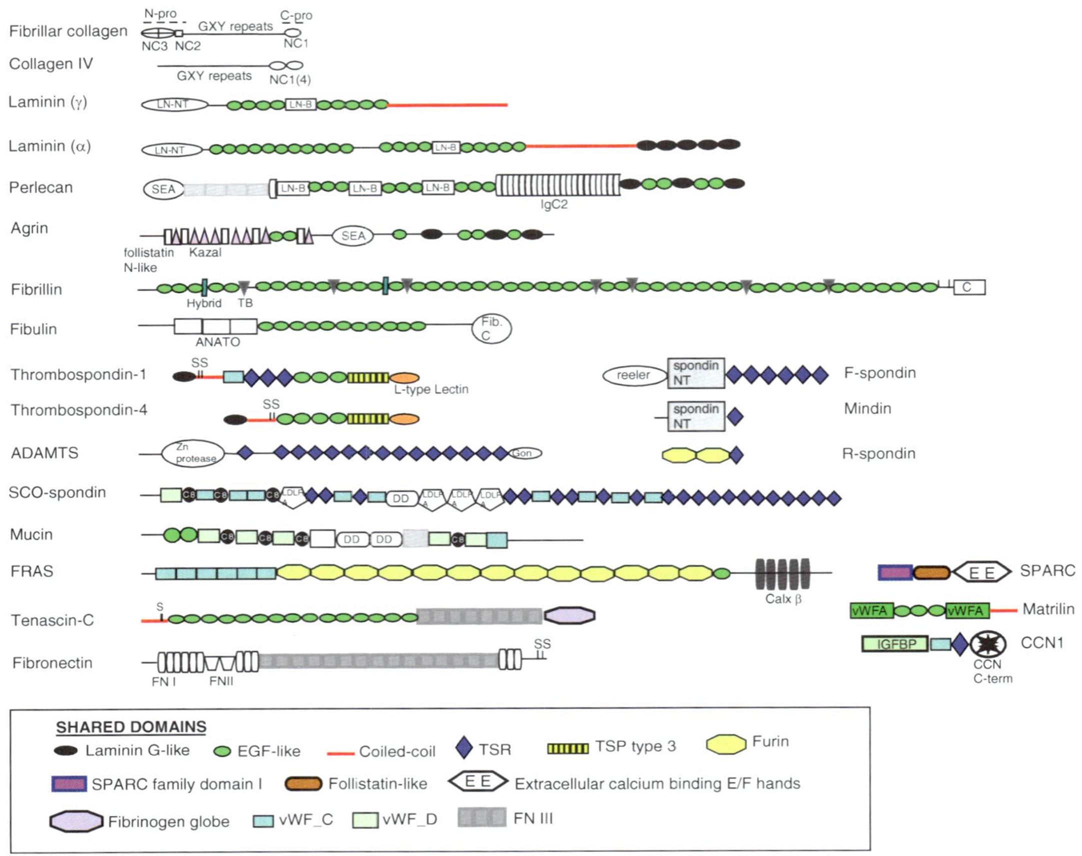

Fig. 1.2 Schematic diagrams of the domain relationships of the major extracellular matrix proteins and proteoglycans discussed in this chapter

domains are found in prokaryotes and eukaryotes and thus clearly evolved long before the ECM components themselves (Beck and Brodsky 1998). Domain shuffling is an important general mechanism for evolution of new genes (Babushok et al. 2007), and many of the domains found in ECM proteins are not exclusive to the metazoa and are also present in many proteins with unrelated functions. Examples which are frequently present in ECM proteins as tandem-repeated domains include epidermal growth factor (EGF)-like domains, thrombospondin type 1 (TSR) or properdin domains, the von Willebrand Factor A domain (vWF_A) and the fibronectin type III (FNIII) domain (Adams and Tucker 2000; Whittaker and Hynes 2002; Wouters et al. 2005; Hynes and Naba 2012). Globular domains found in many ECM components are also present outside the metazoa: the C-propeptide non-collagenous (NC1) domain of collagens; the laminin N-terminal (NT) and laminin G domains; the discoidin domain; and C-type and L-type lectin domains (Dodd and Drickamer 2001; Kiedzierska et al. 2007; King et al. 2008; Exposito et al. 2010). The multidomain nature of ECM proteins and the wide evolutionary conservation of many of their constituent domains mean that special care is needed in applying bioinformatics to analyse the conservation of ECM proteins between different phyla (Adams and Engel 2007).

The occurrence of domains found in metazoan ECM proteins outside the metazoa provides clear evidence that domain shuffling of pre-existing domains contributed to the evolution of metazoan ECM proteins (King et al. 2008). This point is also well-exemplified by the identification of protein-coding sequences in choanoflagellates that contain a motif characteristic of metazoan collagens, the repeated GXY triple helical motif. However, the choanoflagellate proteins are unlikely to have fibril-forming capacity because they lack N- or C-propeptide domains, and so the mechanisms for registration of single molecules for triple helix assembly are lacking. Several other proteins of choanoflagellates do contain a C-propeptide-like domain, leading to a clear scenario of how domain shuffling by exon shuffling or fusion might have led to the origin of fibril-forming collagens (Exposito et al. 2008, 2010).

1.3 ECM in the Basal Metazoa

The question of ECM evolution needs to be placed in context of the evolution of the metazoa themselves. In recent years, evolutionary models based on the fossil record and the application of traditional biological classifications, regarding numbers of differentiated cell types, body plan features and life cycle, have been supplemented, revised and even revolutionised by molecular phylogenetic and phylogenomic analyses. Nevertheless, the origin of animals remains a strongly contested topic of study, due to major difficulties in assigning relationships between the early diverging or basal metazoa (sponges, Ctenophores, Placozoa and Cnidaria) and bilaterians (animals with bilateral symmetry and three embryonic epithelial cell layers) (Table 1.1). In the last 5 years, multiple distinct phylogenetic models have

Table 1.1 Taxonomical classifications of the early diverging metazoa and other unikonts referred to in this chapter

Eukaryotic group/ (phylum)	Notes	Species referred to in the chapter
Apusozoa	Sister group to opisthokonts	*Amastigomonas sp.* (unicellular)
Opisthokonts		
Ichthyosporea	Independent opisthokont, sister group to the holozoa (choanoflagellates and metazoa)	*Capsaspora owczarzaki* (unicellular)
Choanoflagellates	Sister group to metazoa	*Monosiga brevicollis* (unicellular) *Salpingoeca rosetta* (unicellular and colonial)
Metazoa		
(Porifera)	Traditionally considered the earliest-diverging metazoa Four classes based on morphology and spicule features: Demospongiae, Calcarea, Hexactinellida and Homoscleromorpha. Only Homoscleromorpha contain morphologically recognisable basement membranes	Demosponges *Amphimedon queenslandica* *Suberites domuncula* *Lubomirskia baicalensis* Homoscleromorph sponges *Oscarella carmela* *Oscarella lobularis*
(Placozoa)	The simplest metazoa in terms of morphology. Uncertainty over its placement in metazoan phylogeny	*Trichoplax adhaerens*
(Cnidaria)	Radial symmetry, ECM layer has features of basement membranes and connective ECM	*Hydra magnipapillata* *Nematostella vectensis*
(Nematodes)	Bilaterians	*Caenorhabditis elegans*
(Arthropods)	Bilaterians	*Drosophila melanogaster*

been proposed that place either sponges, Ctenophores or Placozoa, respectively, as the earliest-diverging metazoans (Dunn et al. 2008; Srivastava et al. 2008, 2010; Ruiz-Trillo et al. 2008; Schierwater et al. 2009; Pick et al. 2010). In part, these difficulties may be due to under-representation of species from the early diverging phyla in sequencing projects. A study which sampled a wider range of species from these phyla via expressed sequence tags (ESTs) returned a metazoan phylogeny closer to traditional views, with sponges placed as the earliest-diverging branch and Ctenophores as a sister group to Cnidaria (Philippe et al. 2009). The same placement of sponges was obtained with an alternative study method based on analysis of conserved proteins (Torruella et al. 2012). I adopt a "sponges-first" view for the discussions of evolution of ECM protein families in this chapter.

At the morphological level the early diverging metazoan phyla have very different attributes in terms of ECM. Sponges comprise a diverse group characterised by sessile adults that are specialised for flow of water through the body cavity for feeding and removal of waste products. Early sponge fossils date from c. 580 million years ago (Li et al. 1998). Adult sponges consist of two layers of cells with epithelial features supported by a central "mesohyl" consisting of rigid, mineralised spicules and fibrillar material; this corresponds to the ECM. Current data indicate rapid, convergent evolution of biomineralisation in the metazoa and different classes of sponges form spicules from silica, calcium carbonate or on a spongin base; the latter being a sponge-specific family of short-chain collagens (Exposito et al. 1991; Aouacheria et al. 2006; Jackson et al. 2011; Murdock and Donoghue 2011). Collagen fibrils are recognised as major components of the fibrillar material of the mesohyl (e.g. Heinemann et al. 2007; Exposito et al. 2008). Most sponges lack a basement membrane; however, the Homoscleromorpha sponges have a morphologically recognisable basement membrane and cell–cell junctions and encode a type IV collagen (Boute et al. 1996; Ereskovsky et al. 2009). It remains under discussion if sponges are a monophyletic group (Cárdenas et al. 2012; Wörheide et al. 2012), yet the presence of both fibrillar and basement-membrane collagens in homoscleromorph sponges implies an origin for these complex ECM assemblies on the metazoan stem lineage.

Placozoa are the simplest known animals in terms of their multicellular organisation. They are flat, disclike organisms, a few hundred microns in diameter, which comprise two layers of epithelial cells and three additional cell types. Placozoa are found in aquatic environments around the world and move by means of cilia or through amoeboid-like motion. In electron microscopy preparations, the adult organisms contain no morphologically recognisable ECM between the cell layers. It has therefore been considered that Placozoa lack an ECM (Schierwater et al. 2009). In view of evolutionary scenarios that place sponges as the earliest-diverging metazoa, the apparent lack of ECM in adult Placozoa may represent a secondary loss rather than a basal condition (Philippe et al. 2009).

Cnidaria and Ctenophores have considerably more complex body organisations than sponges and Placozoa; both possess nerve and muscle cells and a digestive system. The Cnidaria are widely recognised as the most basal metazoa to have a well-structured ECM with clear features in common with the ECM of bilaterians (Table 1.1). The bodies of Cnidaria consist of two cell layers separated by a continuous ECM sheet, the mesoglea, that is made up of two basement membrane-like sheets connected by a central fibrillar ECM. Mesoglea can be isolated from Cnidaria such as *Hydra magnipapillata* as an acellular ECM; the basement membrane layers contain laminin and collagen IV as major components, and fibrillar collagen is present within the central connective layer (Deutzmann et al. 2000; Fowler et al. 2000; Zhang et al. 2002; Shimizu et al. 2008). Ctenophores are small marine plankton organisms that swim by beating of rows of cilia (Pang and Martindale 2008). Like Cnidaria, these organisms have a morphologically distinct ECM, the mesoglea, that separates the ectoderm and endoderm cell layers and contains fibrillar material as well as basement membrane. Ctenophores are less studied than Cnidaria, and so there is less information on the protein composition of their mesoglea.

1.4 Evolution of ECM Components: The View from Comparative Genomics

Just as comparative genomics has expanded views on metazoan evolution, it has advanced our knowledge of the evolution of metazoan ECM components. With the exception of Ctenophores, genome sequences have now been published for at least one representative species of each metazoan phylum. It is now possible to survey the encoding of known ECM components across these species and draw up models of their points of origin in relation to branch points within the metazoan evolutionary tree. If encoded throughout the metazoa, an origin on the metazoan stem lineage is indicated. Adopting the view that ECM constitutes a dynamic, multi-protein system, these analyses also need to consider other elements that are essential for a functional ECM "adhesome", for example, collagen-processing enzymes, extracellular proteases, cross-linking enzymes and ECM adhesion receptors (Ozbek et al. 2010; Hynes and Naba 2012).

Recent phylogenomic assessments of the evolution of the ECM as a supersystem of network-forming multimolecular assemblies have made it clear that many of the central components of basement membranes and interstitial ECM are present throughout the metazoa (Ozbek et al. 2010; Srivastava et al. 2010) (Fig. 1.1). This leaves the representation of specific molecules within the early diverging metazoa as a question of key interest for understanding fundamental aspects of how ECM is built. A limitation here, at this time, is that the number of species with sequenced genomes or large-scale EST projects is low for these phyla. For example, most knowledge of sponge gene inventories comes from sponges of the Demospongiae class: *Amphimedon queenslandica* (Srivastava et al. 2010), *Suberites domuncula* and *Lubomirskia baicalensis* (Harcet et al. 2010) (Table 1.1). Genome sequences or global EST projects are needed for representatives of other classes of sponges, in particular the Homoscleromorpha that are known by morphological criteria to form basement membranes (see section above). *Oscarella lobularis* is being developed as a homoscleromorph model (Ereskovsky et al. 2009). Until genomic information is available from additional species that represent more fully the different classes of sponges, the absence of an encoding sequence for an otherwise well-conserved molecule in a single species should be taken as provisional for the phylum. With this caveat in mind, the origins of major components of interstitial ECM and basement membranes are discussed in brief in the following sections.

1.5 Interstitial/Connective ECM

1.5.1 Fibril-Forming Collagens

Fibril-forming collagens are conserved throughout the metazoa and represent the most abundant fibril systems of the ECM. As discussed above, the identification of important domains of collagens in separate protein-coding sequences of choanoflagellates has given rise to the model that domain shuffling or fusion events on the metazoan stem lineage brought together GXY triple helix repeats with N- and C-non-collagenous domains into a single open reading frame (Exposito et al. 2008, 2010). Given the presence of genes for both a fibrillar collagen and a basement membrane collagen in a homoscleromorph sponge, the putative collagen "founder gene" was likely duplicated rapidly. *Hydra magnipapillata* fibrillar collagen 1 is a homo-trimer (Deutzmann et al. 2000), but many fibrillar collagens are assembled as heterotrimers from the products of two or three genes, and radiation of collagens with distinct *N*-propeptides appears to have been an early event in the metazoa (Exposito et al. 2010). In addition to their roles in soft tissues, fibrillar collagens have evolved as templates for biomineralisation in contexts such as the avian eggshell (Carrino et al. 1996) and the vertebrate endoskeleton and teeth (Kawasaki et al. 2009).

1.5.2 Fibrillins

Fibrillins are large proteins characterised by 46–47 EGF-like domains, seven interspersed TB (transforming growth factor β binding protein) domains and one or two so-called hybrid domains in which the N-terminus of a TB domain is fused with the C-terminus of a calcium-binding EGF domain (Fig. 1.2). Fibrillins polymerise through intermolecular interactions of their N-terminal regions to form morphologically distinctive 10 nm beaded microfibrils in connective ECM. The microfibrils interact with other ECM glycoproteins and proteoglycans and have important roles in binding and sequestering TGFβ and BMP (bone morphogenetic protein) growth factors, thus contributing to tissue homeostasis of TGFβ activity (Ramirez and Sakai 2010; Massam-Wu et al. 2010). In vertebrates, fibrillins have an additional important role through the binding of elastin (an evolutionary novelty in vertebrates; Sage and Gray 1979) to form elastic fibres that contribute to tissue elasticity in connective tissues of the skin, musculoskeletal system and the eye, and are central to functioning of a high-pressure, pulsatile blood circulation system (Wagenseil and Mecham 2009).

Fibrillin is one of the most ancient fibril-forming proteins in the ECM. Definitive ECM beaded microfibrils and the encoding of fibrillins is apparent in Cnidaria, and fibrillins are conserved throughout most eumetazoa with the exceptions of insects and nematodes (Reber-Müller et al. 1995; Robertson et al. 2011). The TB domains are specific to eumetazoa; however, fibrillin-related proteins containing numerous EGF-like domains with high homology to the EGF domains of fibrillins are encoded in *Trichoplax adhaerens* (XP_002110968), the sponge *Amphimedon queenslandica* (e.g. XP_003390605 and others) and in choanoflagellates (*Salpingoeca rosetta* EGD76297, EGD76295; *Monosiga brevicollis* XP_001748060). These proteins lack TB or hybrid domains and have distinct sequence in their N-terminal regions. Whether these proteins represent convergent evolution of EGF domains or true fibrillin relatives, and whether they have fibril-forming capacity, is currently unknown. The transition from proteins with long runs of contiguous fibrillin-like EGF domains to a fibrillin with interspersed, individual TB domains would have required multiple domain-shuffling events.

1.5.3 Fibulins

Fibulins (a.k.a. hemicentins) are proteins of between 600 and 1,100 amino acids, characterised by variable length N-terminal domains, multiple tandem EGF-like domains and a C-terminal fibulin FC domain (Fig. 1.2). The fibulins of vertebrates contribute, with fibrillins, to the assembly of elastic fibres (Yanagisawa and Davis 2010). Also similar to fibrillins, fibulins are of much earlier evolutionary origin than elastin and are recognised from Cnidaria to humans (Segade 2010). Database searches for predicted fibulin-related proteins in choanoflagellates, sponges and *Trichoplax adhaerens* return the same EGF domain containing polypeptides as those identified as potential fibrillin relatives (see fibrillin section above). Thus, fibrillins and fibulins may have diverged from a common ancestor that originated on the Holozoan stem lineage.

1.5.4 Secreted ECM Proteoglycans

The ECM of vertebrates includes many secreted proteoglycans that contribute to ECM network formation and ECM homeostasis through specific protein–protein interactions of their core proteins and charge-based interactions of their glycosaminoglycan (GAG) chains with other ECM components or growth factors. Vertebrate connective ECM includes proteoglycans such as decorin, biglycan, aggrecan and versican; brain ECM is enriched in brevican and the cornea in lumican (Sarrazin et al. 2011; Iozzo and Sanderson 2011). Strikingly, none of these core proteins can be identified by basic local alignment search tool (BLAST) searches outside the deuterostomes (Ozbek et al. 2010). It is possible that related core proteins are present but are not susceptible to identification due to very low sequence conservation of regions other than the short motifs for GAG addition sites. For example, it is only possible to identify the transmembrane proteoglycan syndecan in early diverging metazoa by using the highly conserved cytoplasmic domain as the database search tool (Ozbek et al. 2010). Chondroitin sulphate and heparan sulphate GAGs are present throughout invertebrates (Cássaro and Dietrich 1977; Yamada et al. 2007), leaving open the question of whether Cnidaria encode lineage-specific proteoglycan core proteins. Protein-based methods will be needed to search for these.

1.6 Basement Membranes

1.6.1 Laminin

Laminin is a large, αβγ heterotrimeric ECM glycoprotein that has a central role in assembling the basement membrane network by interacting with cell-surface integrins, dystroglycan complex and collagen IV molecules (Tzu and Marinkovich 2008). The laminin α, β and γ polypeptides are the products of different genes yet have commonalities of domain architecture that suggest they originated from a common ancestor. All contain a laminin N-terminal domain, a series of repeated EGF-like domains and an extended heptad repeat coiled-coil region that is responsible for heterotrimerisation. Repeated laminin G domains are important for interactions with integrins and the dystroglycan complex, and these are found only in the C-terminal region of the laminin α subunit (see examples in Fig. 1.2). According to the criterion of encoding of multiple laminin chains (i.e. competence for heterotrimer assembly), laminin is found throughout the metazoa, with expansion to a large gene family encoding at least 16 distinct and in many cases, tissue-specific, heterotrimers in vertebrates (Tzu and Marinkovich 2008). Although basement membranes are not evident in demosponges, the genome of the demosponge *Amphimedon queenslandica* encodes multiple laminin-like polypeptides (Srivastava et al. 2010). Four contain all three of the domain hallmarks of laminin subunits: three of these are laminin β– (XP_003387987) or γ-like (XP_003390574, XP_003386652) and one is laminin α-like in that it contains laminin G domains (XP_003382371). All have distinctions of domain organisation compared to the characterised basement membrane forming laminins of Cnidaria and bilaterians (Srivastava et al. 2010). Experimental analysis is needed to establish whether an *Amphimedon queenslandica* laminin has the capacity to form a polymeric network at cell surfaces.

Hydra laminin is well established to be a component of the basement membrane layers of mesoglea (Zhang et al. 2002) and the domain architectures of *Hydra magnipapillata* laminin αβγ polypeptides correspond to those of bilaterians (Chapman et al. 2010). Similar laminin subunits are encoded in another Cnidarian, *Nematostella vectensis* (Putnam et al. 2007; Ozbek et al. 2010). Despite its lack of morphologically recognisable ECM, *Trichoplax adhaerens* (Placozoa) encodes three laminin polypeptides, equivalent to the αβγ subunits of bilaterians (Srivastava et al. 2008; Ozbek et al. 2010). Collectively, the encoding of laminin polypeptides in early diverging metazoa, including several that lack overt basement membrane structures, indicates an origin of laminin on the metazoan stem lineage. Indeed, a large protein containing a laminin NT domain, four tandem EGF-like domains and many furin repeats is encoded in the choanoflagellate *Salpingoeca rosetta* (EGD73449), providing a tantalising indication of how molecular innovations on the metazoan stem lineage might have built on a domain combination pre-existing in the Choanoflagellate/metazoan ancestor.

1.6.2 Collagen IV/Spongins

Collagen IV molecules assemble as homo-trimers (e.g. *Hydra magnipapillata* collagen IV; Fowler et al. 2000) or heterotrimers (Ricard-Blum and Ruggiero 2005). Dimers of these trimers form the building blocks for assembly of the thin sheetlike networks that are essential elements of basement membranes. Collagen IV polypeptides are encoded in all metazoa that are known by morphological criteria to assemble basement membranes, from Homoscleromorpha sponges to mammals (Boute et al. 1996). Curiously, collagen IV α1- and IVα2-like polypeptides are also encoded in *Trichoplax adhaerens* (Placozoa) that is considered to lack an ECM (Srivastava et al. 2008; Ozbek et al. 2010). Spongins define a distinct family of short-chain collagens that are conserved in demosponges, Cnidaria, protostomes (with the exception of Ecdysozoa) and the basal chordates *Ciona intestinalis* and *Branchiostoma floridae*. In demosponges, the spongins provide a template for spicule formation, form part of the adhesive contact between the sponge and its environment, and assemble into spongin microfibrils in the mesohyl (Aouacheria et al. 2006). The NC1 (non-collagenous domain 1) domains of collagen IV and spongin short-chain collagens are similar in structure, leading to the model that these distinct protein families arose from a common ancestor on the metazoan stem lineage (Aouacheria et al. 2006). Given the essential role of collagen IV NC1 domains in the assembly of collagen IV trimers and networks, spongin NC1 domains might be important for spongin microfibril assembly. As yet, nothing is known about the roles of spongin short-chain collagens in organisms with type IV collagen-containing basement membranes.

1.6.3 Basement Membrane Proteoglycans: Perlecan and Agrin

Two proteoglycan core proteins that are readily identifiable across the metazoa are perlecan, an important component of basement membranes, and the related heparan sulphate proteoglycan, agrin. Perlecan acts as a scaffold in basement membranes through its interactions with laminin and collagen IV and is also present in the cartilage of vertebrates. Perlecan is especially important for the mechanical strength and functional integrity of basement membranes and skeletal development (Olsen 1999). Perlecan is identifiable from genome sequences or ESTs of species from all metazoan phyla, including homoscleromorph sponges, Trichoplax and Cnidaria (Ozbek et al. 2010; the identifications in *Oscarella* are from ESTs and thus from incomplete sequences). Agrin exists in mammals in secreted and transmembrane forms. Secreted agrin is a component of many basement membranes, whereas membrane-bound agrin is highly expressed in the brain and contributes to neuronal synapse development and plasticity (Gautam et al. 1996; Daniels 2012). Despite the lack of nerve cells in sponges and Placozoa, agrin-related sequences are both encoded in a homoscleromorph sponge and *Trichoplax adhaerens*; paradoxically, the Cnidarians *Hydra magnipapillata* and *Nematostella vectensis* encode only perlecan-like sequences (Ozbek et al. 2010). No distinct homologies (except for the above-mentioned laminin-like sequence) are identifiable in the predicted proteomes of choanoflagellates. The implication is that perlecan and agrin may have both originated on the metazoan stem lineage and that agrin has been lost from the Cnidarian lineage.

1.6.4 Nidogen/Entactin

A smaller ECM component, nidogen/entactin, acts as another scaffold between laminin and collagen IV in basement membranes (Fig. 1.2). Nidogen is not present in choanoflagellates or Porifera but is represented in *Trichoplax adhaerens*, Cnidaria and other metazoa (Srivastava et al. 2008, 2010; Ozbek et al. 2010). It would be interesting to know if a distinct protein takes the role of nidogen in homoscleromorph sponges. Mammals encode two nidogens, and these make varied contributions to the structure of basement membranes in different tissues: nidogen-1 and nidogen-2 double knockout mice are impaired for basement membrane development in the lung and heart yet other basement membranes remain intact (Bader et al. 2005). A study of supramolecular assemblies isolated from epidermal basement membrane noted weak connections between nidogen and the central complexes, suggesting that the main role of nidogen is in providing stabilisation or additional binding capacities (Behrens et al. 2012).

1.7 Other ECM Components Conserved Throughout the Metazoa

1.7.1 Thrombospondins

TSPs are multidomain, multifunctional, calcium-binding glycoproteins that interact with cell surfaces, other ECM components, growth factors, cytokines and proteases. In vertebrates, TSPs form a gene family with members located in basement membranes or interstitial ECM. Muscle–tendon attachments are a major site of expression in *Drosophila*; currently, little is known about TSP localisation in other invertebrates (Adams and Lawler 2011). TSPs have varied domain organisations in their N-terminal regions and are characterised by a hallmark domain architecture of their C-terminal regions, consisting of contiguous EGF domains, TSP type 3 repeats and an L-type lectin-like domain at the C-terminus (Fig. 1.2). The TSPs of vertebrates fall into two subgroups according to their assembly as trimers or pentamers. TSPs are exclusive to the metazoa and no related proteins are apparent in choanoflagellates. However, many of their component domains (laminin G-like,

TSR, EGF-like and L-type lectin) have pre-metazoan origins, and the evolution of TSPs themselves predates that of many of the TSP ligands that have been identified in mammals. Ancestral functions may include binding to GAGs, fibrillar collagens or certain integrins (Bentley and Adams 2010). Early diverging metazoa and protostomes encode a single TSP gene: exceptions are the sea anemone *Nematostella vectensis* that encodes four TSPs and nematode and planarian worms in which the TSP gene has been lost. Most TSPs of invertebrates have a domain organisation corresponding to the pentameric TSPs, albeit with varied numbers of EGF domains. Distinct domain architectures are evident in *Amphimedon queenslandica* TSP (XP_003389217) and TSPs from shrimps and prawns (Bentley and Adams 2010).

All deuterostomes encode at least 3 TSPs per genome, leading to the inference that gene duplications and domain-shuffling events took place on the deuterostome stem lineage. Subsequent divergence led to novel forms of TSPs: the TSP-A domain architecture, which includes the TSR and is conserved in the trimeric TSP-1 and TSP-2 of vertebrates, and TSP-DD, a monomeric form with an N-terminal discoidin-like domain that is present only in early diverging deuterostomes and basal chordates (Bentley and Adams 2010). TSP evolution in vertebrates involved further gene duplications, likely resulting from the large-scale genome duplications that occurred early in the vertebrate lineage (Dehal and Boore 2005; Meyer and Van de Peer 2005), plus subsequent gene losses, resulting in a total of five TSPs in modern tetrapods and up to seven TSPs in bony fish (McKenzie et al. 2006). Genes encoding the pentamerising TSP-3, TSP-4 and TSP 5 are located within paralogous regions of the human genome (defined as regions that arose from the same large-scale genome duplication event; McLysaght et al. 2002), indicating their evolution as duplicated genes within the vertebrate lineage (McKenzie et al. 2006). Many nonlethal mutations have been identified in the coding sequence of *THBS5* that are causal for pseudoachondroplasia and some forms of multiple epiphyseal dysplasia (Posey and Hecht 2008). In general, newly duplicated genes are under-reduced selection pressure and often accumulate deleterious mutations to eventually be lost or become pseudogenes (Lynch and Conery 2000). It is possible that *THBS5* is in the evolutionary process of becoming silenced.

1.7.2 SPARC Family

SPARC/osteonectin (Fig. 1.2) is the prototypic member of this family, which is represented by four categories of family members in vertebrates: SPARC and SPARCL1/hevin, SMOC1 and SMOC2, testican-1 to testican-3 and follistatin-related protein 1. All SPARC family proteins are characterised by three domains: an acidic, low-affinity calcium-binding domain I, a follistatin-like domain II and high-affinity calcium-binding domain III that contains two EF hand motifs. SPARC is a vital component of basement membranes in *Drosophila melanogaster* and *Caenorhabditis elegans*; embryonic lethality of loss of function mutants in

Drosophila is accompanied by overt morphological perturbation of basement membranes (Schwarzbauer and Spencer 1993; Fitzgerald and Schwarzbauer 1998; Martinek et al. 2002, 2008). A functional role of SPARC in basement membranes is conserved in vertebrates, and the protein has evolved additional functions as a major component of bone and connective tissues such as dermis (Yan et al. 2003; Rentz et al. 2007). Thus, the evolution of SPARC has included adaptation from a collagen IV interacting protein to additional roles in binding fibrillar collagens.

SPARC-like proteins are present throughout the metazoa, although not identified to date in Placozoa (Ozbek et al. 2010). Domain I is the least well-conserved domain between invertebrate SPARCs and is absent from Cnidarian SPARCs (Koehler et al. 2009). The family as a whole shows a complex evolution. Testican and follistatin-related protein 1 are present in Cnidaria and arthropods as well as deuterostomes and thus appear to have originated on the bilaterian stem. SMOC-like proteins are present in insects and deuterostomes but not in Cnidaria. SPARCL1/hevin originated through a gene duplication of the ancestral vertebrate SPARC on the amniote lineage (Kawasaki et al. 2004; reviewed by Mosher and Adams 2012).

1.7.3 Mucins

Mucins are highly glycosylated transmembrane or secreted proteins that protect epithelia and assist in preventing invasion by bacterial pathogens. They are characterised by a PTS (proline, serine and threonine-rich) domain and multiple vWF_D domains; other domains are varied (example in Fig. 1.2). Transmembrane mucins are restricted to vertebrates. The secreted mucins that form extracellular gels are of much earlier evolutionary origin: these are present in protostomes (e.g. *Drosophila* hemolectin) and Cnidaria as well as throughout the chordates (Lang et al. 2007). Since the study of Lang et al., secreted mucins have been identified in *Hydra magnipapillata* (XP_002157307) and *Trichoplax adhaerens* (XP_02110756).

1.7.4 Spondins

Spondins are so-named because of the presence of one or more TSR domains that relate these proteins to the subgroup A TSPs. In other regards, these proteins have quite distinct domain architectures (Fig. 1.2). Spondins include F-spondin (also known as spondin-1), a protein enriched in the floor plate of the neural tube and its smaller relative mindin/spondin-2 (Feinstein and Klar 2004); SCO-spondin, a very large protein secreted by the sub-commissural organ, that forms part of Reissner's fibre in cerebrospinal fluid (Meiniel et al. 2008); and R-spondins 1–4, named for the localisation of the founding family member in the roof-plate of the developing nervous system. Members of the vertebrate R-spondin family bind to leucine-rich repeat-containing G-protein-coupled receptors (Lgr), Lgr4, Lgr5 and Lgr6 and act as agonists of canonical Wnt signalling that activate proliferation of adult stem cells (de Lau et al. 2012).

The F-spondin/mindin domain architecture is the most ancient: these proteins appear to have originated early in the metazoan lineage, because single gene products similar to both F-spondin and mindin are encoded in several early diverging metazoa (*Trichoplax adhaerens* XP_002117808, *Nematostella vectensis* XP_001641958) as well as in nematodes, insects and deuterostomes. In contrast, SCO-spondin is deuterostome-specific (present in acorn worm, *Ciona intestinalis* and vertebrates) and R-spondins are chordate-specific. The N-terminal region of SCO-spondin has some domain similarities to the secreted mucins (Fig. 1.2).

1.7.5 Matrix Proteases

A fundamental property of ECM is its capacity to be remodelled during tissue growth, regeneration, repair and ageing and asexual budding of sponges and Cnidaria, or to enable movement of immune cells to sites of injury or infection. A central mechanism for ECM remodelling is the cleavage of ECM components by extracellular proteases, often leading to uptake and endocytic degradation of cell-surface bound ECM components or fragments. Given the origin of many ECM components on the metazoan stem lineage, it is not surprising that the fundamental metazoan adhesome extends to matrix metalloproteinases (MMP) and ADAMTS-like proteases; both families are conserved in sponges and throughout other metazoa (Nicholson et al. 2005; Fanjul-Fernández et al. 2010; Ozbek et al. 2010; also *Amphimedon queenslandica* XP_003388992; Srivastava et al. 2010). In common with the ECM proteins themselves, MMPs and ADAMTSs bear evidence in their domain architecture of domain shuffling of pre-existing domains: both families contain zinc metalloprotease domains as their catalytic domains, and ADAMTS proteases are characterised by the presence of TSR domains (Fig. 1.2). Although definite homologues of MMP or ADAMTS cannot be identified in *Monosiga brevicollis* or *Salpingoeca rosetta*, both species encode at least one intriguing polypeptide related to ADAMTSs. The relationship is clearest in *Salpingoeca rosetta* which encodes a secreted protein containing a zinc metalloprotease domain, several TSR and two SH3 domains (EGD75659). A related protein of *Monosiga brevicollis* includes a zinc metalloprotease domain and SH3 domains (XP_001743602). Remarkably, related zinc metalloprotease domain proteins are also encoded by the unicellular protozoan *Capsaspora owczarzaki* (EFW46674, EFW46669; N.B., these are mis-annotated as "thrombospondins"). Thus, the ADAMTS proteases of metazoa may have diverged from a secreted zinc metalloprotease that originated on the Holozoan stem lineage.

1.8 Chordate and Vertebrate ECM Innovations

It has long been noted that vertebrates encode a wider repertoire of ECM components and adhesion receptors than protostomes and basal metazoa. Analyses of the adhesome and ECM-encoding repertoires of early diverging deuterostomes, the echinoderm *Strongylocentrotus purpuratus* (sea urchin) and the urochordate *Ciona intestinalis* (sea squirt), have investigated this point in depth. With relatively few exceptions (e.g. the expansion of the TSP gene family; see TSPs section), the sea urchin adhesome is similar to that of other invertebrates, with small laminin and collagen gene families and widely conserved ECM components represented mostly as single genes (Whittaker et al. 2006). In *Ciona intestinalis*, most ECM components are also represented at low gene copy number and a relatively small number of chordate-specific or taxon-specific innovations can be recognised. Huxley-Jones et al. (2007) identified from the *Ciona* genome a complement of 60 ECM genes and around 20 enzymes needed for collagen secretion and processing. They also identified a relative enrichment of ECM-coding capacity in the vertebrate genome, with many components encoded as single genes in *Ciona intestinalis* represented by 2–4 genes in humans. In most cases, these genes are present in paralogous regions of the human genome. These findings demonstrate that much of the expansion of adhesome complexity in vertebrates is accounted for by duplication of genes already present on the deuterostome stem lineage. It is likely that these events occurred during the two rounds of genome duplication early on the vertebrate lineage (Dehal and Boore 2005). In general, many newly duplicated genes are subsequently lost, yet the *Ciona*/human comparison indicates a preferential retention of ECM-encoding genes in vertebrates (Huxley-Jones et al. 2007).

The genomes of modern vertebrates also bear witness to numerous ECM innovations. Novel domain combinations, resulting from domain shuffling, have been identified as one source of innovation that increased the repertoire of ECM protein domain architectures in vertebrates (Kawashima et al. 2009). The origin of elastin and its coordination with fibrillin microfibrils to assemble novel elastic fibres was discussed above. Additional protein innovations related to vascular functions or vascular cell-ECM adhesion that originated on the chordate stem include vitronectin, von Willebrand Factor, fibrinogen and the blood-clotting system (Huxley-Jones et al. 2007; Doolittle et al. 2008; Doolittle 2012). Important innovations for basement membrane complexity include the deuterostome-specific Fras (Fraser syndrome protein) and the FREM-1, -2 (Fraser syndrome related extracellular matrix) proteins of urochordates and chordates. In mammals, these proteins form a distinct supramolecular assembly with a scaffold protein, Grip1, that contributes to anchoring of basement membrane sheets to the underlying interstitial ECM. Mutations in proteins of this complex give rise to the autosomal recessive Fraser syndrome (characterised by embryonic skin blistering, cryptophthalmos, syndactyly and renal agenesis) (Pavlakis et al. 2011). Mutations in FREM1 also underlie the less severe MOTO and BNAR syndromes (Slavotinek et al. 2011).

Other deuterostome or chordate innovations have resulted in entirely novel fibril systems within vertebrate connective ECM. Hyaluronan is an extracellular GAG that is synthesised by cell-surface hyaluronan synthase (HS) and forms massive, water-retaining polymers in association with the secreted proteoglycan aggrecan in cartilage and brain. The origin of this system lies on the gnathostome stem lineage: hyaluronan synthase is chordate-specific and aggrecan is conserved in cartilaginous fish and vertebrates (Weigel and DeAngelis 2007; Ozbek et al. 2010). An important fibril system of vertebrates is assembled by the glycoprotein fibronectin, a chordate-specific ECM protein characterised by repeated fibronectin (FN) type I, II and III domains (Fig. 1.2). Fibronectin dimers assemble into fibrils through a complex process involving fibronectin–integrin interaction, dimer unfolding under the action of cytoskeletal tension and fibril assembly through intermolecular interactions of FNI domains (Singh et al. 2010). Fibronectin fibrils interact with multiple ECM components including collagens, proteoglycans and tenascins and are vital for vertebrate life. Fibronectin-like proteins are encoded in *Ciona* species; however, due to differences in domain organisation to vertebrate fibronectins, it is not obvious whether these proteins have fibril-forming capacity. True fibronectin homologues are apparent through the gnathostomes (these points are discussed in more detail in Ozbek et al. 2010).

Vertebrate ECM complexity has also expanded through the evolution of molecules with novel domain architectures that can be categorised broadly as adaptors within the ECM: i.e. proteins that interact with network-building ECM components, growth factors or morphogens and cell surfaces and provide new links between existing ECM components. Examples include the CCN family, matrilins and tenascins (Tucker et al. 2006; Chen and Lau 2009; Klatt et al. 2011) (Fig. 1.2). The tenascins are of particular evolutionary interest because of their structural relationship with fibronectin through the presence of many tandem FNIII domains. Tenascin appears to have evolved at the same time as, or even before, fibronectin because clear tenascin homologues are present in *Ciona* species and amphioxus (Tucker and Chiquet-Ehrismann 2009). It is speculated that tenascin and fibronectin may have shared a common ancestor because the FNIII domains of *Ciona* tenascin and *Ciona* fibronectin-like are highly related to each other and to the FNIII domains of vertebrate fibronectin (Tucker and Chiquet-Ehrismann 2009; e.g. BLASTP search of the *Ciona intestinalis* proteome with human fibronectin returns *Ciona* tenascin, XP_002124787 as the top hit). In addition, several tenascin family members associate with fibronectin fibrils and are linked functionally with fibronectin. However, the varied functions and ECM interactions of vertebrate tenascins leave ambiguity as to whether this reflects an ancestral relationship of fibronectin and tenascin as co-evolving binding partners or a functional specialisation in vertebrates (discussion of this point can be found in Mosher and Adams 2012).

All of these processes of gene duplication, functional diversification of retained gene products and innovations through domain shuffling contributed to the richness of vertebrate ECM and its potential for complexity of ECM signalling and tissue-specific ECM micro-environments. It is thought that these factors acted as drivers for the major innovations of body plan, developmental processes and tissue specialisation that are evident in vertebrates: the elaborate central nervous system and diversity of basement membranes and the innovations of the notochord, pharyngeal arches and neural crest cells, high-pressure vascular system and cartilage and bone (Huxley-Jones et al. 2007; Jeffery et al. 2008; Kawashima et al. 2009; Wada 2010).

1.9 Concluding Remarks

Recent genome sequencing and EST projects for metazoan species and representatives of metazoan sister groups are illuminating our understanding of the evolution of the metazoan ECM. Doubtless further insights and surprises will arise as the density of sequenced genomes increases for early diverging metazoa and unikont protozoa, yet the current picture is extremely interesting. It includes some enigmas in which ECM protein-coding capacity cannot be reconciled easily with previous morphological observations. This is particularly the case for the Placozoan, *Trichoplax adhaerens*, for which no ECM has been observed in the adult organism, yet its repertoire of encoded ECM components is similar to that of the Cnidarians that have far more complex body plans. Is an ECM only assembled at certain lifecycle stages? What might the roles of these proteins be in *Trichoplax*?

Comparative genomics has confirmed the particular complexity of the ECM of vertebrates and provides a clearer view on the origins of this complexity. Phylogenomics has also shed light on the origins of central components of basement membranes and connective ECM and presents a view of an organised ECM inherent from the origin of the metazoa. It appears that a major wave of innovation took place on the metazoan stem lineage, leading to the appearance of collagens, laminin, TSP and others. It is fascinating to consider the considerable adaptations of domain shuffling, posttranslational subunit assembly and the capacities of the secretory pathway that must have been involved in the origin of these complex, multidomain, oligomeric proteins and their functional ability to form stable networks. These events remain a “black box”, yet genome sequences from choanoflagellate species, representing the sister group to metazoa, are bringing tantalising views of the encoding of secreted proteins that share not one but several features of domain organisation with metazoan ECM proteins. Proteins with some similarities to laminin, ADAMTS proteases and possibly fibrillin have been discussed in this chapter. The functions of these proteins are currently unknown and clearly must be adapted to the ecology and life cycle of the choanoflagellates. Nevertheless, identification of these proteins gives an indication of how “pre-adapted” domain combinations that originated on the Holozoan stem might have provided building blocks for the evolution of the large and complex ECM proteins of metazoa.

References

Abedin M, King N (2008) The premetazoan ancestry of cadherins. Science 319:946–948

Adams JC, Tucker RP (2000) The thrombospondin type 1 repeat (TSR) superfamily: diverse proteins with related roles in neuronal development. Dev Dyn 218:280–299

Adams JC, Engel J (2007) Bioinformatic analysis of adhesion proteins. Methods Mol Biol 370:147–171

Adams JC, Lawler J (2011) The thrombospondins. Cold Spring Harb Perspect Biol 3(10):a009712

Aouacheria A, Geourjon C, Aghajari N, Navratil V, Deléage G, Lethias C, Exposito JY (2006) Insights into early extracellular matrix evolution: spongin short chain collagen-related proteins are homologous to basement membrane type IV collagens and form a novel family widely distributed in invertebrates. Mol Biol Evol 23:2288–2302

Babushok DV, Ostertag EM, Kazazian HH (2007) Current topics in genome evolution: molecular mechanisms of new gene formation. Cell Mol Life Sci 64:542–554

Bader BL, Smyth N, Nedbal S, Miosge N, Baranowsky A, Mokkapati S, Murshed M, Nischt R (2005) Compound genetic ablation of nidogen 1 and 2 causes basement membrane defects and perinatal lethality in mice. Mol Cell Biol 25:6846–6856

Beck K, Brodsky B (1998) Supercoiled protein motifs: the collagen triple-helix and the alpha-helical coiled coil. J Struct Biol 122:17–29

Behrens DT, Villone D, Koch M, Brunner G, Sorokin L, Robenek H, Bruckner-Tuderman L, Bruckner P, Hansen U (2012) The epidermal basement membrane is a composite of separate laminin- or collagen IV-containing networks connected by aggregated perlecan, but not by nidogens. J Biol Chem 287:18700–18709

Bentley AA, Adams JC (2010) The evolution of thrombospondins and their ligand-binding activities. Mol Biol Evol 27:2187–2197

Boute N, Exposito JY, Boury-Esnault N, Vacelet J, Noro N, Miyazaki K, Yoshizato K, Garrone R (1996) Type IV collagen in sponges, the missing link in basement membrane ubiquity. Biol Cell 88:37–44

Cárdenas P, Pérez T, Boury-Esnault N (2012) Sponge systematics facing new challenges. Adv Mar Biol 61:79–209

Carrino DA, Dennis JE, Wu TM, Arias JL, Fernandez MS, Rodriguez JP, Fink DJ, Heuer AH, Caplan AI (1996) The avian eggshell extracellular matrix as a model for biomineralization. Connect Tissue Res 35:325–329

Cássaro CM, Dietrich CP (1977) Distribution of sulfated mucopolysaccharides in invertebrates. J Biol Chem 252:2254–2261

Chapman JA, Kirkness EF, Simakov O, Hampson SE, Mitros T, Weinmaier T, Rattei T, Balasubramanian PG, Borman J, Busam D, Disbennett K, Pfannkoch C, Sumin N, Sutton GG, Viswanathan LD, Walenz B, Goodstein DM, Hellsten U, Kawashima T, Prochnik SE, Putnam NH, Shu S, Blumberg B, Dana CE, Gee L, Kibler DF, Law L, Lindgens D, Martinez DE, Peng J, Wigge PA, Bertulat B, Guder C, Nakamura Y, Ozbek S, Watanabe H, Khalturin K, Hemmrich G, Franke A, Augustin R, Fraune S, Hayakawa E, Hayakawa S, Hirose M, Hwang JS, Ikeo K, Nishimiya-Fujisawa C, Ogura A, Takahashi T, Steinmetz PR, Zhang X, Aufschnaiter R, Eder MK, Gorny AK, Salvenmoser W, Heimberg AM, Wheeler BM, Peterson KJ, Böttger A, Tischler P, Wolf A, Gojobori T, Remington KA, Strausberg RL, Venter JC, Technau U, Hobmayer B, Bosch TC, Holstein TW, Fujisawa T, Bode HR, David CN, Rokhsar DS, Steele RE (2010) The dynamic genome of Hydra. Nature 464:592–596

Chen CC, Lau LF (2009) Functions and mechanisms of action of CCN matricellular proteins. Int J Biochem Cell Biol 41:771–783

Daniels MP (2012) The role of agrin in synaptic development, plasticity and signaling in the central nervous system. Neurochem Int 61(6):848–853

Dehal P, Boore JL (2005) Two rounds of whole genome duplication in the ancestral vertebrate. PLoS Biol 3:e314

Deutzmann R, Fowler S, Zhang X, Boone K, Dexter S, Boot-Handford RP, Rachel R, Sarras MP Jr (2000) Molecular, biochemical and functional analysis of a novel and developmentally important fibrillar collagen (Hcol-I) in hydra. Development 127:4669–4680

Dodd RB, Drickamer K (2001) Lectin-like proteins in model organisms: implications for evolution of carbohydrate-binding activity. Glycobiology 11:71R–79R

Doolittle RF (2012) The protochordate Ciona intestinalis has a protein like full-length vertebrate fibrinogen. J Innate Immun 4:219–222

Doolittle RF, Jiang Y, Nand J (2008) Genomic evidence for a simpler clotting scheme in jawless vertebrates. J Mol Evol 66:185–196

Dunn CW, Hejnol A, Matus DQ, Pang K, Browne WE, Smith SA, Seaver E, Rouse GW, Obst M, Edgecombe GD, Sørensen MV, Haddock SH, Schmidt-Rhaesa A, Okusu A, Kristensen RM, Wheeler WC, Martindale MQ, Giribet G (2008) Broad phylogenomic sampling improves resolution of the animal tree of life. Nature 452:745–749

Ereskovsky AV, Borchiellini C, Gazave E, Ivanisevic J, Lapébie P, Perez T, Renard E, Vacelet J (2009) The Homoscleromorph sponge Oscarella lobularis, a promising sponge model in evolutionary and developmental biology: model sponge Oscarella lobularis. Bioessays 31:89–97

Exposito JY, Le Guellec D, Lu Q, Garrone R (1991) Short chain collagens in sponges are encoded by a family of closely related genes. J Biol Chem 266:21923–21928

Exposito JY, Larroux C, Cluzel C, Valcourt U, Lethias C, Degnan BM (2008) Demosponge and sea anemone fibrillar collagen diversity reveals the early emergence of A/C clades and the maintenance of the modular structure of type V/XI collagens from sponge to human. J Biol Chem 283:28226–28235

Exposito JY, Valcourt U, Cluzel C, Lethias C (2010) The fibrillar collagen family. Int J Mol Sci 11:407–426

Fanjul-Fernández M, Folgueras AR, Cabrera S, López-Otín C (2010) Matrix metalloproteinases: evolution, gene regulation and functional analysis in mouse models. Biochim Biophys Acta 1803:3–19

Feinstein Y, Klar A (2004) The neuronal class 2 TSR proteins F-spondin and Mindin: a small family with divergent biological activities. Int J Biochem Cell Biol 36:975–980

Fitzgerald MC, Schwarzbauer JE (1998) Importance of the basement membrane protein SPARC for viability and fertility in Caenorhabditis elegans. Curr Biol 8:1285–1288

Fowler SJ, Jose S, Zhang X, Deutzmann R, Sarras MP Jr, Boot-Handford RP (2000) Characterization of hydra type IV collagen. Type IV collagen is essential for head regeneration and its expression is up-regulated upon exposure to glucose. J Biol Chem 275:39589–39599

Gautam M, Noakes PG, Moscoso L, Rupp F, Scheller RH, Merlie JP, Sanes JR (1996) Defective neuromuscular synaptogenesis in agrin-deficient mutant mice. Cell 85:525–535

Harcet M, Roller M, Cetković H, Perina D, Wiens M, Müller WE, Vlahovicek K (2010) Desmosponge EST sequencing reveals a complex genetic toolkit of the simplest metazoans. Mol Biol Evol 27:2747–2756

Heinemann S, Ehrlich H, Douglas T, Heinemann C, Worch H, Schatton W, Hanke T (2007) Ultrastructural studies on the collagen of the marine sponge Chondrosia reniformis Nardo. Biomacromolecules 8:3452–3457

Hulpiau P, van Roy F (2009) Molecular evolution of the cadherin superfamily. Int J Biochem Cell Biol 41:349–369

Huxley-Jones J, Robertson DL, Boot-Handford RP (2007) On the origins of the extracellular matrix in vertebrates. Matrix Biol 26:2–11

Hynes RO, Naba A (2012) Overview of the matrisome—an inventory of extracellular matrix constituents and functions. Cold Spring Harb Perspect Biol 4(1):a004903

Iozzo RV, Sanderson RD (2011) Proteoglycans in cancer biology, tumour microenvironment and angiogenesis. J Cell Mol Med 15:1013–1031

Jackson DJ, Macis L, Reitner J, Wörheide G (2011) A horizontal gene transfer supported the evolution of an early metazoan biomineralization strategy. BMC Evol Biol 11:238

Jeffery WR, Chiba T, Krajka FR, Deyts C, Satoh N, Joly JS (2008) Trunk lateral cells are neural crest-like cells in the ascidian Ciona intestinalis: insights into the ancestry and evolution of the neural crest. Dev Biol 324:152–160
Kawashima T, Kawashima S, Tanaka C, Murai M, Yoneda M, Putnam NH, Rokhsar DS, Kanehisa M, Satoh N, Wada H (2009) Domain shuffling and the evolution of vertebrates. Genome Res 19:1393–1403
Kawasaki K, Suzuki T, Weiss KM (2004) Genetic basis for the evolution of vertebrate mineralized tissue. Proc Natl Acad Sci U S A 101:11356–113561
Kawasaki K, Buchanan AV, Weiss KM (2009) Biomineralization in humans: making the hard choices in life. Annu Rev Genet 43:119–142
Kiedzierska A, Smietana K, Czepczynska H, Otlewski J (2007) Structural similarities and functional diversity of eukaryotic discoidin-like domains. Biochim Biophys Acta 1774:1069–1078
King N, Westbrook MJ, Young SL, Kuo A, Abedin M, Chapman J, Fairclough S, Hellsten U, Isogai Y, Letunic I, Marr M, Pincus D, Putnam N, Rokas A, Wright KJ, Zuzow R, Dirks W, Good M, Goodstein D, Lemons D, Li W, Lyons JB, Morris A, Nichols S, Richter DJ, Salamov A, Sequencing JG, Bork P, Lim WA, Manning G, Miller WT, McGinnis W, Shapiro H, Tjian R, Grigoriev IV, Rokhsar D (2008) The genome of the choanoflagellate Monosiga brevicollis and the origin of metazoans. Nature 451:783–788
Klatt AR, Becker AK, Neacsu CD, Paulsson M, Wagener R (2011) The matrilins: modulators of extracellular matrix assembly. Int J Biochem Cell Biol 43:320–330
Knoll AH, Javaux EJ, Hewitt D, Cohen P (2006) Eukaryotic organisms in Proterozoic oceans. Philos Trans R Soc Lond B Biol Sci 361:1023–1038
Koehler A, Desser S, Chang B, MacDonald J, Tepass U, Ringuette M (2009) Molecular evolution of SPARC: absence of the acidic module and expression in the endoderm of the starlet sea anemone, Nematostella vectensis. Dev Genes Evol 219:509–521
Lang T, Hansson GC, Samuelsson T (2007) Gel-forming mucins appeared early in metazoan evolution. Proc Natl Acad Sci U S A 104:16209–16214
de Lau WB, Snel B, Clevers HC (2012) The R-spondin protein family. Genome Biol 13:242
Li C-W, Chen J-Y, Hua T-E (1998) Precambrian sponges with cellular structures. Science 279:879–882
Lynch M, Conery JS (2000) The evolutionary fate and consequences of duplicate genes. Science 290:1151–1155
Marshall CR, Valentine JW (2010) The importance of preadapted genomes in the origin of the animal bodyplans and the Cambrian explosion. Evolution 64:1189–1201
Martinek N, Zou R, Berg M, Sodek J, Ringuette M (2002) Evolutionary conservation and association of SPARC with the basal lamina in Drosophila. Dev Genes Evol 212:124–133
Martinek N, Shahab J, Saathoff M, Ringuette M (2008) Haemocyte-derived SPARC is required for collagen-IV-dependent stability of basal laminae in Drosophila embryos. J Cell Sci 121:1671–1680
Massam-Wu T, Chiu M, Choudhury R, Chaudhry SS, Baldwin AK, McGovern A, Baldock C, Shuttleworth CA, Kielty CM (2010) Assembly of fibrillin microfibrils governs extracellular deposition of latent TGF beta. J Cell Sci 123:3006–3018
McLysaght A, Hokamp K, Wolfe KH (2002) Extensive genomic duplication during early chordate evolution. Nat Genet 31:200–204
McKenzie P, Chadalavada SC, Bohrer J, Adams JC (2006) Phylogenomic analysis of vertebrate thrombospondins reveals fish-specific paralogues, ancestral gene relationships and a tetrapod innovation. BMC Evol Biol 6:33
Meiniel O, Meiniel R, Lalloué F, Didier R, Jauberteau MO, Meiniel A, Petit D (2008) The lengthening of a giant protein: when, how, and why? J Mol Evol 66:1–10
Meyer A, Van de Peer Y (2005) From 2R to 3R: evidence for a fish-specific genome duplication (FSGD). Bioessays 27:937–945
Mosher DF, Adams JC (2012) Adhesion-modulating/matricellular ECM protein families: a structural, functional and evolutionary appraisal. Matrix Biol 31:155–161

Murdock DJ, Donoghue PC (2011) Evolutionary origins of animal skeletal biomineralization. Cells Tissues Organs 194:98–102

Nichols SA, Roberts BW, Richter DJ, Fairclough SR, King N (2012) Origin of metazoan cadherin diversity and the antiquity of the classical cadherin/β-catenin complex. Proc Natl Acad Sci U S A 109(32):13046–13051

Nicholson AC, Malik SB, Logsdon JM Jr, Van Meir EG (2005) Functional evolution of ADAMTS genes: evidence from analyses of phylogeny and gene organization. BMC Evol Biol 5:11

Olsen BR (1999) Life without perlecan has its problems. J Cell Biol 147:909–912

Ozbek S, Balasubramanian PG, Chiquet-Ehrismann R, Tucker RP, Adams JC (2010) The evolution of extracellular matrix. Mol Biol Cell 21:4300–4305

Pang K, Martindale MQ (2008) Comb jellies (ctenophora): a model for Basal metazoan evolution and development. CSH Protoc: doi:10.110/pdb.emo106

Pavlakis E, Chiotaki R, Chalepakis G (2011) The role of Fras1/Frem proteins in the structure and function of basement membrane. Int J Biochem Cell Biol 43:487–495

Philippe H, Derelle R, Lopez P, Pick K, Borchiellini C, Boury-Esnault N, Vacelet J, Renard E, Houliston E, Quéinnec E, Da Silva C, Wincker P, Le Guyader H, Leys S, Jackson DJ, Schreiber F, Erpenbeck D, Morgenstern B, Wörheide G, Manuel M (2009) Phylogenomics revives traditional views on deep animal relationships. Curr Biol 19:706–712

Pick KS, Philippe H, Schreiber F, Erpenbeck D, Jackson DJ, Wrede P, Wiens M, Alié A, Morgenstern B, Manuel M, Wörheide G (2010) Improved phylogenomic taxon sampling noticeably affects nonbilaterian relationships. Mol Biol Evol 27:1983–1987

Posey KL, Hecht JT (2008) The role of cartilage oligomeric matrix protein (COMP) in skeletal disease. Curr Drug Targets 9:869–877

Prochnik SE, Umen J, Nedelcu AM, Hallmann A, Miller SM, Nishii I, Ferris P, Kuo A, Mitros T, Fritz-Laylin LK, Hellsten U, Chapman J, Simakov O, Rensing SA, Terry A, Pangilinan J, Kapitonov V, Jurka J, Salamov A, Shapiro H, Schmutz J, Grimwood J, Lindquist E, Lucas S, Grigoriev IV, Schmitt R, Kirk D, Rokhsar DS (2010) Genomic analysis of organismal complexity in the multicellular green alga Volvox carteri. Science 329:223–226

Putnam NH, Srivastava M, Hellsten U, Dirks B, Chapman J, Salamov A, Terry A, Shapiro H, Lindquist E, Kapitonov VV, Jurka J, Genikhovich G, Grigoriev IV, Lucas SM, Steele RE, Finnerty JR, Technau U, Martindale MQ, Rokhsar DS (2007) Sea anemone genome reveals ancestral eumetazoan gene repertoire and genomic organization. Science 317:86–94

Ramirez F, Sakai LY (2010) Biogenesis and function of fibrillin assemblies. Cell Tissue Res 339:71–82

Reber-Müller S, Spissinger T, Schuchert P, Spring J, Schmid V (1995) An extracellular matrix protein of jellyfish homologous to mammalian fibrillins forms different fibrils depending on the life stage of the animal. Dev Biol 169:662–672

Rentz TJ, Poobalarahi F, Bornstein P, Sage EH, Bradshaw AD (2007) SPARC regulates processing of procollagen I and collagen fibrillogenesis in dermal fibroblasts. J Biol Chem 282:22062–22071

Ricard-Blum S, Ruggiero F (2005) The collagen superfamily: from the extracellular matrix to the cell membrane. Pathol Biol (Paris) 53:430–442

Robertson I, Jensen S, Handford P (2011) TB domain proteins: evolutionary insights into the multifaceted roles of fibrillins and LTBPs. Biochem J 433:263–276

Rokas A (2008) The origins of multicellularity and the early history of the genetic toolkit for animal development. Annu Rev Genet 42:235–251

Ruiz-Trillo I, Roger AJ, Burger G, Gray MW, Lang BF (2008) A phylogenomic investigation into the origin of metazoa. Mol Biol Evol 25:664–672

Sage H, Gray WR (1979) Studies on the evolution of elastin—I. Phylogenetic distribution. Comp Biochem Physiol B 64:313–327

Sarrazin S, Lamanna WC, Esko JD (2011) Heparan sulfate proteoglycans. Cold Spring Harb Perspect Biol 3(7). pii:a004952

Sebé-Pedrós A, Roger AJ, Lang FB, King N, Ruiz-Trillo I (2010) Ancient origin of the integrin-mediated adhesion and signaling machinery. Proc Natl Acad Sci U S A 107:10142–10147
Schierwater B, de Jong D, Desalle R (2009) Placozoa and the evolution of Metazoa and intrasomatic cell differentiation. Int J Biochem Cell Biol 41:370–379
Schwarzbauer JE, Spencer CS (1993) The Caenorhabditis elegans homologue of the extracellular calcium binding protein SPARC/osteonectin affects nematode body morphology and mobility. Mol Biol Cell 4:941–952
Segade F (2010) Molecular evolution of the fibulins: implications on the functionality of the elastic fibulins. Gene 464:17–31
Shimizu H, Aufschnaiter R, Li L, Sarras MP Jr, Borza DB, Abrahamson DR, Sado Y, Zhang X (2008) The extracellular matrix of hydra is a porous sheet and contains type IV collagen. Zoology (Jena) 111:410–418
Singh P, Carraher C, Schwarzbauer JE (2010) Assembly of fibronectin extracellular matrix. Annu Rev Cell Dev Biol 26:397–419
Slavotinek AM, Baranzini SE, Schanze D, Labelle-Dumais C, Short KM, Chao R, Yahyavi M, Bijlsma EK, Chu C, Musone S, Wheatley A, Kwok PY, Marles S, Fryns JP, Maga AM, Hassan MG, Gould DB, Madireddy L, Li C, Cox TC, Smyth I, Chudley AE, Zenker M (2011) Manitoba-oculo-tricho-anal (MOTA) syndrome is caused by mutations in FREM1. J Med Genet 48:375–382
Srivastava M, Begovic E, Chapman J, Putnam NH, Hellsten U, Kawashima T, Kuo A, Mitros T, Salamov A, Carpenter ML, Signorovitch AY, Moreno MA, Kamm K, Grimwood J, Schmutz J, Shapiro H, Grigoriev IV, Buss LW, Schierwater B, Dellaporta SL, Rokhsar DS (2008) The Trichoplax genome and the nature of placozoans. Nature 454:955–960
Srivastava M, Simakov O, Chapman J, Fahey B, Gauthier ME, Mitros T, Richards GS, Conaco C, Dacre M, Hellsten U, Larroux C, Putnam NH, Stanke M, Adamska M, Darling A, Degnan SM, Oakley TH, Plachetzki DC, Zhai Y, Adamski M, Calcino A, Cummins SF, Goodstein DM, Harris C, Jackson DJ, Leys SP, Shu S, Woodcroft BJ, Vervoort M, Kosik KS, Manning G, Degnan BM, Rokhsar DS (2010) The Amphimedon queenslandica genome and the evolution of animal complexity. Nature 466:720–726
Tomitani A, Knoll AH, Cavanaugh CM, Ohno T (2006) The evolutionary diversification of cyanobacteria: molecular–phylogenetic and paleontological perspectives. PNAS 103:5442–5447
Torruella G, Derelle R, Paps J, Lang BF, Roger AJ, Shalchian-Tabrizi K, Ruiz-Trillo I (2012) Phylogenetic relationships within the Opisthokonta based on phylogenomic analyses of conserved single-copy protein domains. Mol Biol Evol 29:531–544
Tucker RP, Hess J, Drabikowski K, Ferralli J, Chiquet-Ehrismann R, Adams JC (2006) Phylogenetic analysis of the tenascin gene family: evidence of origin early in the chordate lineage. BMC Evol Biol 6:60
Tucker RP, Chiquet-Ehrismann R (2009) Evidence for the evolution of tenascin and fibronectin early in the chordate lineage. Int J Biochem Cell Biol 41:424–434
Tzu J, Marinkovich MP (2008) Bridging structure with function: structural, regulatory, and developmental role of laminins. Int J Biochem Cell Biol 40:199–214
Wada H (2010) Origin and genetic evolution of the vertebrate skeleton. Zoolog Sci 27:119–123
Wagenseil JE, Mecham RP (2009) Vascular extracellular matrix and arterial mechanics. Physiol Rev 89:957–989
Weigel PH, DeAngelis PL (2007) Hyaluronan synthases: a decade-plus of novel glycosyltransferases. J Biol Chem 282:36777–36781
Whittaker CA, Hynes RO (2002) Distribution and evolution of von Willebrand/integrin A domains: widely dispersed domains with roles in cell adhesion and elsewhere. Mol Biol Cell 13:3369–3387
Whittaker CA, Bergeron KF, Whittle J, Brandhorst BP, Burke RD, Hynes RO (2006) The echinoderm adhesome. Dev Biol 300:252–266

Wickström SA, Radovanac K, Fässler R (2011) Genetic analyses of integrin signaling. Cold Spring Harb Perspect Biol 3(2)

Wörheide G, Dohrmann M, Erpenbeck D, Larroux C, Maldonado M, Voigt O, Borchiellini C, Lavrov DV (2012) Deep phylogeny and evolution of sponges (phylum porifera). Adv Mar Biol 61:1–78

Wouters MA, Rigoutsos I, Chu CK, Feng LL, Sparrow DB, Dunwoodie SL (2005) Evolution of distinct EGF domains with specific functions. Protein Sci 14:1091–1103

Yamada S, Morimoto H, Fujisawa T, Sugahara K (2007) Glycosaminoglycans in Hydra magnipapillata (Hydrozoa, Cnidaria): demonstration of chondroitin in the developing nematocyst, the sting organelle, and structural characterization of glycosaminoglycans. Glycobiology 17:886–894

Yan Q, Blake D, Clark JI, Sage EH (2003) Expression of the matricellular protein SPARC in murine lens: SPARC is necessary for the structural integrity of the capsular basement membrane. J Histochem Cytochem 51:503–511

Yanagisawa H, Davis EC (2010) Unraveling the mechanism of elastic fiber assembly: the roles of short fibulins. Int J Biochem Cell Biol 42:1084–1093

Zhang X, Fei K, Agbas A, Yan L, Zhang J, O'Reilly B, Deutzmann R, Sarras MP Jr (2002) Structure and function of an early divergent form of laminin in hydra: a structurally conserved ECM component that is essential for epithelial morphogenesis. Dev Genes Evol 212:159–172

Chapter 2
Domain Shuffling and the Evolution of Vertebrate Extracellular Matrix

Hiroshi Wada

Abstract Contribution of domain shuffling events in the evolution of the vertebrate extracellular matrix (ECM) is reviewed. In some cases, domain shuffling contributed to the evolution of molecules via minor modification of existed molecules by fusing a domain to existing proteins. In contrast, some ECM molecules such as aggrecan and fibronetin seem to be created rather de novo via domain shuffling. In the case of occludin, a domain of completely different function (ELL: elongation of RNA synthesis) was recruited for use in cell adhesion molecules. These cases are spectacular examples of molecular tinkering.

2.1 Introduction

Evolution of the extracellular matrix (ECM) has played a major role in the evolution of multicellularity of animals. Many components of the ECM, including laminin and type IV collagen, were acquired in common ancestors of metazoans (Patthy 2003). Since then ECMs have continued to evolve. The evolution of ECM molecules contributed to the diversification of various metazoans (Patthy 1999, 2003).

Because most ECM molecules show multidomain architectures, the evolution of ECM molecules likely progressed through two steps, namely, fixation of functional amino acid domains (domain emergence) and subsequent shuffling of functional domains (domain shuffling). These mechanisms have generated the variety of functional ECM molecules found today. Ekman et al. (2007) indicated that the evolutionary rate of domain emergence was relatively constant during eukaryotic evolution or possibly relatively high in eukaryotic ancestors prior to the divergence

H. Wada (✉)
Graduate School of Life and Environmental Sciences, University of Tsukuba, Tsukuba, Japan
e-mail: hwada@biol.tsukuba.ac.jp

F.W. Keeley and R.P. Mecham (eds.), *Evolution of Extracellular Matrix*,
Biology of Extracellular Matrix, DOI 10.1007/978-3-642-36002-2_2,

of plants, fungi, and metazoans. However, domain shuffling has accelerated in metazoan lineages.

Duplication divergence is one of the most widely accepted mechanisms for the molecular evolution of domain emergence. According to this theory, novel functional sequences were established by gene duplication followed by divergence of amino acid sequences. In cases where one of the duplicated genes performs an ancestral function, the other copy is typically under less functional constraint and thus free to fix distinct amino acid sequences. On the other hand, the recent accumulation of genomic data may increase our understanding of the other aspect of gene evolution. The genome sequences analyzed to date contain a large number of genes that show no homology to any genomes of other species. Some authors such as Schmidt and Davies (2007) and Tautz and Domazet-Loso (2011) proposed that these sequences emerged from noncoding genome sequences and that novel functional amino acid sequences continue to emerge.

In this chapter we focus on the second step of ECM molecular evolution, namely, domain shuffling. Our previous research identified several novel ECM molecules produced via domain shuffling in vertebrate ancestors (Kawashima et al. 2009). Our basic strategy to evaluate domain shuffling events was as follows. First, novel domain architectures in gene models were evaluated by listing domain pairs found in single-gene models of several representative genomes. Using this method, we identified 47 vertebrate genes created by domain shuffling. These genes possessed domain pairs conserved in all vertebrate species but were not found in any invertebrate genomes. We found that most of the domain shuffling events were likely due to exon shuffling, because most of the new domain pairs were split by intron(s) (Kawashima et al. 2009), and nice correlation was observed between exon and domain boundaries (Liu and Grigoriev 2004). Among the exons involved in domain shuffling, a number were split by phase 1 introns at both the 5′ and 3′ termini (exons of intron phase type 1-1), while 0-0 phase exons were abundant in all exons (Patthy 1999; Kawashima et al. 2009). The mechanism behind this type of shuffling remains unclear but Patthy (1999) suggested that it is due to the larger number of exons encoding extracellular protein domains encoded by 1-1 phase exons. Below we review how these novel genes were created by domain shuffling and how they contributed to evolution of the vertebrate body.

2.2 Lecticans and Evolution of Vertebrate Cartilage

Cartilage, one of the most important features of vertebrates, was established through ECM evolution. The most abundant component of vertebrate cartilage is fibrillar collagen, and gene duplication of fibrillar collagens has contributed to the formation of various vertebrate skeletal structures (Wada et al. 2006). An interesting exception is agnathan cartilage, whose main component is elastin-like molecules (Wright and Youson 1983; Robson et al. 1993, 1997); therefore, the evolutionary relationship between agnathan and gnathostomes skeletons remains

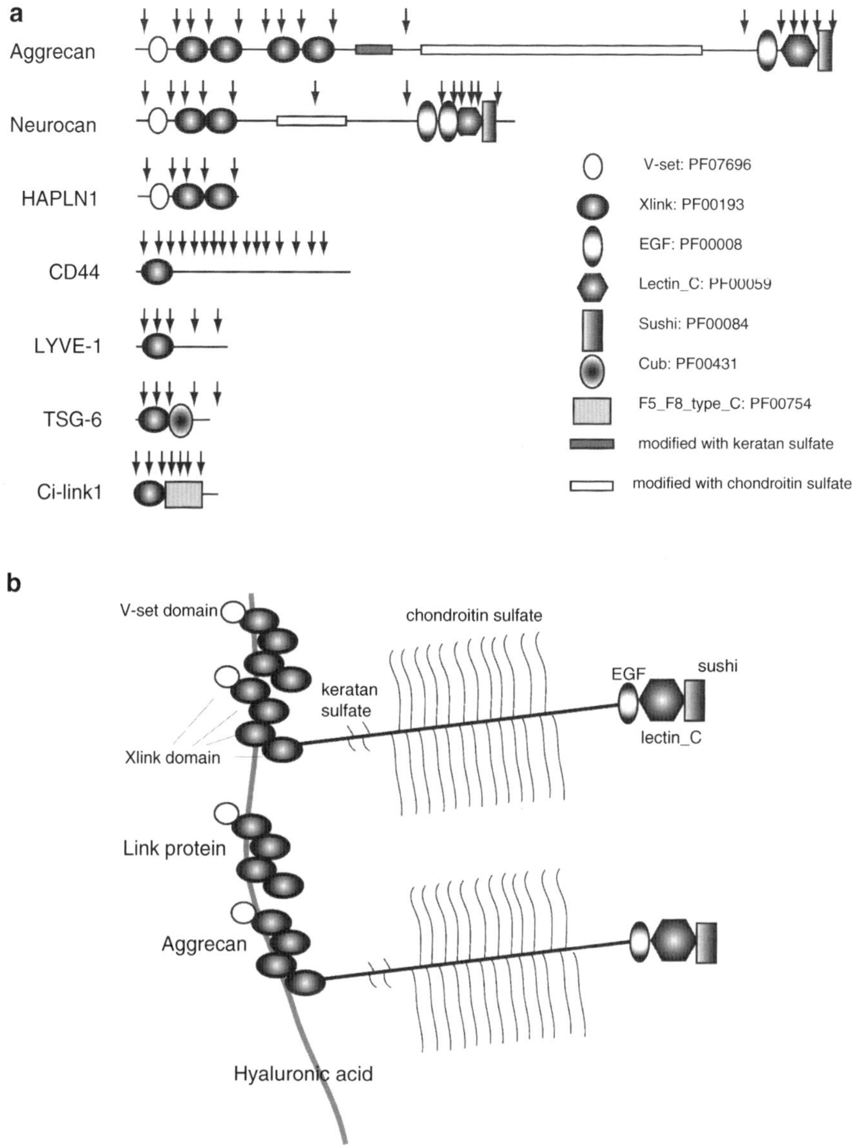

Fig. 2.1 Schematic illustration of the domain architecture of aggrecan. (**a**) Comparison of the domain architecture to proteins with Xlink domain. (**b**) Schematic illustration of the domain architecture of aggrecan and HAPLN and their interactions with hyaluronic acid

unclear (McCauley and Bronner-Fraser 2006; Zhang et al. 2006; Ohtani et al. 2008). Aggrecan is the most abundant non-collagenous protein component of gnathostome cartilage. The aggrecan domain architecture is shown in Fig. 2.1a; the domain pair V-set and Xlink is novel in vertebrates. This domain pair is also found in other lecticans, such as neurocan and bevican, as well as in HAPLNs (Fig. 2.1a). By examining each domain in the amphioxus and ascidian genomes, we found that these domain architectures were established de novo. The V-set, Lectin C, EGF, and Sushi domains were abundant in both genomes. For example, V-set domains were found in 269 and 26 gene models of the amphioxus and ascidian genomes, respectively, and Sushi domains were found in 365 and 61 gene models. In contrast, 20 amphioxus gene models and 2 ascidian gene models contain Xlink domains. One of the ascidian gene models (*Ci-link1*) consisted of a single Xlink domain and the F5/8 type C domain (Fig. 2.1a). In vertebrates, the Xlink domain was also found in CD44, TSG-6, LYVE-1, and stabilin-1. These genes contain single Xlink domains with architectures similar to Ci-link1 (Fig. 2.1a).

The vertebrate molecules with single Xlink domains are involved in lymphocyte migration (Prevo et al. 2001; Cichy and Pure 2003; Ponta et al. 2003; Lesley et al. 2004), and Ci-link1 is expressed in some juvenile blood cells (Kawashima et al. 2009; Yoneda et al. 2010). These observations suggest that the Xlink domain is a component of the surface molecules of blood cells in protochordate ancestors. The Xlink domain then combined with another component, such as the V-set domain in a vertebrate ancestor, and was recruited as a component of cartilage, a novel vertebrate structure (Kawashima et al. 2009; Yoneda et al. 2010).

This scenario introduced other aspects of the evolution of the Xlink domain. The Xlink domain functions with the aggrecan molecule as the interface with hyaluronan (HA) (Fig. 2.1b). The complex consisting of HA, link protein/HAPLN1, and aggrecan provides tensile strength for cartilage to absorb shock and resist compression in the joint. Mutations in aggrecan or the link protein/HAPLN1 cause severe defects in cartilage and result in dwarfism (Watanabe et al. 1994; Watanabe and Yamada 1999). Although Xlink is present in various metazoans, including nematodes, no HA had been identified in invertebrates (Yoneda et al. 2010). Indeed, hyaluronan synthase (HAS) was believed to have been acquired in vertebrate ancestors via horizontal gene transfer (Salzberg et al. 2001). Our recent studies of HAS have revealed that amphioxus also possesses HAS genes, and thus HAS was perhaps acquired in the common ancestors of chordates (Yoneda et al. 2010). On the other hand, ascidians do not possess HA, and the ascidian Xlink domain binds heparin (Yoneda et al. 2010).

The above observations reveal rather dramatic evolutionary events during the establishment of the molecular complex of Aggrecan and HA for important cartilage components (Yoneda et al. 2010). Chordate ancestors acquired HAS possibly by horizontal gene transfer, but their Xlink domain was not utilized for HA binding, instead functioning as a heparan-binding domain. In the common ancestors of vertebrates, Xlink acquired binding specificity against HA. Subsequently, through domain shuffling, genes encoding lecticans were established de novo. After gene duplications in the common ancestors of vertebrates (perhaps two rounds of genome duplications), one of the lectican genes is utilized as a cartilage matrix component.

Another novel molecule, cartilage acidic protein 1 precursor, was recruited as cartilage matrix (Kawashima et al. 2009). This molecule was established via minor

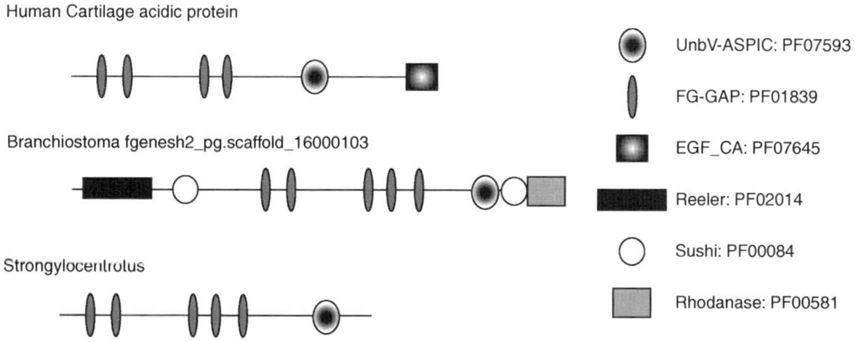

Fig. 2.2 Schematic illustration of the domain architecture of the cartilage acidic protein. Amphioxus (*Branchiostoma*) and sea urchin (*Strongylocentrotus*) possess genes with similar domain architecture, but they lack EGF_CA domain in their C-terminus

modifications of the domain architecture. Amphioxus and sea urchins possess genes of similar domain architectures with FG-GAP domain and UnbV-ASPIC repeats but lack EGF_CA (Fig. 2.2). The combination of the UnbV_ASPIC and FG-GAP domains may have occurred in common ancestors of bilaterians, and subsequently this gene acquired the EGF_CA domain in ancestral vertebrates, which allowed it to function as the cartilage acidic protein.

2.3 Domain Shuffling and Evolution of the Tight Junction

The tight junction exists only in vertebrates and ascidians, and the major component of the tight junction, claudin, is found in these species (Kollmar et al. 2001). We previously demonstrated that the other tight junction component, occludin (Tsukita et al. 2001), is found only in vertebrates (Kawashima et al. 2009). The combination of the MARVEL domain and occluding ELL domains is specific to vertebrates (Fig. 2.3a). Amphioxus and ascidian possess the occludin_ELL domain in RNA polymerase II elongation factor ELL (Fig. 2.3b). Thus, it is likely that, after genome duplication in the vertebrate lineage, one of the paralogs was combined with the MARVEL domain and the occludin protein was created. Seventeen amphioxus genes and four ascidian genes contain the MARVEL domain, and all of the genes from amphioxus and ascidians contain only the MARVEL domain (Fig. 2.3a, b). The structures of protochordate genes containing the MARVEL domain are more similar to MAL and physins involved in vesicle trafficking, in which the MARVEL domain is thought to be involved in cholesterol-rich membrane apposition events (Sánchez-Pulido et al. 2002) (Fig. 2.3a). The protochordate genes containing the MARVEL domain might perform roles similar to MAL or physins. Subsequently, in an ancestral vertebrate, a gene acquired the occludin_ELL domain and was committed to a novel function as a component of the tight junction.

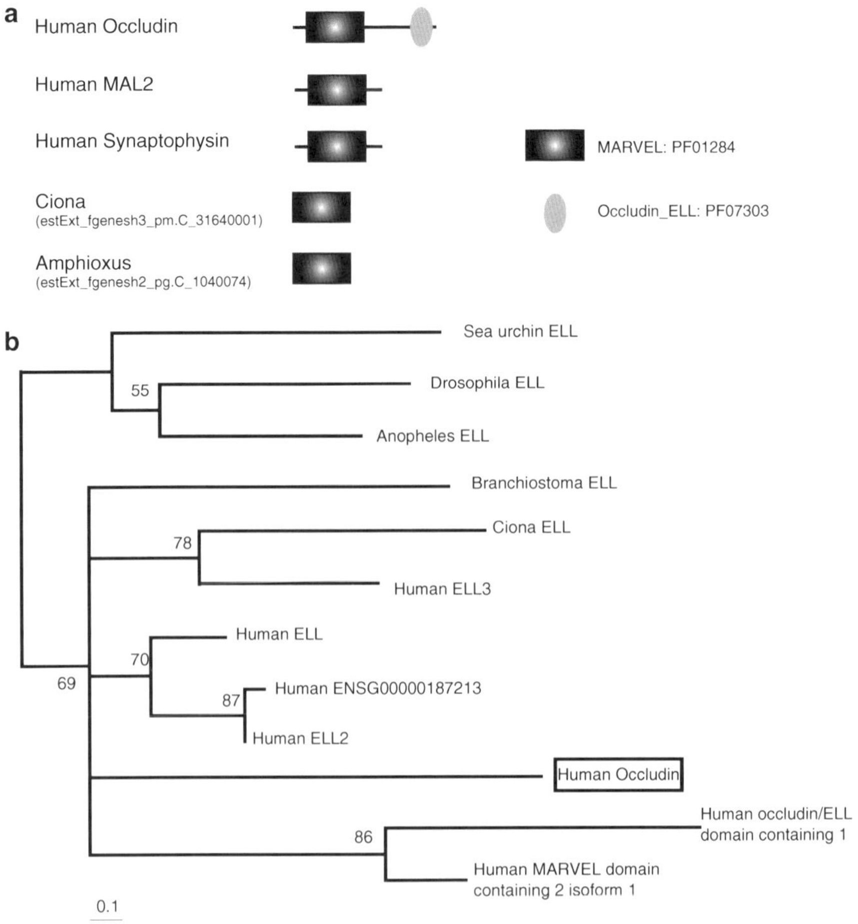

Fig. 2.3 Domain architecture of occludin and evolutionary origin of ELL domain. (**a**) Comparison of domain architecture of occludin to some proteins with MARVEL domain. (**b**) Maximum likelihood phylogenetic tree of occludin_ELL domains. Numbers on the nodes show bootstrap values

Occludin is not an essential component of the tight junction because occludin-deficient knockout mice lack any detectable barrier defects (Saitou et al. 2000; Schulzke et al. 2005). Rather, occludin is thought to be involved in a regulatory aspect of the tight junction such as mediating extracellular signals for cell adhesion (Saitou et al. 2000). Because the *Ciona* genome lacks occludin, it is likely that the tight junction did not contain occludin when it originated in the common ancestor of ascidians and vertebrates. Upon acquiring occludin as a novel component of the tight junction, vertebrates may have used the junction in a more flexible manner.

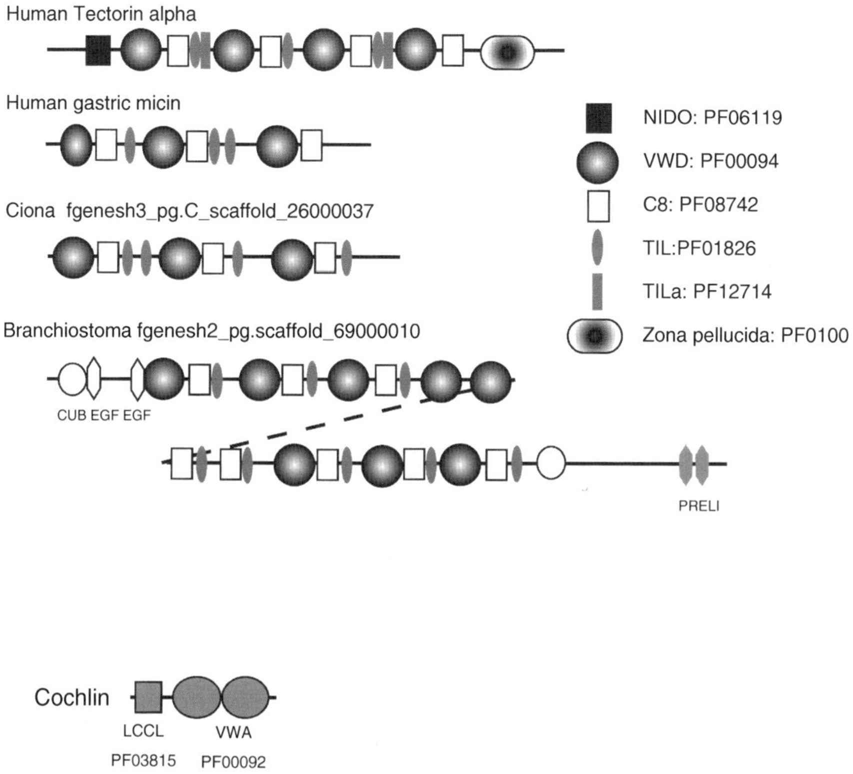

Fig. 2.4 Domain architecture of tectorin-alpha and cochlin. Ascidian (*Ciona*) and amphioxus (*Branchiostoma*) also possess proteins with similar domain architecture of VWD-C8-TIL repeats

2.4 Domain Shuffling and Evolution of the Vertebrate Auditory System

Two new genes established by domain shuffling contributed to the evolution of the vertebrate auditory system (Kawashima et al. 2009). Tectorin-alpha is one of the major components of the tectorial membrane in the mammalian inner ear. In humans, a missense mutation in tectorin-alpha causes hearing loss of about 50–80 dB (Verhoeve et al. 1998). Tectorin-alpha consists of three or four repeats of vWD-C8-TIL domains with N-terminal NIDO domain and C-terminal zona_pellucida-like domain (Fig. 2.4). Domain combinations between the zona pellucida-like and NIDO domains and between the NIDO and TIL are only observed in vertebrate genomes. The repeats of vWD-C8-TIL domains were found in mucins, and protochordates possessed genes with similar domain architectures (Fig. 2.4). Therefore, ancestral molecules may have functioned as mucin, and the novel molecule recruited as a component of the tectorial membrane fused with NIDO and zona pellucida domains.

Cochlin is also an essential component of the vertebrate auditory system, and a mutation in *cochlin* causes deafness (Robertson et al. 2006). Cochlin shows domain architecture of the LCCL with the repeat of VWA domain, whose combinations are found only in vertebrates (Fig. 2.4) (Kawashima et al. 2009). Thus, this molecule was apparently established de novo. These observations clearly indicate that domain shuffling played critical roles in the evolution of vertebrate auditory systems.

2.5 Other ECM Molecules Established in Vertebrates

Fibronectin and vitronectin also show domain architectures unique to vertebrates (Kawashima et al. 2009). Fibronectins consist of three types of domain repeats: FN1, FN2, and FN3. FN3 repeats were found in the tyrosine phosphatase gene of various metazoans including *Drosophila* or *Hirudo*. FN1 and FN2 repeats were also found in respective amphioxus genes (Fig. 2.5). Therefore, fusions of these repeats in the vertebrate ancestors established fibronectin as a novel ECM molecule.

Vitronectin consists of Hemopexin domain repeats with an N-terminal somatomedin domain (Fig. 2.6). Hemopexin domain repeats were found in several metalloproteinases, including MMP9 (Fig. 2.6). Vitronectin may have originated from one of these metalloproteinases genes by fusing with the somatomedin domain.

2.6 Future Prospects

We previously performed a comprehensive survey of molecules established by domain shuffling in the common ancestors of vertebrates (Kawashima et al. 2009). However, this list may remain incomplete. As more genome sequences are determined, we will identify other ECM molecules or domain pairs thought to be exclusive to vertebrates in invertebrate species. Indeed, our further analyses revealed two ECM molecules (fibulin and nidogen) thought to be exclusive to vertebrates in other invertebrates. There are an unexpected number of genes found in vertebrates and cnidarians, but not in genomes of *Drosophila* or *C. elegans*, indicating that multiple gene loss occurs frequently (Putnam et al. 2007). Also, incomplete gene model prediction may hinder the identification of true domain pairs. Prediction of gene models remains one of the most technically difficult issues in genome analyses. Although several algorithms for gene model prediction are available, gene model prediction is dependent on transcriptome data. Thus, the list of domain pairs in single-gene models is not complete.

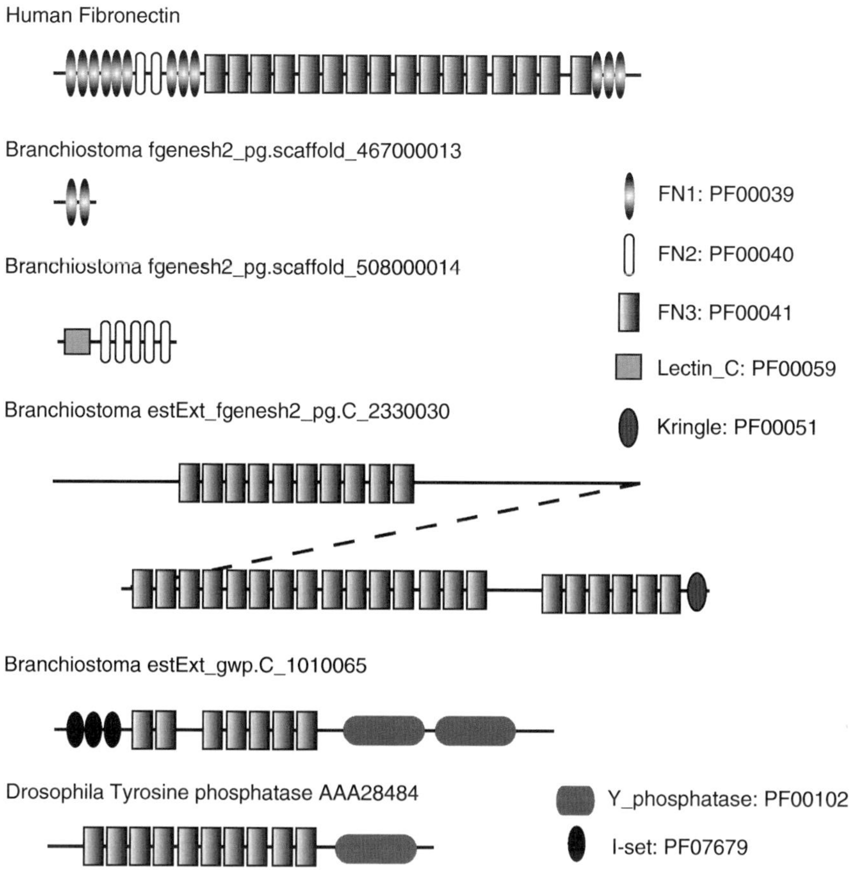

Fig. 2.5 Domain architecture of fibronectin. Amphioxus (*Branchiostoma*) possess FN1 repeat, FN2 repeat, and FN3 repeat in different proteins

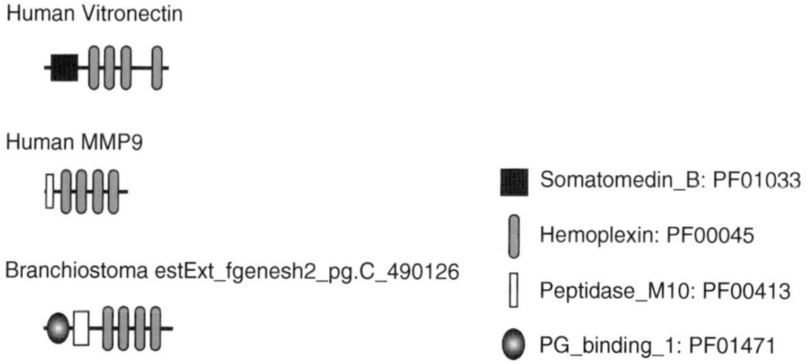

Fig. 2.6 Domain architecture of vitronectin. The repeat of hemopexin was found in MMP9, and similar domain architecture is found in some amphioxus genes

While keeping these incompleteness in mind, we found that domain shuffling contributed to the evolution of molecules, not only via minor modification of molecules by fusing a domain to existing proteins, but also by producing a novel molecule de novo, such as aggrecan and fibronectin. In the case of occludin, a domain of completely different function (ELL: elongation of RNA synthesis) was recruited for use in cell adhesion molecules. This may be a spectacular example of molecular tinkering.

It is also important to understand how novel protein domains are established during evolution. Unique domains continue to emerge, and there are 278 domains identified to be unique to vertebrates (Ekman et al. 2007). It was widely believed that most of these new domains were the result of replication divergence. However, recent reviews by Schmidt and Davies (2007) and Tautz and Domazet-Loso (2011) make the point that novel sequence may have emerged from noncoding genomic sequences. Further analyses of the mechanistic aspects of gene evolution may reveal dynamic aspects of molecular evolution.

References

Cichy J, Pure E (2003) The liberation of CD44. J Cell Biol 161:839–843

Ekman D, Bjorkund A, Elofsson A (2007) Quantification of the elevated rate of domain rearrangements in Metazoa. J Mol Biol 372:1337–1348

Kawashima T, Kawashima S, Tanaka C, Murai M, Yoneda M, Putnam NH, Rokhsar DS, Kanehisa M, Satoh N, Wada H (2009) Domain shuffling and the evolution of vertebrates. Genome Res 19:1393–1403

Kollmar R, Nakamura SK, Kappler JA, Hudespeth AJ (2001) Expression and phylogeny of claudins in vertebrate primordia. Proc Natl Acad Sci USA 98:10196–10201

Lesley J, Gal I, Mahoney DJ, Cordell MR, Rugg MS, Hyman R, Day AJ, Mikecz K (2004) TSG-6 modulates the interaction between hyaluronan and cell surface CD44. J Biol Chem 279:25745–25754

Liu M, Grigoriev A (2004) Protein domains correlate strongly with exons in multiple eukaryotic genomes—evidence of exon shuffling? Trends Genet 20:399403

McCauley DW, Bronner-Fraser M (2006) Importance of SoxE in neural crest development and the evolution of the pharynx. Nature 441:750–752

Ohtani K, Yao T, Kobayashi M, Kusakabe R, Kuratani S, Wada H (2008) Expression of Sox and fibrillar collagen genes in lamprey larval chondrogenesis with implications for the evolution of vertebrate cartilage. J Exp Zool 310B:596–607

Patthy L (1999) Genome evolution and the evolution of exon-shuffling—a review. Gene 238:103–114

Patthy L (2003) Modular assembly of genes and the evolution of new functions. Genetica 118:217–231

Ponta H, Sherman L, Herrlich PA (2003) CD44: from adhesion molecules to signalling regulators. Nat Rev Mol Cell Biol 4:33–45

Prevo R, Banerji S, Ferguson DJP, Clasper S, Jackson DG (2001) Mouse LYVE-1 is an endocytic receptor for hyaluronan in lymphatic endothelium. J Biol Chem 276:19420–19430

Putnam NH, Srivastava M, Hellsten U, Dirks B, Chapman J, Salamov A, Terry A, Shapiro H, Lindquist E, Kapitonov VV, Jurka J, Genikhovich G, Grigoriev IV, Lucas SM, Steele RE, Finnerty JR, Technau U, Martindale MQ, Rokhsar DS (2007) Sea anemone genome reveals ancestral eumetazoan gene repertoire and genomic organization. Science 317:86–94

Robertson N, Hamaker S, Patriub V, Aster J, Morton C (2006) Subcellular localisation, secretion, and post-translational processing of normal cochlin, and of mutants causing the sensorineural deafness and vestibular disorder, DFNA9. J Med Genet 40:479–486
Robson P, Wright G, Sitarz E, Maiti A, Rawat M, Youson J, Keeley F (1993) Characterization of lamprin, an unusual matrix protein from lamprey cartilage. Implications for evolution, structure, and assembly of elastin and other fibrillar proteins. J Biol Chem 268:1440–1447
Robson P, Wright GM, Youson JH, Keely FW (1997) A family of non-collagen-based cartilages in the skeleton of the sea lamprey, *Petromyzon marinus*. Comp Biochem Physiol 118B:71–78
Saitou M, Furuse M, Sasaki H, Schulzke J-D, Fromm M, Takano H, Noda T, Tsukita S (2000) Complex phenotype of mice lacking occludin, a component of tight junction strands. Mol Biol Cell 11:4131–4142
Salzberg SL, White O, Peterson J, Eisen JA (2001) Microbial genes in the human genome: lateral transfer or gene loss. Science 292:1903–1906
Sánchez-Pulido L, Martín-Belmonte F, Valencia A, Alonso MA (2002) MARVEL: a conserved domain involved in membrane apposition events. Trends Biochem Sci 27:599–601
Schmidt EE, Davies CJ (2007) The origin of polypeptide domains. Bioessays 29:262–270
Schulzke JD, Gitter AH, Mankertz J, Spiegel S, Seidler U, Amasheh S, Saitou M, Tsukita S, Fromm M (2005) Epithelial transport and barrier function in occludin-deficient mice. Biochim Biophys Acta 1669:34–42
Tautz D, Domazet-Loso T (2011) The evolutionary origin of orphan genes. Nat Rev Genet 12 (10):692–702
Tsukita S, Furuse M, Itoh M (2001) Multifunctional strands in tight junctions. Nat Rev Mol Cell Biol 2:285–293
Verhoeve K, Van Laer L, Kirschhofer K, Legan P, Hughes D, Schatteman I, Verstreken M, Van Hauwe P, Coucke P, Chen A, Smith R, Somers T, Offeciers F, Van de Heyning P, Richardson G, Wachtler F, Kimberling W, Willems P, Govaerts P, Van Camp G (1998) Mutations in the human alpha-tectorin gene cause autosomal dominant non-syndromic hearing impairment. Nat Genet 19:60–62
Wada H, Okuyama M, Satoh N, Zhang S (2006) Molecular evolution of fibrillar collagen in chordates, with implications for the evolution of vertebrate skeletons and chordate phylogeny. Evol Dev 8:370–377
Watanabe H, Yamada Y (1999) Mice lacking link protein develop dwarfism and craniofacial abnormalities. Nat Genet 21:225–229
Watanabe H, Kimata K, Line S, Strong D, Gao L, Kozak C, Yamada Y (1994) Mouse cartilage matrix deficiency (cmd) caused by a 7 bp deletion in the aggrecan gene. Nat Genet 7:154–157
Wright GM, Youson JH (1983) Ultrastructure of cartilage from young adult sea lamprey, Petromyzon marinus L: a new type of vertebrate cartilage. Am J Anat 167:59–70
Yoneda M, Nakamura T, Murai M, Wada H (2010) Evidence for the heparin-binding ability of the ascidian Xlink domain and insight into the evolution of the Xlink domain in chordates. J Mol Evol 71:51–59
Zhang G, Miyamoto MM, Cohn MJ (2006) Lamprey type II collagen and *Sox9* reveal an ancient origin of the vertebrate collagenous skeleton. Proc Natl Acad Sci USA 103:3180–3185

Chapter 3
Invertebrate and Vertebrate Collagens

Jean-Yves Exposito and Claire Lethias

Abstract Collagens form a family of extracellular matrix proteins that is often associated with metazoan multicellularity and evolution. These modular proteins elaborate supramolecular networks that contribute to the physical and biological properties of tissues. Among the different collagen families characterized in vertebrates, three of them represent ubiquitous components of extracellular matrices (fibrillar collagens), and basement membranes (type IV collagens and multiplexins). Interestingly, these three collagen families are already present in basal animals, sponges, and cnidarians. Other metazoan collagen families have a phylum-specific origin or have emerged during chordate evolution. Collagen-like proteins that are not involved in animal extracellular matrix structure have been discovered in vertebrates and bacterial pathogens. Among these, vertebrate defense collagens of the innate immune system provide protection against pathogens while bacterial collagens seem to be involved in pathogen virulence.

3.1 Introduction

Metazoa are one of the eukaryotic kingdoms, and their evolution is often associated with collagen-based extracellular matrices (Heino et al. 2009). Collagens represent a family of extracellular matrix proteins. They contain a triple helical structure and participate in the formation of supramolecular aggregates. Collagens form homotrimeric (three identical α chains) or heterotrimeric molecules

J.-Y. Exposito (✉)
Institut de Biologie et Chimie des Protéines, UMR 5305, CNRS Université Lyon 1,
SFR BioSciences Gerland-Lyon Sud, 7 Passage du Vercors, 69367, Lyon, France
e-mail: jy.exposito@ibcp.fr

C. Lethias
Institut de Biologie et Chimie des Protéines, UMR 5305, Université Lyon 1, SFR BioSciences
Gerland-Lyon Sud, 7 Passage du Vercors, 69367, Lyon, France
e-mail: c.lethias@ibcp.fr

F.W. Keeley and R.P. Mecham (eds.), *Evolution of Extracellular Matrix*,
Biology of Extracellular Matrix, DOI 10.1007/978-3-642-36002-2_3,

(two or three distinct polypeptides chains). Each α chain has at least one collagenous domain characterized by a repetition of Gly–Xaa–Yaa triplets, where Xaa and Yaa are often proline and hydroxyproline residues, respectively. The collagenous segments of the three α chains participate in the formation of the characteristic right-handed super helix. Collagens form supramolecular networks that often are intimately linked to the structural integrity of tissues, organs, and ultimately animals. For example, these structural functions include corneal transparency, tensile strength of the dermis, resilience of tendons, or cohesion of the sponge skeleton. Although collagens are well known for their structural functions, they are also involved in the regulation of cell behavior (migration, adhesion, differentiation, and spreading) during development, tissue formation, and healing. These biological functions involve collagen receptors such as integrins and discoidin domain receptors (Heino et al. 2009). In vertebrates, 28 types of collagens have been described (types I–XXVIII; Gordon and Hahn 2010; Ricard-Blum 2011). They have been subdivided into several families according to their primary structure and their supramolecular organization. Among these families, the fibrillar or fibril-forming collagens, the basement membrane type IV collagen and the multiplexins (for multiple triple-helix domains and interruptions) have been characterized from the first multicellular animals (sponges and cnidarians) to human, and will constitute the major topics of this review. The other vertebrate collagen and collagen-like families will be briefly presented with regard to invertebrate data, as some "versatile" collagens (Engel 1997) that seem to be specific for an invertebrate phylum. Although considered as a metazoan hallmark, collagen-like proteins have been described in several bacterial species (Rasmussen et al. 2003). Their potential role at the surface of pathogenic bacteria will also be discussed.

3.2 Fibrillar Collagen

3.2.1 Modular Structure of Mammalian Fibrillar Collagen α Chains

Striated fibril-forming (or fibrillar) collagens are the most abundant extracellular matrix glycoproteins. They were first described from the glue-like properties of gelatin (boiled collagens), and have been used in food and other industries (cosmetic, pharmaceutical, photography, leather, etc.). Fibrillar procollagen chains are the primary elements of striated fibrils and are built with the same architecture, i.e., a major central triple helix or collagenous domain (approximately 338 Gly–Xaa–Yaa triplets) flanked by N- and C-terminal non-collagenous domains, the N- and the C-propeptides (Gordon and Hahn 2010). These chains are involved in the formation of homo- or heterotrimeric procollagen molecules. The major steps leading to the formation of a striated fibril from three individual α chains are illustrated in Fig. 3.1 (Kadler et al. 1996; Exposito et al. 2010). A single cell

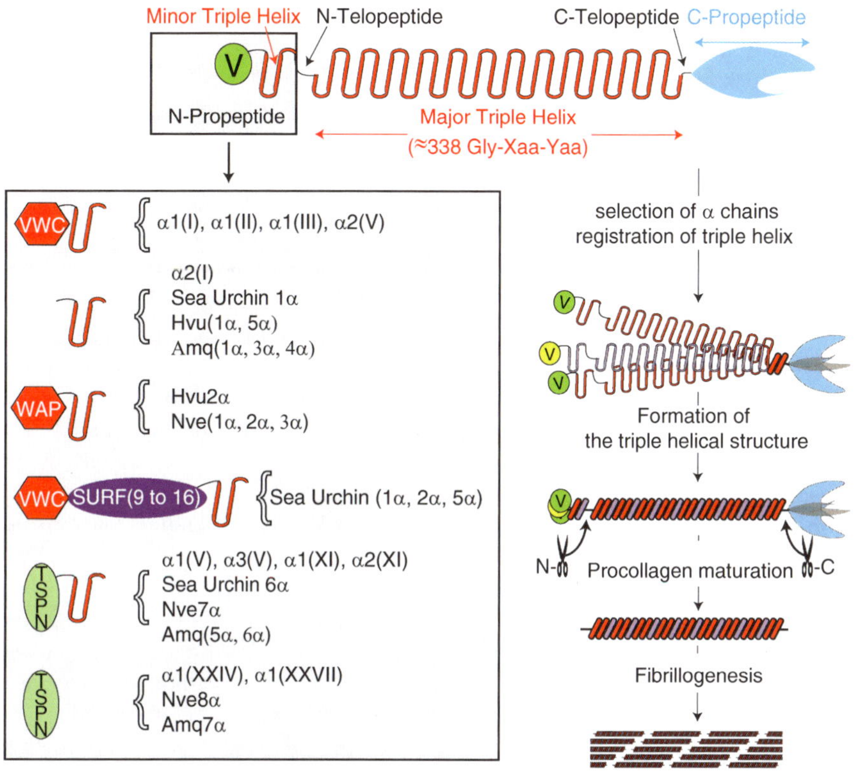

Fig. 3.1 Fibrillar collagens. The different domains of a prototype fibrillar procollagen chain are depicted at the top. The major steps leading to formation of a striated fibril are illustrated on the right, and the different N-propeptide modular structures on the left. Note: hydra N-propeptide harboring copies of VWA and WAP and described by Zhang et al. (2007) are not drawn. SURF, sea urchin fibrillar module; TSPN, thrombospondin N-terminal-like domain; V, variable region; WAP, whey acidic protein; VWA, von Willebrand factor A domain; VWC, von Willebrand factor-type C

being able to synthesize different fibrillar collagen molecules, the first step is the selection of three specific α chains via their chain recognition sequence located in the C-propeptide domain (Lees et al. 1997). Selection and association of the C-propeptides are the main mechanisms contributing to the alignment and the registration of the major triple helical sequences. After nucleation, the super helix elongates in a zipper-like fashion towards the N-terminus. These procollagen molecules are generally processed by specific proteases, and the resultant collagen molecules are incorporated into the well-known striated fibrils.

In humans, seven types (I–III, V, XI, XXIV, and XXVII) of fibrillar collagens have been described (Boot-Handford et al. 2003; Koch et al. 2003; Pace et al. 2003). Types II and III, and probably types XXIV and XXVII form homotrimeric molecules. Type I is generally a heterotrimer composed of two proα(1) chains and one proα2(I) chain, but homotrimers of the proα1(I) chain have been detected.

Types V and XI collagens have multiple isoforms, and heterotypic molecules made of types V and XI chains have been described (Kleman et al. 1992; Mayne et al., 1993; Wu et al. 2009). Types V and XI are now described as a single collagen type (Hoffman et al. 2010). Although human α chains display the general modular structures presented in Fig. 3.1, they can be divided into different groups according to features in their N-propeptides, their major triple helices, and their C-propeptides. Hence, with the exception of proα1(XXIV) and proα1(XXVII), the other fibrillar procollagen chains have an uninterrupted major triple helix and a short triple helix in their N-propeptide that are separated by a non-collagenous segment, the N-telopeptide. For types XXIV and XXVII collagens, the major triple helix is shorter (997 residues), and contains two glycine substitutions in the N-terminal region, and a Gly–Xaa–Yaa–Zaa imperfection localized in the C-terminal region. A similar imperfection has been previously determined in the C-terminal region of the major triple helix of a sponge fibrillar α chain (Exposito and Garrone 1990). Most of the chain recognition sequence is absent in the C-propeptide domain of types XXIV and XXVII collagens, and in invertebrate fibrillar chains (Boot-Handford and Tuckwell 2003). The last feature concerns the non-collagenous part of the N-propeptide domain. A cysteine-rich module, VWC, is present for the proα1(I), the proα1(II), the proα1(III), and proα2(V) chains, while a TSPN module flanked at its C-terminus by a variable region is observed in the proα1(V), proα(V), proα1(XI), proα2(XI), proα(XXIV), and proα(XXVII) chains. For the last fibrillar chain, proα(I), the N-propeptide region is limited to the short triple helix. As shown in Fig. 3.1, some invertebrate α chains present some particularities in their N-propeptides. Several hydra α chains have a WAP module or a series of WAP and VWA modules (not illustrated in Fig. 3.1) located in their N-propeptide (Zhang et al. 2007). In sea urchin, three fibrillar α chains harbor a series of SURF modules between VWC and the minor triple helix of the N-propeptide (Exposito et al. 1992a, 2010; Cluzel et al. 2004). The major form of the sea urchin 1α chain presents a modular architecture similar to that of the vertebrate proα2(I) chain (Exposito et al. 1992b).

3.2.2 Ancestral Gene Structure

As first described by electron microscopy and R-loop analysis (hybridization between genomic and cDNA; Ohkubo et al. 1980), fibrillar collagen genes contain more than 50 exons and in humans have a length comprised between 17 kb to more than 230 kb. From the partial characterization of the genomic sequence encoding the major triple helix of the chicken α2(I) collagen chain, Yamada et al. (1980) have revealed that in this region exons begin with an intact glycine codon, end by an intact Yaa codon, and have a length of 54 bp or a multiple of 54−9 bp (99 bp). From this observation, they suggested that the primordial unit of a fibrillar collagen gene is an exon of 54 bp flanked by these intronic sequences. In their evolutionary scheme leading to the formation of an ancestral fibrillar collagen gene, the first steps were successive rounds of duplication of the 54 bp primordial unit by recombination within introns. The formation of 54−9 bp-related exons arose by unequal crossing-over. Subsequently, the sequencing of the genes encoding the types I–III fibrillar collagen chains (Chu et al. 1984; Upholt et al. 1985) revealed that in the region encoding the major triple helix, half of the exons are 54 bp long whereas the others are multiples of 54 bp (108 and 162) or multiples of 54 bp minus 9 bp (45 and 99). Invertebrate genomic data have confirmed the pattern of introns and exons in the genomic region encoding the major triple helix of their fibrillar α chains. Hence, the study of a freshwater demosponge α chain (Emf1α from *Ephydatia mülleri*) reveals a compact gene with small intronic sequences, most of them having a length comprised between 50 and 100 bp, and exons with a length related to 54 or 54–9 bp (Exposito and Garrone 1990; Exposito et al. 1993). After alignment with the available mammalian type I-III fibrillar collagen genes, and by taking into account all the exon–intron junctions, we have been able to suggest the exon/intron distribution of an ancestral gene, or more precisely the length pattern of the exons encoding its major triple helix. From this structure, it has been possible to suggest a new model of evolution (Fig. 3.2). As in the Yamada's model, the first step was the multiple rounds of duplication of the 54 bp primordial genetic sequence. In a second step, an unequal crossing-over leads to the formation of a 45 bp exon, which in our model is included in a new genetic unit with a 54 bp exon. The multiple rounds of duplication of this new genetic unit might explain the distribution of stretches of 54 bp and (54 and 45 bp) exons in fibrillar collagen genes. The "perfect" relationship of exon organization between the metazoan fibrillar collagen genes has been confirmed (Takahara et al. 1995; Vuoristo et al. 1995; Exposito et al. 2000, 2008; Aouacheria et al. 2004), and, as detailed below, has been a useful tool to investigate the evolution of this family of genes. In the model presented in Fig. 3.2, we have chosen to arbitrarily indicate that the primordial 54 bp exon might have encoded 6 Gly–Pro–Pro triplets, this suggestion being in agreement to Benveniste-Schrode et al. (1985), who postulated that the primordial ancestor of the type I fibrillar collagen gene is a 9 bp sequence encoding the tripeptide Gly–Pro–Pro.

3.2.3 Evolution of the Fibrillar Collagen Family

Until recently, before the availability of numerous metazoan genomes, the evolution of fibrillar collagen chains has been more focused on mammalian data. From genomic data (Takahara et al. 1995; Vuoristo et al. 1995), and phylogenic analysis (Sicot et al. 1997), the fibrillar collagen chains have been divided into two clades. The A clade includes types I–III and the pro-α2(V) chains, while the pro-α1(V), pro-α3(V), and type XI chains belong to the B clade. This subdivision is in good agreement with the N-propeptide configuration of these chains presented in Fig. 3.1, with the exception of the pro-α2(I) chain for which the N-propeptide is reduced to the short triple helical region. More recently, two new members of the

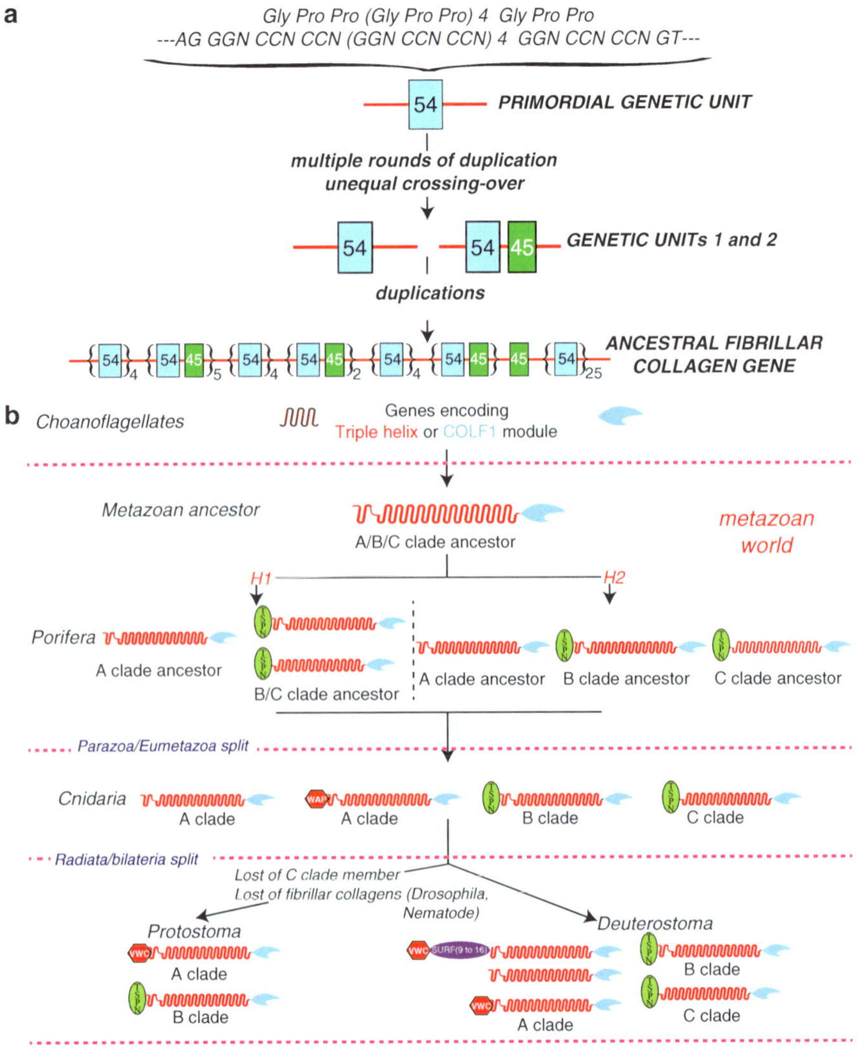

Fig. 3.2 Fibrillar collagen evolution. (**a**) Steps leading to the formation of a putative ancestral fibrillar collagen gene from two primordial genetic units. Open boxes represent exons, and Arabic numbers indicate lengths of exons in bp. (**b**) Early diversification of the fibrillar collagen family. Genes encoding collagenous or COLF1 domains are present in choanoflagellates, the sister group to Metazoa (King et al. 2008). The formation of an ancestral fibrillar collagen gene arose before the metazoan radiation. Furthermore, the divergence of A and B/C clade progenitors occurred at the very dawn of the metazoan world before the poriferan radiation. The formation of B and C clade fibrillar collagen genes from a B/C clade gene (hypotheses 1 or 2) predated the Parazoa/Eumetazoa split. In this illustration, the B/C clade ancestral chain contains a TSPN module in its N-propeptide. We have previously suggested a common origin of WAP and VWC modules (Exposito et al. 2008). In this scenario, the inclusion of a VWC module in the N-propeptide of an A clade chain predated the Cnidaria/Bilateria split. In this model, the fibrillar collagen gene diversification that predates the vertebrate radiation (see text), and the steps leading to acquisition of a complete

fibrillar collagen family have been characterized, and phylogenic analysis has permitted to place these two new chains (α1(XXIV) and α1(XXVII)) in a new fibrillar collagen clade C (Boot-Handford and Tuckwell 2003; Boot-Handford et al. 2003). Moreover, this study has shown that C clade members are closely related to B clade members, these two clades arising from the duplication of an ancestral B/C clade gene. In these studies, the available invertebrate chains used in phylogenetic analyses seem to be unrelated to any of the vertebrate clades. The fact that some of invertebrate fibrillar α chains have an N-propeptide including a VWC module and that none of them contain a TSPN module have suggested that the three clades retrieved in vertebrates result from duplication events from an ancestral A/B/C fibrillar collagen gene at the very dawn of the vertebrate world, this primordial unit encoding an α chain possessing a VWC module (Boot-Handford et al. 2003). The sequence data available at the time of writing are in good agreement with this proposed scheme of evolution of the collagen family. However, gene structure (Exposito and Garrone, 1990), biochemical (Miura and Kimura 1985; Tillet et al. 1996) and ultrastructural studies (Garrone et al., 1993) have previously led several authors to suggest the presence of type V/XI collagens in invertebrates. Fishing of fibrillar collagen genes in *Apis mellifera* and *Anopheles gambiae* (honeybee and mosquito, protostomes) and *Strongylocentrotus purpuratus* (sea urchin, a deuterostome) reveal that these invertebrates can encode fibrillar collagen chains presenting a modular structure related to the A and B clades (insects) and the three clades in sea urchin (Aouacheria et al. 2004). However, phylogenetic analysis using the C-propeptide sequences fails to relate any invertebrate α chains to a vertebrate clade. The C-propeptide has been considered to be the most conserved region of the fibrillar collagen chains. However, while this domain harbors highly conserved segments, several regions of the C-propeptide are highly variable in length and sequence (Dion and Myers 1987). As indicated in the previous paragraph, the exon–intron organization in the region encoding the major triple helix is conserved from sponges to human, and recent studies have demonstrated the powerful use of exon–intron junctions in evolutionary studies (Roy and Gilbert 2005; Yandell et al. 2006). It has been shown that gene structure might be an efficient tool to confirm or to correct multiple alignments (Irimia and Roy 2008). In agreement with this statement, we have been able to use the major triple helical sequences to decipher the evolution of this family of proteins after alignment of all the metazoan fibrillar collagen genes. A first study has made it possible to relate invertebrate sequences to the three fibrillar collagen clades defined in vertebrates (Aouacheria et al. 2004). Two more recent studies (Zhang et al. 2007; Exposito et al., 2008) have resulted in a better understanding of the first steps in fibrillar

Fig. 3.2 (continued) COLF1 recognition sequence in A and B clade vertebrate progenitors are not represented. Boot-Handford and Tuckwell (2003) have suggested that the acquisition of such a sequence occurred at the very dawn of the vertebrate world. A more recent study has suggested that the convergent acquisition of a long-chain selection sequence in an A or B clade gene occurred before or after the ascidian/vertebrate split, respectively (Wada et al. 2006)

collagen evolution (Fig. 3.2). In this model, the formation of an ancestral fibrillar collagen gene predated the Choanoflagellata/Metazoa split. The formation of A clade and B/C clade progenitors preceded metazoan cladogenesis, while the divergence of B and C clade members from a B/C clade ancestor arose either before or after the Parazoa/Eumetazoa split. As previously indicated (Zhang et al. 2007), the addition of a VWC module in the N-propeptide domain of an A clade chain occurred before the bilaterian radiation. The WASP module is either specific to cnidarian A clade chains or is evolutionarily related to the VWC module (Exposito et al. 2008). From this model, C clade collagens might have been lost in protostomes while fibrillar collagen genes are absent in drosophila and nematodes (Hutter et al. 2000).

3.2.4 Fibrillar Collagen Gene Expansion in Vertebrates

During the study of Hox cluster evolution, Bailey et al. (1997) have shown that A clade members are associated with Hox clusters. The expansion of the fibrillar collagen genes in vertebrates has to be related to the two rounds of whole genome duplication or 2R hypothesis (Ohno 1970). Hox cluster/A clade synteny has also been shown in sea urchin (Cameron et al. 2006). From this study, and also available genomic data (http://genome.ucsc.edu/cgi-bin/hgGateway; scaffold69286), the three A clade fibrillar collagen genes from *Strongylocentrotus purpuratus* are arranged in tandem at the vicinity of the Hox cluster. A comparable situation has been shown for B clade collagen genes that are located in synteny to *C* (complement), *NOTCH*, and *TN* (tenascin) genes (Katsanis et al. 1996; Olinski et al. 2006; Exposito et al. 2010). In agreement with their close evolutionary relationships with B clade genes, the two C clade genes are located in these clusters of genes or paralogons (Popovici et al. 2001). In humans, *COL5A1* and *COL27A1* are located in synteny to *NOTCH1*, *C5*, and *TNC* (9q32 to 9q34.3), while similar gene families were observed in chromosome 1 (*COL11A1*, *COL24A1*, *NOTCH2*, *C3*, *TNR*, *TNN*), chromosome 6 (*COL11A2*, *NOTCH4*, *C4A*, *C4B*, *TNXA*, *TNXB*), and chromosome 19 (*COL5A3*, *NOTCH3*, *C3*). It has been shown that the two rounds of duplication occurred at the very dawn of vertebrates, before the divergence of Agnatha and Gnathostoma (Kuraku 2008; Kuraku et al. 2009). True orthologs of *COL2A1* have been characterized in both vertebrate classes Agnatha and Gnathostoma. *COL2A1* encodes the proα1(II) chain, a cartilage-specific collagen in Gnathostoma (jawed vertebrates). Several studies have shown that cartilage in agnathans (hagfish and lamprey), previously described as a non-collagenous tissue, possesses type II collagen, suggesting that all extant vertebrates have a type II collagen-based cartilage (Ohtani et al. 2008; Wright et al. 2001; Zhang et al. 2006; Zhang and Cohn 2006). Cartilage-like tissues that are rich in A clade fibrillar collagens have also been described in invertebrates (hemichordates and cephalochordates) (Rychel et al. 2006; Rychel and Swalla 2007; Wada et al. 2006; Zhang and Cohn 2006). These studies have made it possible to describe pre-type II evolution, with the formation of collagenous cartilage-like tissues near the time of deuterostome diversification (Rychel and Swalla 2007), notochord expression in lancelet (cephalochordate) of the pre-2R A clade gene (Zhang and Cohn 2006) suggesting that vertebrate type II collagen expressing chondrocytes may have evolved from notochordal cells.

3.2.5 Molecules and Fibrils in Invertebrates

The melting temperatures (T_m) of invertebrate fibrillar collagen molecules are often lower than those estimated for human collagens, and have been related to the environmental temperature of animals (Rigby and Robinson 1975). At the primary structure level, glycine substitutions, Gly–Gly–Yaa and Gly–Xaa–Gly triplets, and low proline content have been described as factors lowering the T_m of some invertebrate and lower vertebrate collagens (Exposito et al. 1992b; Saito et al. 2001; Su et al. 1991; Yoneda et al. 1999). Triple helical peptide studies or use of recombinant collagens have made it possible to describe the destabilizing effects of Gly–Gly containing triplets (Shah et al. 1997) and the effect of glycine substitution on triple-helix folding and stability (Xiao et al. 2011; Cheng et al. 2011). Habitat temperature might act as an evolutionary pressure to maintain an "adequate" primary structure and the melting temperature of fibrillar collagens. Hence, a collagen chain from a deep-sea thermophilic worm is richer in stabilizing triplets (e.g., Gly–Pro–Pro) than the orthologous chain of a coastal worm living in colder conditions (Sicot et al. 2000). Another special feature has been characterized in a "fibrillar-like" collagen of the glass-sponge *Hyalonema sieboldi* (Ehrlich et al. 2010). In this study, the authors have shown the presence in the organic fraction of the anchoring spicules of a collagen rich in unusual (Gly–3Hyp–4Hyp) triplets. This hydroxylated collagen may serve as a template for biosilification. While the presence of 4Hyp residues in the Yaa position is an important stabilizing factor for the collagen triple helix, proline residues in position Xaa are usually unmodified. Recently, a small number of 3Hyp residues have been identified in mammalian fibrillar collagens, and it has been suggested that these might play a function in fibril supramolecular assembly (Weis et al. 2010).

Depending on the tissue, and the collagen type involved, fibrils might present different morphological aspects, from orthogonal layers of small-diameter fibrils in cornea to large fibrils in tendon. These different packing arrangements of fibrils seem to play important structural functions, for example the corneal transparency or tendon elasticity and resistance. Fibrillar collagen networks also play important functions in cell behavior and during development. In mammals, fibrils are generally heterotypic (one or two quantitatively major collagens (I, II, and III) and one minor collagen (V or XI)), and can be divided into cartilage (types II and XI) and non-cartilage fibrils (I, III, and V). In articular cartilage, Wu et al. (2010) have shown that type III collagen molecules with unprocessed N-propeptides form covalently linked polymeric filaments. These type III collagen polymers are cross-linked with type II collagen fibrils, and may increase the stability of this network. Several studies have demonstrated that in heterotypic fibrils, types V/XI play a pivotal function, as fibril nucleators and in the regulation of fibril diameter. Type V collagen induces the formation of smaller diameter fibrils than type I collagen in an in vitro self-assembly assay (Birk et al. 1990). Moreover, the mean diameter of type I collagen fibrils was reduced in the presence of type V molecules. Collagen V and XI-knockout mouse models have shown the depletion or drastic diminution of type I and type II fibrils, respectively (Li et al. 1995; Wenstrup et al., 2004, 2011; Segev et al. 2006). In a Cre-*lox*P mouse model leading to the collagen type V ablation in cornea, the fibrils present in this tissue are less abundant, present a larger diameter, and have lost their orthogonal organization (Sun et al. 2011). This corneal depletion of type V collagen results in a loss of transparency. Retention of the N-propeptides in types V or XI has also been defined as one of the factors modulating the diameter of the fibrils (Thom and Morris 1991; Linsenmayer et al. 1993; Gregory et al. 2000). Linsenmayer et al. (1993) have suggested that in types I/V fibrils, the presence of the type V N-propeptides on their surface inhibits further accretion of collagen molecules. Due to the presence of glycine substitutions, triple helical imperfections, and shorter collagenous domains, the implication of types XXIV and XXVII in the formation of classical collagen fibrils is questionable. Recently, it has been shown that collagen XXVII can form thin 10 nm unstriated fibril-like structures (Plumb et al. 2007), but it can also be detected in filamentous meshworks (Hjorten et al. 2007). As already described, the three fibrillar collagen clades seem to have arisen before the poriferan/cnidarian split, this result suggesting that invertebrate B clade collagens might also play a pivotal role in fibril nucleation and in the regulation of fibril diameter. In sea urchin, N-propeptide retention has been visualized in rotary shadowing images of fibrils from plutei (Lethias et al. 1997). These thin fibrils harbor large N-propeptide projections consisting of VWC and SURF modules, and have a brush-bottle aspect. In sea urchin adult tissues, heterotypic fibrils have been described (Cluzel et al. 2004), indicating that the formation of heterotypic fibrils arose also early during metazoan evolution.

3.3 Basement Membrane Type IV Collagen

3.3.1 Type IV Collagen Network

Basement membranes are acellular-specialized structures that delineate epithelial, endothelial, muscle, nerve, and fat cells from the surrounding connective tissues (LeBleu et al. 2007; Yurchenco 2011). These sheet-like structures anchor cells to their adjacent extracellular matrix, and might act as filtration barriers for macromolecules and as reservoirs of growth factors, for example. Basement membranes are also important in the structural integrity and/or differentiation of

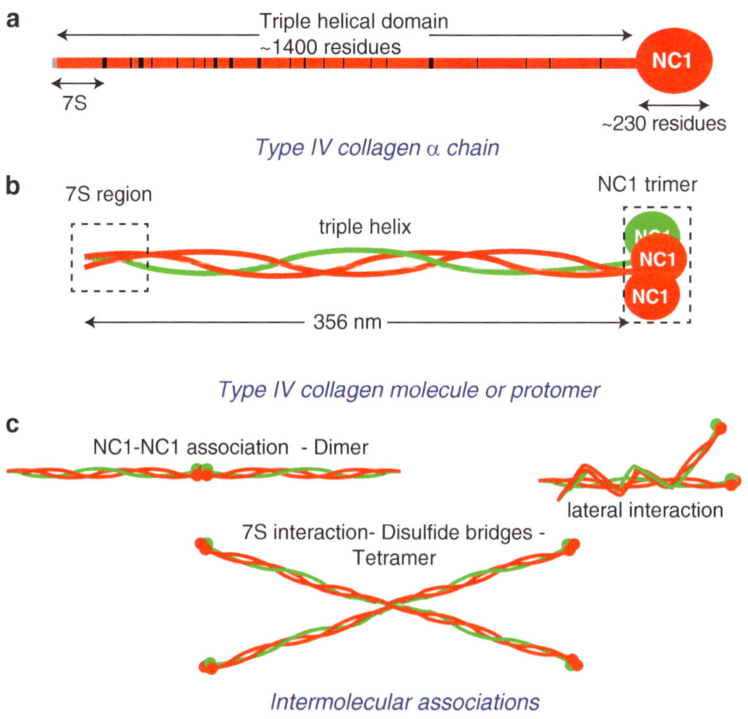

Fig. 3.3 Type IV collagen. (**a**) Prototype type IV collagen chain. Vertical black lines in the triple helical region indicate the approximate locations of the 21 Gly–Xaa–Yaa interruptions present in the human α1(IV) collagen chain. (**b**) Type IV collagen protomer. (**c**) Intermolecular associations at the base of type IV collagen network

tissues. They are made of network forming components including extracellular matrix glycoproteins (such as type IV collagen and laminin) and proteoglycans (such as perlecan).

Type IV collagen is a metazoan extracellular matrix glycoprotein hallmark, and represents one of the major elements of basement membranes. In mammals, six chains (α1–α6) are involved in the formation of heterotrimeric molecules (Khoshnoodi et al. 2008). Among these, we can cite the major and ubiquitous form which is made of two α1(IV) and one α2(IV) chains, the other isoforms (α3α4α5 and $(\alpha5)_2\alpha6$) showing tissue-specific distributions. Type IV collagen chains are made of three major domains, i.e., a large interrupted triple helix flanked by a short non-collagenous segment and the NC1 domain at the N- and C-termini, respectively. At the molecular level, these three regions are involved in the formation of the type IV collagen network illustrated in Fig. 3.3c (Timpl et al. 1981; Yurchenco and Ruben 1987).

The short N-terminal non-collagenous region is rich in cysteine and lysine residues, and is included in a cross-linking region mostly made of a collagenous sequence, the 7S domain (Risteli et al. 1980). Interchain disulfide bonds and lysine-related cross-links are implicated in the covalent association of the 7S domain of four molecules or protomers (Timpl et al. 1981). An N-glycosylation site (NXS/T) located at the C-terminus of the 7S domain is present in almost all mammalian type IV chains with the notable exception of the human α3(IV) chain. The conservation of this carbohydrate attachment site in hydra, nematode, mollusk, and sea urchin type IV α chains (Fowler et al. 2000; Guo et al. 1991; Fleury et al. 2011; Exposito et al. 1994) suggests its evolutionary and structural importance, in the protection against proteases, in 7S tetramer assembly and/or during cellular secretion (Timpl et al. 1981; Langeveld et al. 1991; Nayak and Spiro 1991).

The C-terminal non-collagenous NC1 domain is made of approximately 230 residues and can be divided into two similar subdomains, NC1-A and NC1-B, sharing 35 % amino acid identity (Netzer et al. 1998). During the formation of type IV collagen molecules, the NC1 domain is involved in the molecular selection and association of three specific α chains, and in the nucleation and propagation of the major triple helix (Boutaud et al. 2000; Borza et al. 2001; Söder and Pöschl 2004; Khoshnoodi et al. 2006). Two adjacent type IV collagen molecules can form a dimer through interactions of their NC1 trimers. The three dimensional structure of this NC1 hexamer has been determined (Sundaramoorthy et al. 2002; Than et al. 2002). Each NC1 unit is rich in ß-sheets, the NC1 trimer being organized by interactions through a domain swapping mechanism. The formation of the NC1 hexamer from two NC1 trimers is mediated by hydrophobic and hydrophilic interactions at their interface, the stabilization of this structure being reinforced by a covalent cross-link involving a modified lysine residue (S-hydroxylysyl) and a methionine residue from two adjacent trimers (Than et al. 2002, 2005; Vanacore et al. 2005). More recently, Vanacore et al. (2009) have revealed that the cross-link between these two residues is a sulfilimine bond. From multiple alignment studies, it has been reported that all bilaterian animals and the cniderian *Nematostella vectensis* type IV chains possess these two residues, which are absent in sponge, placozoa, the cnidarian *Hydra magnipapillata*, and the flatworm *Schistosoma japonicum* (Aouacheria et al. 2006; Vanacore et al. 2009). Hence, the stabilization of the type IV collagen network via sulfilimine bonds between NC1 trimers must have arisen before the cnidarian radiation (Vanacore et al. 2009).

In addition to these two intermolecular associations, lateral protomer interactions are supported by the collagen domain (Yurchenco and Ruben 1987), which corresponds to a large triple helical region of approximately 1400 residues, the Gly–Xaa–Yaa repeats being interrupted by numerous short non-collagenous sequences (21 of 2–11 residues long in the case of the human α1(IV) chain, for example) that might promote lateral triple-helix associations (Hwang et al. 2010a). These interruptions might also participate to the flexibility of type IV molecules, as shown by rotary shadowing images (Timpl et al. 1981). The three inter-protomer type IV collagen interactions (Fig. 3.3c) represent the backbone of the type IV network observed in basement membranes.

3.3.2 Type IV Collagen Evolution

In vertebrates, type IV collagen chains have been divided into two families, i.e., the α1(IV)-like (α1, α3, and α5) and the α2(IV)-like (α2, α4, and α6) chains (Mariyama et al. 1992; Zhou et al. 1994; Leinonen et al. 1994; Netzer et al. 1998). Furthermore, genes encoding the mammalian type IV collagen chains (*COL4A1* to *COL4A6*) are arranged pairwise in a head-to-head orientation (*COL4A1/COL4A2*; *COL4A3/COL4A4*; *COL4A5/COL4A6*), these gene pairs reflecting the two type IV collagen chain families or clades.

As previously indicated, type IV collagen has been identified from sponges to human. Among the four sponge lineages, basement membrane-like structures have only been detected in Homoscleromorpha (Garrone 1984). This observation is in agreement with the presence of a type IV collagen chain in the sponge *Pseudocorticium jarrei* belonging to Homoscleromorpha (Boute et al. 1996). However, it has recently been shown that despite the lack of basement-membrane structures in Calcarea, sponges belonging to this lineage possess at least two type IV collagen genes (Leys and Riesgo 2012). Based upon evolutionary studies, these type IV collagen genes might be related to the *COL4A1*-like and *COL4A2*-like subfamilies. From these results, as well as evolutionary data suggesting either monophyly (Philippe et al. 2009) or paraphyly (Sperling et al. 2009) of the four major sponge lineages, the first steps in type IV collagen gene evolution could follow two scenarios (Fig. 3.4). In the first, representing the paraphyletic hypothesis, the formation of an ancestral type IV collagen gene predated the divergence of Calcisponges/Homoscleromorphs lineages and Eumetazoa, although the loss of type IV collagen genes in the demosponge-hexactinellid ancestor may have been an alternative way of evolution. In the second scenario, favoring sponge monophyly, the formation of an ancestral type IV collagen gene has predated the Porifera-Eumetazoa split, and this gene has been lost in some sponge lineages. The next step has been the divergence of the *COL4A1-like* and *COL4A2-like* genes from an ancestral type IV gene, possibly before the poriferan-eumetazoan divergence (Leys and Riesgo 2012) or the cnidarian–bilaterian split (as illustrated in Fig. 3.4). The presence of type IV collagen gene pairs in a head-to-head fashion in drosophila (Yasothornsrikul et al. 1997), and in invertebrate species such as the sea urchin *Strongylocentrotus purpuratus* (see http://www.spbase.org/ and Scaffold559), argued that the formation of this bigene occurred during the duplication event leading to the formation of the genes encoding the α1- and α2-like chains. However, it should be noted that in nematodes the type IV collagen genes are localized on different chromosomes (Guo and Kramer 1989). The formation of the three type IV collagen gene pairs may have arisen via the two "whole" genome duplication events at the dawn of vertebrate evolution (Ohno 1970), the bigenes *COL4A1/COL4A2* and *COL4A5/COL4A6* being more closely related than the gene pair *COL4A3/COL4A4*.

Despite the presence of type IV α chains, calcarean sponges lack basement membrane-like structures (Leys and Riesgo 2012). This situation should be related

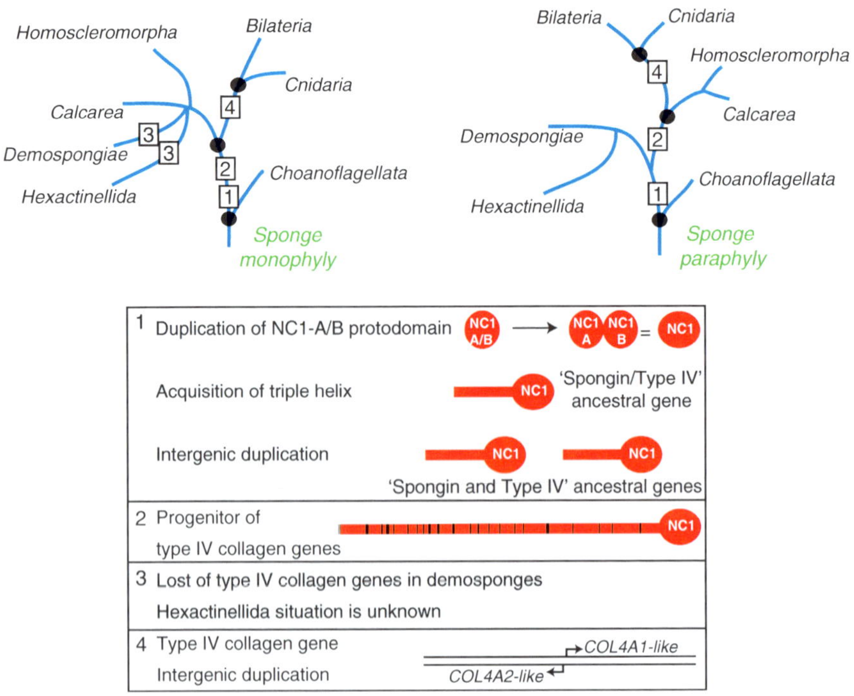

Fig. 3.4 Type IV collagen evolution. The four main steps (*numbers* included in *open boxes*) in the evolution of type IV collagen genes are presented along simplified trees (*top* of the figure) reflecting sponge monophyly (Philippe et al. 2009) or sponge paraphyly (Sperling et al. 2009). Step 1 predates the metazoan radiation, and consists of (i) duplication of a primordial NC1-AB sequence (PFAM name is C4) allowing the formation of an ancestral NC1 domain; (ii) acquisition of a triple-helix sequence leading to the formation of an ancestral spongin short-chain/type IV collagen gene; (iii) intergenic duplication of this ancestral gene. Step 2 predates either the poriferan/cnidarian split (sponge monophyly hypothesis) or calcarean+homoscleromorphan/cnidarian split (sponge paraphyly hypothesis). A progenitor of all extant type IV collagen genes was formed during this step. Step 3 concerns the putative loss of type IV collagen genes in some sponge lineages. Step 4 has been suggested to predate the bilaterian radiation (Aouacheria et al. 2006), and possibly the cnidarian/bilaterian split as shown in the simplified trees. During this step, a head-to-head bigene encoding α1-like and α2-like type IV collagen chains was obtained from an intergenic duplication event. The type IV collagen gene expansion predating the vertebrate radiation is not included is this illustration

to the absence of this collagen in some tissues of the nematode *Caenorhabditis elegans* (Graham et al. 1997), or the presence of basement structures in mouse embryos depleted of α1(IV) and α2(IV) chains (Pöschl et al. 2004). Hence, type IV collagen is not an essential element in the formation of basement membranes, but plays an important function in the conservation of this structure. This structural function of type IV collagens has been observed in different animal models. In nematode null mutants for type IV collagens, muscle detachment from the body wall was observed after contraction (Gupta et al. 1997). In Drosophila, collagen IV determines organ shape and animal form, in association with the proteoglycan perlecan (Pastor-Pareja and Xu 2011).

3.3.3 Spongin Short-Chain Collagens

In demosponges, basement membrane-like structures have not been identified, while collagen IV genes are absent in the newly characterized genome of the marine demosponge *Amphimedon queenslandica* (Srivastava et al. 2010). In addition to fibrillar collagens, genes encoding spongin short-chain collagens have first been described in the freshwater demosponge *Ephydatia mülleri* (Exposito et al. 1990). These collagens contain two uninterrupted collagenous domains of 66 and 171 residues (COL2 and COL1, respectively) that are interspersed with three non-collagenous sequences (NC3–NC1, from the N- to the C-termini). Spongin short-chain collagens present several similarities with nematode cuticular collagens. A similar distribution of cysteine residues has been pointed out between the spongin short-chain collagen NC3 and NC2 domains and comparable domains of cuticular collagens (Exposito et al. 2002). Like cuticular collagens, spongin short-chain collagens are encoded by a large gene family (Exposito et al. 1991; Cox et al. 1984). Furthermore, we have suggested that these spongin short-chain collagens might be considered as an exocollagen like the nematode cuticular collagens. Hence, spongin short-chain collagen genes are expressed by cells of the basal epithelial layer and cells neighboring the inorganic siliceous skeleton (Exposito et al. 1991). These cells express spongin, a collagenous protein surrounding the silicate spicules forming the sponge skeleton (note that in the bath sponge, the skeleton is free of inorganic spicules and consists mainly of spongin) and also helping to attach the animal to its substratum. Recently, Pozzolini et al. (2012) have shown in another demosponge free of silica spicules, *Chondrosia reniformis*, that short-chain collagen genes are expressed in the cortical part (the ectosome), and could play an important role in the attachment of this sponge to its substratum.

As represented in Fig. 3.4, spongin short-chain collagens might be considered as a sister family of basement membrane type IV collagen (Aouacheria et al. 2006). The relationships between these two collagen families consist of the presence of a collagenous region and a structurally and evolutionarily related non-collagenous C-terminal NC1 domain corresponding to the duplication of an ancestral NC1-AB (PFAM domain C4). This common evolutionary origin is also accompanied by functional analogy. Spongin short-chain collagen attaches the sponge to its substratum, and the type IV collagen network is an important factor in basement membrane and tissue integrity, and plays a pivotal role in cell attachment mediated by integrin and discoidin domain receptors (Leitinger and Hohenester 2007).

```
Nematode-Cle-1    DRVIHMIALSQPFSGNLHGLRGADLQCYREARAAGYTTTFRAMLSSNVQDLVRIVHSVDF
Sea anemone-Col   VILLRLVALNAPVTGELGSLAGVDHQCWRQSRQAGLKGTFRAFLSNRHQHLYSVVPREER
Drosophila-Dmp    PRMLRVAALNEPSTGDLQGIRGADFACYRQGRRAGLLGTFKAFLSSRVQNLDTIVRPADR
Human-XVIII       QPVLHLVALNSPLSGGMRGIRGADFQCFQQARAVGLAGTFRAFLSSRLQDLYSIVRRADR
Human-XV          KPALHLAALNMPVSGDIR----ADFQCFKQARAAGLLSTYRAFLSSHLQDLSTIVRKAER

-DTTVVNVAGHHLFPSWRSFVNGAQ--MNPHAKLFSFDRHDVLNDSRWPDKRVWHGSKDGGIR-AEQYCDGWRRAD
-DLPVVNLKGDQLFTSWRNMFSKNYG-FNPDIPIYTFDGRDVIKDEKWPSKYFWHGSRGDGGSKYPDNCRSWQSRK
-DLPVVNTRGDVLFNSWKGIFNGQGGFFSQAPRIYSFSGKNVMTDSTWPMKMVWHGSLPNGERSMDTYCDAWHSGD
AAVPIVNLKDELLFPSWEALFSGSEGPLKPGARIFSFDGKDVLRHPTWPQKSVWHGSDPNGRRLTESYCETWRTEA
YSLPIVNLKGQVLFNNWDSIFSGHGGQFNMHIPIYSFDGRDIMTDPSWPQKVIWHGSSPHGVRLVDNYCEAWRTAD

SSLTSLAGHISSNTSIFQSSGSEKCENKLVVLCVENM
PLDMSTAAAIDGSTAMFAEN-VFTCDNRFSILCIQIS
HLKGSFASNLDGHK-LLEQK-RQSCDSKLIILCVEAL
PSATGQASSLLGGR-LLGQS-AASCHHAYIVLCIENS
TAVTGLASPLSTGK-ILDQK-AYSCANRLIVLCIENS
```

Fig. 3.5 Representative multiple alignments of multiplexin endostatin domains. Accession numbers of the sequences are ACZ94635 (drosophila Dmp), AAC78500 (human type XV collagen), AAR83296 (human type XVIII collagen), and NP_492358 (nematode Cle-1). The sea anemone Col sequence was collected via the StellaBase web site (http://nematostella.bu.edu/stellabase/; c401101236.Contig1)

3.4 Multiplexins, a Basement Membrane Collagen Family

Collagens XV and XVIII have been called multiplexins (Oh et al. 1994). Multiplexin collagens represent with basement membrane type IV and fibrillar collagens one of the most widespread collagen families in Metazoa. Collagen XV/XVIII has been described in drosophila (Meyer and Moussian 2009) and nematode (Ackley et al. 2001), while a genomic study has pointed out that the sea anemone *N. vectensis* (Cnidaria: Anthozoa) might potentially encode a type XV/XVIII collagen (Reitzel et al. 2008; Fahey and Degnan 2010). The general modular structure of multiplexins is conserved, with a highly interrupted collagenous domain flanked in its N-terminus by a TSPN module. At the C-terminus of these collagens, a trimerization domain (for which the crystal structure has been described by Boudko et al. (2009)) links the collagenous region to an endostatin module. The triple helix of vertebrate type XVIII collagen is longer than in type XV, and the interruptions separating the sub-collagenous domains are shorter. These differences might explain why type XVIII molecules appear to be less flexible than type XV monomers in rotary shadowing electron microscopy images (Marneros et al. 2004; Myers et al. 2007). The collagenous region of drosophila Dpm and nematode Cle-1 collagen XV/XVIII is shorter than vertebrate multiplexins. The endostatin domain is the most conserved region of metazoan multiplexins (see Fig. 3.5) with an average sequence identity superior to 40 %. It should be noted that different isoforms resulting from alternative promotors or splicing have been described for nematode, drosophila, and vertebrate type XVIII multiplexins (Ackley et al. 2001; Meyer and Moussian 2009; Oh et al. 1994; Suzuki et al. 2002).

Multiplexins are localized in basement membranes. Collagen XVIII is a heparan sulfate proteoglycan (Halfter et al. 1998) found ubiquitously in vascular and epithelial basement membranes, and seems to be polarized with its C-terminus in the lamina densa and its N-terminus in the sublamina densa (Marneros et al., 2004). Mutations of the corresponding gene have been identified in patients suffering from Knobloch syndrome. This disease is characterized by eye defects including a high myopia, while mice deleted for the type XVIII collagen gene present several ocular abnormalities (Sertié et al. 2000; Fukai et al. 2002). Type XV collagen gene-knockout mice have skeletal myopathy and cardiovascular defects (Eklund et al. 2001). Multiplexin functions are often associated with their endostatin domain. The biological relevance of this domain in angiogenesis, tumor growth, and apoptosis, is well described in a recent review (Seppinen and Pihlajaniemi 2011).

3.5 "Versatile" Invertebrate Collagens

3.5.1 Cuticular Nematode Collagen

In nematode, the cuticle is considered as an exoskeleton, and acts as an environmental barrier. A large gene family of more than 170 units encodes these special collagens that constitute the major component of the basal layer of cuticle or endocuticle, and are also present in the medial layer. Cuticular collagens have been divided into several subfamilies from their pattern of cysteine residues (Johnstone 2000), and contain two collagenous domains interspaced by three non-collagenous regions. In addition to some similarities with spongin short-chain collagens, the distribution of cysteine residues at the junction of the collagenous and C-terminal non-collagenous domains is similar to that observed in FACIT (fibril-associated collagens with interrupted triple helices) collagens. In the cuticle, the collagens form multiple highly cross-linked layers, and mutations in their corresponding genes are often associated with abnormal morphologies. Plant-parasitic nematodes can be infected by *Pasteuria*, this event beginning by adhesion of bacterial endospores to the surface of cuticle. In animal-parasitic *Bacillus*, the presence of the collagen BclA has been demonstrated at the surface of the endospores (Kailas et al. 2011). For this reason, Davies and Curtis (2011) have suggested that *Pasteuria* collagens might play an important function during animal infection.

3.5.2 Minicollagens of the Cnidarian Nematocyst

In Cnidaria, specialized organelles are used for prey capture and defense, the nematocysts (Ozbek 2011). A family of fibrous minicollagens have been described in these organelles (Holstein et al., 1994), most of them being localized in the capsule wall. The collagenous domain is made of 14–27 Gly–Xaa–Yaa triplets, and is flanked by cysteine-rich domains or CRD that are involved in the supramolecular assembly of minicollagens. A disulfide bond-swapping model has been proposed in the formation and stabilization of the mini-collagen network (Ozbek et al. 2002). The high tensile strength of the nematocyst wall is mediated by a continuous suprastructure involving intermolecular disulfide bonds between CRDs of minicollagens and NOWA, a nematocyst outer wall antigen (Meier et al. 2007). Another cnidarian-specific collagen type is present in the tubule of nematocysts, and consists of a protein made of collagenous and galectin sugar-binding domains (Hwang et al. 2010b). Minicollagens have been characterized in myxozoans, which are aquatic endoparasites. This study has confirmed that myxozoans group within the cnidarians (Holland et al. 2011).

3.5.3 Mussel Byssus and Extensible Collagen

Mussels are attached to the substratum via byssal threads (2–4 cm long with diameters of 10–150 μm) that can be defined as tendon-like structure (Coyne et al. 1997). The foot synthesizes byssal threads which can be divided into an elastic and proximal region and a more rigid and stiff distal portion. Different collagen-like proteins have been described in the byssal threads. These collagen chains form homotrimeric molecules and possess a central interrupted triple helix made of 146–175 Gly–Xaa–Yaa triplets, with His-rich regions at both extremities (Qin et al. 1997; Waite et al. 1998). Between the two His-rich sequences and the central collagenous region, elastic domains sharing similarities with elastin are present in the collagen chain PreCol-P, while silk fibroin domains have been identified in the chain PreCol-D. Interestingly, these two collagens form opposite gradients along the byssal threads, the elastic-containing collagen predominating at the proximal elastic region. The third α chain (Pre-ColNG) is equally distributed along the thread, and contains glycine-rich sequences between His-rich and collagenous domains. Waite et al. (2002) have suggested that the His-rich sequences provide sites for lateral and longitudinal cross-linking.

3.5.4 Partially Characterized Invertebrate Collagens

In annelids, biochemical and electron microscopy (Murray and Tanzer 1983; Gaill et al. 1991) have revealed a collagen involved in the formation of the longest collagen molecule known to date (up to 2,500 nm), 8 times the length of fibrillar collagen molecules. This collagen participates to the formation of large unstriated fibrils in the cuticle. These fibrils form orthogonal layers in a plywood-like structure (Lepescheux 1988). Cuticle collagen analysis from annelids living in different habitats has revealed that collagens extracted from animals living around hydrothermal vents have a melting temperature superior to those from worms collected in shallow sea water (Gaill et al. 1995). The increase in cuticle collagen stability in hydrothermal species has been understood from peptidic sequencing of cuticle collagen from *Riftia pachyptila* (Mann et al. 1996). The frequent presence of glycosylated threonine residues in position Xaa of the collagenous triplets favors triple helical structure stability. Sponge collagens of undetermined function have been characterized by genomic sequencing (Aho et al. 1993).

3.6 Collagens Specific to Chordates

3.6.1 Transmembrane Collagens

In vertebrates, types XIII, XVII, XXIII, and XXV correspond to the type II transmembrane collagens and form homotrimeric molecules. From mutational and computational analyses, it has been suggested that each collagenous domain of these collagens is preceded by an α-helical coiled-coil sequence that participate in the association and the correct folding of the triple helix (Snellman et al. 2007). The role of α-helical coiled-coil sequences in collagen folding has been shown in several collagen types (McAlinden et al. 2003). The extracellular region or ectodomain of transmembrane collagens can be released from the membrane by proteolytic shedding (Franzke et al. 2005; Hashimoto et al. 2002; Väisänen et al. 2006; Veit et al. 2007). Some other transmembrane proteins contain a collagenous domain but are not members of the collagen family. Among these collagen-like proteins, there are the defense collagens (macrophage scavenger receptors, macrophage receptor MARCO, collectins like CL-P1), gliomedin, and ectodysplasin A (Acton et al. 1993; Exposito et al. 2002; Franzke et al. 2005). Most of these transmembrane collagens have not been characterized in non-chordate animals. However, proteins related to gliomedin have been described in drosophila and nematode (Loria et al. 2004). Data mining of extracellular matrix components of the invertebrate chordate *Ciona intestinalis* has revealed the presence of a putative transmembrane collagen chain (Huxley-Jones et al. 2007).

3.6.2 FACIT Collagens

Members of the FACIT collagen family (types IX, XII, XIV, XVI, XIX, XX, XXI, and XXII, in vertebrates) are highly variable in length and in modular composition (Ricard-Blum 2011). FACIT collagens can be associated with the surface of striated collagen fibrils, as shown for type IX molecules (Eyre et al. 1987; van der Rest and Mayne 1988). Genomic analysis and cDNA cloning have determined several FACIT-like proteins in an invertebrate chordate, and one FACIT-like collagen chain in the honeybee *A. melifera* (Huxley-Jones et al. 2007; Vizzini et al. 2002).

3.6.3 Types VIII/X and C1q-Containing Collagens

Types VIII and X collagens form homotrimeric molecules. The C-terminal C1q module of these collagens acts as a trimerization domain (Barber and Kwan 1996). Soluble and transmembrane collagens (complement C1q, adiponectin, …), EMILINs, proteins found in the blood of mammalian hibernators are also included in the C1q-containing collagen family (Exposito et al. 2002). Several collagen-like proteins containing a C1q module have been determined in invertebrate chordates (Carland and Gerwick 2010). Finally, a C1q-containing collagen-like chain has been characterized in the annelid *Hirudo medicinalis* (Tahtouh et al. 2009). As indicated by Carland and Gerwick (2010), more data are needed to determine if the annelid sequence represents the ancestral gene of the mammalian complement C1q.

3.6.4 Other Vertebrate Collagens

Until now, and to our knowledge, there is no clear ortholog of vertebrate types VI, VII, XXVI, and XXVIII collagens in invertebrates. Among the collagen-like proteins, ficolins are soluble defense collagens implicated in innate immunity and acting as pattern recognition molecules against pathogens. These are structurally related to complement C1q, with the replacement of the C1q module by a fibrinogen-like domain. Ficolins have been characterized in invertebrate chordates (Huang et al. 2011; Kenjo et al. 2001).

3.7 Bacterial Collagens

Mimicry is a common strategy used in nature against predators or to bypass host defense. In a mouse model, it has been shown that a collagen-binding protein from *Streptococcus mutans* might increase hemorrhagic stroke by interacting with the collagen present in injured blood vessels. This interaction inhibits platelet aggregation, and affects hemostasis (Nakano et al. 2011). Despite the lack of prolyl hydroxylase, several collagen-like proteins have been characterized in bacteria, and for some of them, it has been demonstrated that the collagenous sequence is involved in the formation of a stable triple helix (Xu et al. 2002), the melting point of this domain being approximately 37 °C. From data mining, Rasmussen et al. (2003) have determined that most of the bacterial collagens display structural motifs of surface or spore-associated proteins. From microscopic and physical studies, Kailas et al. (2011) have presented a model showing that in different *Bacillus* species, the collagen-like protein BclA is attached to the surface of the exosporium (outermost layer of the spore), and projects into the external environment. In this bacterial family, BclA does not seem to be involved in virulence, but might favor spore uptake by macrophages by inhibiting their nonspecific uptake by nonprofessional phagocytic cells (Sylvestre et al. 2002; Bozue et al. 2007). In *Streptococcus pyogenes*, two major collagenous proteins (Scl1 and Scl2) display several interactions with host extracellular matrix, plasma components (Han et al. 2006; Caswell et al. 2010; Reuter et al. 2010), but also integrin cellular receptors (Humtsoe et al. 2005; Caswell et al. 2008; Chen et al. 2010). Other studies have pointed out the importance of these surface collagens in bacterial pathogenicity. Indeed, a recent study has described the importance of Scl1 in the formation and "robustness" of biofilms in the *Streptococcus* family (Oliver-Kozup et al. 2011). Biofilms might be considered as the bacterial equivalent of the metazoan extracellular matrix (Exposito et al. 2002), and are an important determinant of virulence (for a recent review, see Høiby et al. (2011)). Collagen-like proteins have been identified outside the three domains of life (Rasmussen et al. 2003). Interestingly, the seven collagen-related proteins of mimivirus are rich in lysine residues in the Yaa position of the triple helical triplets, and are subject to posttranslational modifications by an enzyme having lysyl hydroxylase and glycosyltransferase functions (Luther et al. 2011).

From genome-based identification of bacterial and viral collagen-like proteins, Rasmussen et al. (2003) suggest that a collagen sequence has been horizontally transferred and presumably from Metazoa to Bacteria or that prokaryotic collagen genes have arisen de novo in bacteria (convergent evolution). An interesting feature of collagens in a bacterial species is that the collagenous domain is often variable in size according to the serotype analyzed, and that their Gly–Xaa–Yaa sequence often consists of multiple copies of one or several triplets. In the crustacean bacteria *Pasteuria ramosa*, more than half of the collagenous segment of a collagen-like protein consists of 25 tandem repeats of the highly conserved sequence GDQ GEA GMP GPK (Mouton et al. 2009). In addition to some amino acid substitutions, the major difference in the sequence of this collagen in two different isolates is the number of copies of this highly conserved 12-residue repeat peptide. Other types of tandem repeats have been recently identified from the characterization of 37 new putative collagen genes in *P. ramosa* (McElroy et al. 2011). The variable length of the collagenous domain, and the presence of different tandem repeats in the Gly–Xaa–Yaa sequence of the collagen-like protein Scl1 (also named SclA) have also been described in different *S. pyogenes* strains (Rasmussen et al. 2000). Layton et al. (2008) have suggested that the collagen gene from the Cyanobacteria *Trichodesmium erythraeum* represents an interdomain transfer event from a devonian or Silurian marine animal, a second and less parsimonious hypothesis consisting of a repeat expansion of a short sequence. The protein encoded by this gene has a large collagenous region (1678 residues) including a Gly–Xaa–Yaa–Zaa imperfection. Careful inspection of the sequence clearly indicates that the collagenous domain of *T. erythraeum* collagen can be divided into 48 tandem copies of a moderately conserved repeat with the consensus sequence GPVGPAGADGVPGLAGPVGPVGPTGAPGPAGPI, this subdivision favoring a repeat expansion hypothesis.

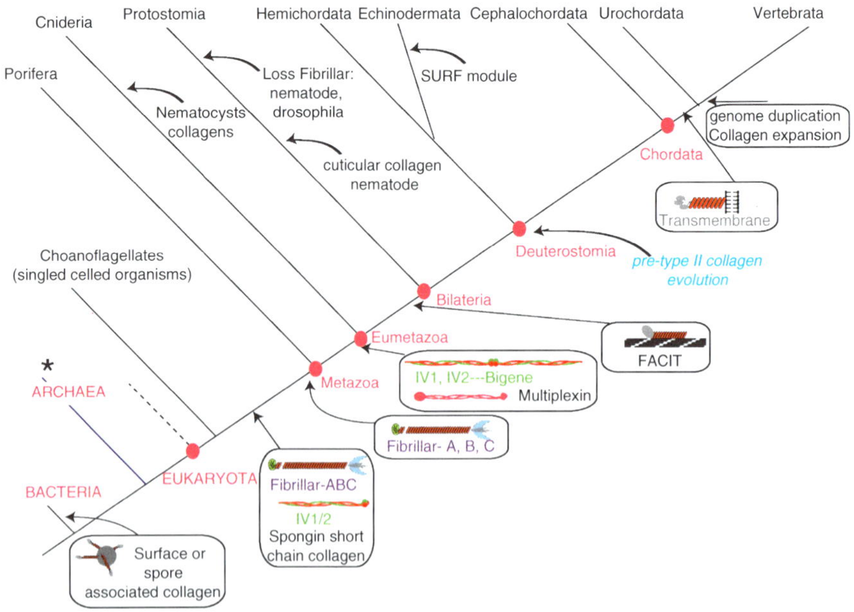

Fig. 3.6 Domains of life and collagens. The first known appearances of collagen families are indicated. This simplified tree of life is based on several studies (Huxley-Jones et al. 2007; Exposito et al. 2010; Ozbek et al. 2010; Hynes 2012). (*asterisk*) Indicates the presence of collagen-like sequence in Archaea (see accession number CP000866, *Nitrosopumilus maritimus* SCM1, marine archaeal group 1 strain)

The presence of sequence repeats and the variability of their copy number in different strains have been associated with membrane-associated proteins (de Castro et al. 2006; Kassai-Jáger et al. 2008), a situation observed for bacterial collagens. As indicated by Kassai-Jáger et al. (2008), tandem repeat polymorphisms play an important function in the variability of membrane-associated proteins in prokaryotes, and might be a useful adaptive mechanism of evolution against environmental constraints.

3.8 Conclusion

Recent reviews on the evolution of metazoan extracellular matrix (Ozbek et al. 2010; Hynes 2012) have pointed out several important concepts. A large number of the gene families encoding extracellular matrix components have emerged early during animal evolution, presumably before the eumetazoan radiation. This extracellular matrix toolkit has evolved with some gene loss, some emergence of phylum-specific genes, and gene expansion at the dawn of vertebrate world. Exon shuffling is an essential mode of evolution that has allowed the genesis of the multimodular extracellular matrix proteins, especially the proteins formed in early vertebrates (Nagy and Patthy 2011). Finally, a few extracellular matrix components have predated the metazoan world, such as integrins, and several putative proteins and modules related to animal matrices have been identified in choanoflagellates. Collagens have been detected in almost all extant animals. Fibrillar, type IV and multiplexin collagens are present, with only a few exceptions, throughout all species. The appearance of the collagen families as well as of proteins specific to a taxon are presented in Fig. 3.6. A next step in the comprehension of collagen evolution will be to perform an exhaustive study of all metazoan genomes available to date, while future genome projects in species belonging to the different sponge groups will be of interest to decipher the early steps of evolution of this family of proteins.

Acknowledgements We are indebted to David J. Hulmes for critically reading the manuscript.

References

Ackley BD, Crew JR, Elamaa H, Pihlajaniemi T, Kuo CJ, Kramer JM (2001) The NC1/endostatin domain of *Caenorhabditis elegans* type XVIII collagen affects cell migration and axon guidance. J Cell Biol 152:1219–1232. doi:10.1083/jcb.152.6.1219

Acton S, Resnick D, Freeman M, Ekkel Y, Ashkenas J, Krieger M (1993) The collagenous domains of macrophage scavenger receptors and complement component C1q mediate their similar, but not identical, binding specificities for polyanionic ligands. J Biol Chem 268:3530–3537

Aho S, Turakainen H, Onnela ML, Boedtker H (1993) Characterization of an intronless collagen gene family in the marine sponge *Microciona prolifera*. Proc Natl Acad Sci U S A 90:7288–7292. doi:10.1073/pnas.90.15.7288

Aouacheria A, Cluzel C, Lethias C, Gouy M, Garrone R, Exposito JY (2004) Invertebrate data predict an early emergence of vertebrate fibrillar collagen clades and an anti-incest model. J Biol Chem 279:47711–47719. doi:10.1074/jbc.M408950200

Aouacheria A, Geourjon C, Aghajari N, Navratil V, Deléage G, Lethias C, Exposito JY (2006) Insights into early extracellular matrix evolution: spongin short chain collagen-related proteins are homologous to basement membrane type IV collagens and form a novel family widely distributed in invertebrates. Mol Biol Evol 23:2288–2302. doi:10.1093/molbev/msl100

Bailey WJ, Kim J, Wagner GP, Ruddle FH (1997) Phylogenetic reconstruction of vertebrate Hox cluster duplications. Mol Biol Evol 14:843–853

Barber RE, Kwan AP (1996) Partial characterization of the C-terminal non-collagenous domain (NC1) of collagen type X. Biochem J 320:479–485

Benveniste-Schrode K, Doering JL, Hauck WW, Schrode J, Kendra KL, Drexler BK (1985) Evolution of chick type I procollagen genes. J Mol Evol 22:209–219. doi:10.1007/BF02099750

Birk DE, Fitch JM, Babiarz JP, Doane KJ, Linsenmayer TF (1990) Collagen fibrillogenesis in vitro: interaction of types I and V collagen regulates fibril diameter. J Cell Sci 95:649–657

Boot-Handford RP, Tuckwell DS (2003) Fibrillar collagen: the key to vertebrate evolution? A tale of molecular incest. Bioessays 25:142–151. doi:10.1002/bies.10230

Boot-Handford RP, Tuckwell DS, Plumb DA, Rock CF, Poulsom R (2003) A novel and highly conserved collagen (pro(α)1(XXVII)) with a unique expression pattern and unusual molecular characteristics establishes a new clade within the vertebrate fibrillar collagen family. J Biol Chem 278:31067–31077. doi:10.1074/jbc.M212889200

Borza DB, Bondar O, Ninomiya Y, Sado Y, Naito I, Todd P, Hudson BG (2001) The NC1 domain of collagen IV encodes a novel network composed of the α1, α2, α5, and α6 chains in smooth muscle basement membranes. J Biol Chem 276:28532–28540. doi:10.1074/jbc.M103690200

Boudko SP, Sasaki T, Engel J, Lerch TF, Nix J, Chapman MS, Bächinger HP (2009) Crystal structure of human collagen XVIII trimerization domain: A novel collagen trimerization Fold. J Mol Biol 392:787–802. doi:10.1016/j.jmb.2009.07.057

Boutaud A, Borza DB, Bondar O, Gunwar S, Netzer KO, Singh N, Ninomiya Y, Sado Y, Noelken ME, Hudson BG (2000) Type IV collagen of the glomerular basement membrane. Evidence that the chain specificity of network assembly is encoded by the noncollagenous NC1 domains. J Biol Chem 275:30716–30724. doi:10.1074/jbc.M004569200

Boute N, Exposito JY, Boury-Esnault N, Vacelet J, Noro N, Miyazaki K, Yoshizato K, Garrone R (1996) Type IV collagen in sponges, the missing link in basement membrane ubiquity. Biol Cell 88:37–44. doi:10.1016/S0248-4900(97)86829-3

Bozue J, Moody KL, Cote CK, Stiles BG, Friedlander AM, Welkos SL, Hale ML (2007) *Bacillus anthracis* spores of the bclA mutant exhibit increased adherence to epithelial cells, fibroblasts, and endothelial cells but not to macrophages. Infect Immun 75:4498–4505. doi:10.1128/IAI.00434-07

Cameron RA, Rowen L, Nesbitt R, Bloom S, Rast JP, Berney K, Arenas-Mena C, Martinez P, Lucas S, Richardson PM, Davidson EH, Peterson KJ, Hood L (2006) Unusual gene order and organization of the sea urchin hox cluster. J Exp Zool B Mol Dev Evol 306:45–58. doi:10.1002/jez.b.21070

Carland TM, Gerwick L (2010) The C1q domain containing proteins: Where do they come from and what do they do? Dev Comp Immunol 34:785–790. doi:10.1016/j.dci.2010.02.014

Caswell CC, Barczyk M, Keene DR, Lukomska E, Gullberg DE, Lukomski S (2008) Identification of the first prokaryotic collagen sequence motif that mediates binding to human collagen receptors, integrins α2ß1 and α11ß1. J Biol Chem 283:36168–36175. doi:10.1074/jbc.M806865200

Caswell CC, Oliver-Kozup H, Han R, Lukomska E, Lukomski S (2010) Scl1, the multifunctional adhesin of group A *Streptococcus*, selectively binds cellular fibronectin and laminin, and mediates pathogen internalization by human cells. FEMS Microbiol Lett 303:61–68. doi:10.1111/j.1574-6968.2009.01864.x

Chen SM, Tsai YS, Wu CM, Liao SK, Wu LC, Chang CS, Liu YH, Tsai PJ (2010) Streptococcal collagen-like surface protein 1 promotes adhesion to the respiratory epithelial cell. BMC Microbiol 10:320. doi:10.1186/1471-2180-10-320

Cheng H, Rashid S, Yu Z, Yoshizumi A, Hwang E, Brodsky B (2011) Location of glycine mutations within a bacterial collagen protein affects degree of disruption of triple-helix folding and conformation. J Biol Chem 286:2041–2046. doi:10.1074/jbc.M110.153965

Chu ML, de Wet W, Bernard M, Ding JF, Morabito M, Myers J, Williams C, Ramirez F (1984) Human proα1(I) collagen gene structure reveals evolutionary conservation of a pattern of introns and exons. Nature 310:337–340. doi:10.1038/310337a0

Cluzel C, Lethias C, Garrone R, Exposito JY (2004) Distinct maturations of N-propeptide domains in fibrillar procollagen molecules involved in the formation of heterotypic fibrils in adult sea urchin collagenous tissues. J Biol Chem 279:9811–9817. doi:10.1074/jbc.M311803200

Cox GN, Kramer JM, Hirsh D (1984) Number and organization of collagen genes in *Caenorhabditis elegans*. Mol Cell Biol 4:2389–2395. doi:10.1128/MCB.4.11.2389

Coyne KJ, Qin XX, Waite JH (1997) Extensible collagen in mussel byssus: a natural block copolymer. Science 277:1830–1832. doi:10.1126/science.277.5333.1830

Davies KG, Curtis RH (2011) Cuticle surface coat of plant-parasitic nematodes. Annu Rev Phytopathol 49:135–156. doi:10.1146/annurev-phyto-121310-111406

de Castro LA, Rodrigues Pedroso T, Kuchiishi SS, Ramenzoni M, Kich JD, Zaha A, Henning Vainstein M, Bunselmeyer Ferreira H (2006) Variable number of tandem aminoacid repeats in adhesion-related CDS products in *Mycoplasma hyopneumoniae* strains. Vet Microbiol 116:258–269. doi:10.1016/j.vetmic.2006.04.022

Dion AS, Myers JC (1987) COOH-terminal propeptides of the major human procollagens. Structural, functional and genetic comparisons. J Mol Biol 193:127–143. doi:10.1016/0022-2836(87)90632-2

Ehrlich H, Deutzmann R, Brunner E, Cappellini E, Koon H, Solazzo C, Yang Y, Ashford D, Thomas-Oates J, Lubeck M, Baessmann C, Langrock T, Hoffmann R, Wörheide G, Reitner J, Simon P, Tsurkan M, Ereskovsky AV, Kurek D, Bazhenov VV, Hunoldt S, Mertig M, Vyalikh DV, Molodtsov SL, Kummer K, Worch H, Smetacek V, Collins MJ (2010) Mineralization of the metre-long biosilica structures of glass sponges is templated on hydroxylated collagen. Nat Chem 2:1084–1088. doi:10.1038/nchem.899

Eklund L, Piuhola J, Komulainen J, Sormunen R, Ongvarrasopone C, Fássler R, Muona A, Ilves M, Ruskoaho H, Takala TE, Pihlajaniemi T (2001) Lack of type XV collagen causes a skeletal myopathy and cardiovascular defects in mice. Proc Natl Acad Sci U S A 98:1194–1199. doi:10.1073/pnas.031444798

Engel J (1997) Versatile collagens in invertebrates. Science 277:1785–1786. doi:10.1126/science.277.5333.1785

Exposito JY, Garrone R (1990) Characterization of a fibrillar collagen gene in sponges reveals the early evolutionary appearance of two collagen gene families. Proc Natl Acad Sci U S A 87:6669–6673. doi:10.1073/pnas.87.17.6669

Exposito JY, Ouazana R, Garrone R (1990) Cloning and sequencing of a Porifera partial cDNA coding for a short-chain collagen. Eur J Biochem 190:401–406. doi:10.1111/j.1432-1033.1990.tb15589.x

Exposito JY, Le Guellec D, Lu Q, Garrone R (1991) Short chain collagens in sponges are encoded by a family of closely related genes. J Biol Chem 266:21923–21928

Exposito JY, D'Alessio M, Ramirez F (1992a) Novel amino-terminal propeptide configuration in a fibrillar procollagen undergoing alternative splicing. J Biol Chem 267:17404–17408

Exposito JY, D'Alessio M, Solursh M, Ramirez F (1992b) Sea urchin collagen evolutionarily homologous to vertebrate pro-α2(I) collagen. J Biol Chem 267:15559–15562

Exposito JY, van der Rest M, Garrone R (1993) The complete intron/exon structure of *Ephydatia mülleri* fibrillar collagen gene suggests a mechanism for the evolution of an ancestral gene module. J Mol Evol 37:254–259

Exposito JY, Suzuki H, Geourjon C, Garrone R, Solursh M, Ramirez F (1994) Identification of a cell lineage-specific gene coding for a sea urchin α2(IV)-like collagen chain. J Biol Chem 269:13167–13171

Exposito JY, Cluzel C, Lethias C, Garrone R (2000) Tracing the evolution of vertebrate fibrillar collagens from an ancestral α chain. Matrix Biol 19:275–279. doi:10.1016/S0945-053X(00)00067-6

Exposito JY, Cluzel C, Garrone R, Lethias C (2002) Evolution of collagens. Anat Rec 268:302–316. doi:10.1002/ar.10162

Exposito JY, Larroux C, Cluzel C, Valcourt U, Lethias C, Degnan BM (2008) Demosponge and sea anemone fibrillar collagen diversity reveals the early emergence of A/C clades and the maintenance of the modular structure of type V/XI collagens from sponge to human. J Biol Chem 283:28226–28235. doi:10.1074/jbc.M804573200

Exposito JY, Valcourt U, Cluzel C, Lethias C (2010) The fibrillar collagen family. Int J Mol Sci 11:407–426. doi:10.3390/ijms11020407

Eyre DR, Apon S, Wu JJ, Ericsson LH, Walsh KA (1987) Collagen type IX: evidence for covalent linkages to type II collagen in cartilage. FEBS Lett 220:337–341. doi:10.1016/0014-5793(87)80842-6

Fahey B, Degnan BM (2010) Origin of animal epithelia: insights from the sponge genome. Evol Dev 12:601–617. doi:10.1111/j.1525-142X.2010.00445.x

Fleury C, Serpentini A, Kypriotou M, Renard E, Galéra P, Lebel JM (2011) Characterization of a non-fibrillar-related collagen in the mollusc *Haliotis tuberculata* and its biological activity on human dermal fibroblasts. Mar Biotechnol 13:1003–1016. doi:10.1007/s10126-011-9364-9

Fowler SJ, Jose S, Zhang X, Deutzmann R, Sarras MP Jr, Boot-Handford RP (2000) Characterization of hydra type IV collagen. Type IV collagen is essential for head regeneration and its expression is up-regulated upon exposure to glucose. J Biol Chem 275:39589–39599. doi:10.1074/jbc.M005871200

Franzke CW, Bruckner P, Bruckner-Tuderman L (2005) Collagenous transmembrane proteins: recent insights into biology and pathology. J Biol Chem 280:4005–4008. doi:10.1074/jbc.R400034200

Fukai N, Eklund L, Marneros AG, Oh SP, Keene DR, Tamarkin L, Niemelä M, Ilves M, Li E, Pihlajaniemi T, Olsen BR (2002) Lack of collagen XVIII/endostatin results in eye abnormalities. EMBO J 21:1535–1544. doi:10.1093/emboj/21.7.1535

Gaill F, Wiedemann H, Mann K, Kühn K, Timpl R, Engel J (1991) Molecular characterization of cuticle and interstitial collagens from worms collected at deep sea hydrothermal vents. J Mol Biol 221:209–223. doi:10.1016/0022-2836(91)80215-G

Gaill F, Mann K, Wiedemann H, Engel J, Timpl R (1995) Structural comparison of cuticle and interstitial collagens from annelids living in shallow sea-water and at deep-sea hydrothermal vents. J Mol Biol 246:284–294. doi:10.1006/jmbi.1994.0084

Garrone R (1984) Formation and involvement of extracellular matrix in the development of sponges, a primitive multicellular system. In: Trelstad RL (ed) The role of extracellular matrix in development. Alan R. Liss, New York, pp 461–477

Garrone R, Exposito JY, Franc JM, Franc S, Humbert-David N, Qin L, Tillet E (1993) Phylogenesis of the extracellular matrix. C R Seances Soc Biol Fil 187:114–123

Gordon MK, Hahn RA (2010) Collagens. Cell Tissue Res 339:247–257. doi:10.1007/s00441-009-0844-4

Graham PL, Johnson JJ, Wang S, Sibley MH, Gupta MC, Kramer JM (1997) Type IV collagen is detectable in most, but not all, basement membranes of *Caenorhabditis elegans* and assembles on tissues that do not express it. J Cell Biol 137:1171–1183. doi:10.1083/jcb.137.5.1171

Gregory KE, Oxford JT, Chen Y, Gambee JE, Gygi SP, Aebersold R, Neame PJ, Mechling DE, Bächinger HP, Morris NP (2000) Structural organization of distinct domains within the non-collagenous N-terminal region of collagen type XI. J Biol Chem 275:11498–11506. doi:10.1074/jbc.275.15.11498

Guo XD, Kramer JM (1989) The two *Caenorhabditis elegans* basement membrane (type IV) collagen genes are located on separate chromosomes. J Biol Chem 264:17574–17582

Guo XD, Johnson JJ, Kramer JM (1991) Embryonic lethality caused by mutations in basement membrane collagen of *C. elegans*. Nature 349:707–709. doi:10.1038/349707a0

Gupta MC, Graham PL, Kramer JM (1997) Characterization of α1(IV) collagen mutations in *Caenorhabditis elegans* and the effects of α1 and α2(IV) mutations on type IV collagen distribution. J Cell Biol 137:1185–1196. doi:10.1083/jcb.137.5.1185

Halfter W, Dong S, Schurer B, Cole GJ (1998) Collagen XVIII is a basement membrane heparan sulfate proteoglycan. J Biol Chem 273:25404–25412. doi:10.1074/jbc.M212244200

Han R, Caswell CC, Lukomska E, Keene DR, Pawlowski M, Bujnicki JM, Kim JK, Lukomski S (2006) Binding of the low-density lipoprotein by streptococcal collagen-like protein Scl1 of *Streptococcus pyogenes*. Mol Microbiol 61:351–367. doi:10.1111/j.1365-2958.2006.05237.x

Hashimoto T, Wakabayashi T, Watanabe A, Kowa H, Hosoda R, Nakamura A, Kanazawa I, Arai T, Takio K, Mann DM, Iwatsubo T (2002) CLAC: a novel Alzheimer amyloid plaque component derived from a transmembrane precursor, CLAC-P/collagen type XXV. EMBO J 21:1524–1534. doi:10.1093/emboj/21.7.1524

Heino J, Huhtala M, Käpylä J, Johnson MS (2009) Evolution of collagen-based adhesion systems. Int J Biochem Cell Biol 41:341–348. doi:10.1016/j.biocel.2008.08.021

Hjorten R, Hansen U, Underwood RA, Telfer HE, Fernandes RJ, Krakow D, Sebald E, Wachsmann-Hogiu S, Bruckner P, Jacquet R, Landis WJ, Byers PH, Pace JM (2007)

Type XXVII collagen at the transition of cartilage to bone during skeletogenesis. Bone 41:535–542. doi:10.1016/j.bone.2007.06.024

Hoffman GG, Branam AM, Huang G, Pelegri F, Cole WG, Wenstrup RM, Greenspan DS (2010) Characterization of the six zebrafish clade B fibrillar procollagen genes, with evidence for evolutionarily conserved alternative splicing within the pro-α1(V) C-propeptide. Matrix Biol 29:261–275. doi:10.1016/j.matbio.2010.01.006

Høiby N, Ciofu O, Johansen HK, Song ZJ, Moser C, Jensen PØ, Molin S, Givskov M, Tolker-Nielsen T, Bjarnsholt T (2011) The clinical impact of bacterial biofilms. Int J Oral Sci 3:55–65. doi:10.4248/IJOS11026

Holland JW, Okamura B, Hartikainen H, Secombes CJ (2011) A novel minicollagen gene links cnidarians and myxozoans. Proc Biol Sci 278:546–553. doi:10.1098/rspb.2010.1301

Holstein TW, Benoit M, Herder GV, David CN, Wanner G, Gaub HE (1994) Fibrous mini-collagens in hydra nematocysts. Science 265:402–404. doi:10.1126/science.265.5170.402

Huang H, Huang S, Yu Y, Yuan S, Li R, Wang X, Zhao H, Yu Y, Li J, Yang M, Xu L, Chen S, Xu A (2011) Functional characterization of a ficolin-mediated complement pathway in amphioxus. J Biol Chem 286:36739–36748. doi:10.1074/jbc.M111.245944

Humtsoe JO, Kim JK, Xu Y, Keene DR, Höök M, Lukomski S, Wary KK (2005) A streptococcal collagen-like protein interacts with the α2β1 integrin and induces intracellular signaling. J Biol Chem 280:13848–13857. doi:10.1074/jbc.M410605200

Hutter H, Vogel BE, Plenefisch JD, Norris CR, Proenca RB, Spieth J, Guo C, Mastwal S, Zhu X, Scheel J, Hedgecock EM (2000) Conservation and novelty in the evolution of cell adhesion and extracellular matrix genes. Science 287:989–994. doi:10.1126/science.287.5455.989

Huxley-Jones J, Robertson DL, Boot-Handford RP (2007) On the origins of the extracellular matrix in vertebrates. Matrix Biol 26:2–11. doi:10.1016/j.matbio.2006.09.008

Hwang ES, Thiagarajan G, Parmar AS, Brodsky B (2010a) Interruptions in the collagen repeating tripeptide pattern can promote supramolecular association. Protein Sci 19:1053–1064. doi:10.1002/pro.383

Hwang JS, Takaku Y, Momose T, Adamczyk P, Özbek S, Ikeo K, Khalturin K, Hemmrich G, Bosch TC, Holstein TW, David CN, Gojobori T (2010b) Nematogalectin, a nematocyst protein with GlyXY and galectin domains, demonstrates nematocyte-specific alternative splicing in Hydra. Proc Natl Acad Sci U S A 107:18539–18544. doi:10.1073/pnas.1003256107

Hynes RO (2012) Evolution: The evolution of metazoan extracellular matrix. J Cell Biol 196:671–679. doi:10.1083/jcb.201109041

Irimia M, Roy SW (2008) Spliceosomal introns as tools for genomic and evolutionary analysis. Nucleic Acids Res 36:1703–1712. doi:10.1093/nar/gkn012

Johnstone IL (2000) Cuticle collagen genes. Expression in Caenorhabditis elegans. Trends Genet 16:21–27. doi:10.1016/S0168-9525(99)01857-0

Kadler KE, Holmes DF, Trotter JA, Chapman JA (1996) Collagen fibril formation. Biochem J 316:1–11

Kailas L, Terry C, Abbott N, Taylor R, Mullin N, Tzokov SB, Todd SJ, Wallace BA, Hobbs JK, Moir A, Bullough PA (2011) Surface architecture of endospores of the *Bacillus cereus/anthracis/thuringiensis* family at the subnanometer scale. Proc Natl Acad Sci U S A 108:16014–16019. doi:10.1073/pnas.1109419108

Kassai-Jáger E, Ortutay C, Tóth G, Vellai T, Gáspári Z (2008) Distribution and evolution of short tandem repeats in closely related bacterial genomes. Gene 410:18–25. doi:10.1016/j.gene.2007.11.006

Katsanis N, Fitzgibbon J, Fisher EM (1996) Paralogy mapping: identification of a region in the human MHC triplicated onto human chromosomes 1 and 9 allows the prediction and isolation of novel *PBX* and *NOTCH* loci. Genomics 35:101–108. doi:10.1006/geno.1996.0328

Kenjo A, Takahashi M, Matsushita M, Endo Y, Nakata M, Mizuochi T, Fujita T (2001) Cloning and characterization of novel ficolins from the solitary ascidian, *Halocynthia roretzi*. J Biol Chem 276:19959–19965. doi:10.1074/jbc.M011723200

Khoshnoodi J, Cartailler JP, Alvares K, Veis A, Hudson BG (2006) Molecular recognition in the assembly of collagens: terminal noncollagenous domains are key recognition modules in the formation of triple helical protomers. J Biol Chem 281:38117–38121. doi:10.1074/jbc.R600025200

Khoshnoodi J, Pedchenko V, Hudson BG (2008) Mammalian collagen IV. Microsc Res Tech 71:357–370. doi:10.1002/jemt.20564

King N, Westbrook MJ, Young SL, Kuo A, Abedin M, Chapman J, Fairclough S, Hellsten U, Isogai Y, Letunic I, Marr M, Pincus D, Putnam N, Rokas A, Wright KJ, Zuzow R, Dirks W, Good M, Goodstein D, Lemons D, Li W, Lyons JB, Morris A, Nichols S, Richter DJ, Salamov A, Sequencing JG, Bork P, Lim WA, Manning G, Miller WT, McGinnis W, Shapiro H, Tjian R, Grigoriev IV, Rokhsar D (2008) The genome of the choanoflagellate *Monosiga brevicollis* and the origin of metazoans. Nature 451:783–788. doi:10.1038/nature06617

Kleman JP, Hartmann DJ, Ramirez F, van der Rest M (1992) The human rhabdomyosarcoma cell line A204 lays down a highly insoluble matrix composed mainly of α1 type-XI and α2 type-V collagen chains. Eur J Biochem 210:329–335. doi:10.1111/j.1432-1033.1992.tb17425.x

Koch M, Laub F, Zhou P, Hahn RA, Tanaka S, Burgeson RE, Gerecke DR, Ramirez F, Gordon MK (2003) Collagen XXIV, a vertebrate fibrillar collagen with structural features of invertebrate collagens: selective expression in developing cornea and bone. J Biol Chem 278:43236–43244. doi:10.1074/jbc.M302112200

Kuraku S (2008) Insights into cyclostome phylogenomics: pre-2R or post-2R. Zoolog Sci 25:960–968. doi:10.2108/zsj.25.960

Kuraku S, Meyer A, Kuratani S (2009) Timing of genome duplications relative to the origin of the vertebrates: did cyclostomes diverge before or after? Mol Biol Evol 26:47–59. doi:10.1093/molbev/msn222

Langeveld JP, Noelken ME, Hård K, Todd P, Vliegenthart JF, Rouse J, Hudson BG (1991) Bovine glomerular basement membrane. Location and structure of the asparagine-linked oligosaccharide units and their potential role in the assembly of the 7 S collagen IV tetramer. J Biol Chem 266:2622–2631

Layton BE, D'Souza AJ, Dampier W, Zeiger A, Sabur A, Jean-Charles J (2008) Collagen's triglycine repeat number and phylogeny suggest an interdomain transfer event from a Devonian or Silurian organism into *Trichodesmium erythraeum*. J Mol Evol 66:539–554. doi:10.1007/s00239-008-9111-7

LeBleu VS, Macdonald B, Kalluri R (2007) Structure and function of basement membranes. Exp Biol Med 232:1121–1129. doi:10.3181/0703-MR-72

Lees JF, Tasab M, Bulleid NJ (1997) Identification of the molecular recognition sequence which determines the type-specific assembly of procollagen. EMBO J 16:908–916. doi:10.1093/emboj/16.5.908

Leinonen A, Mariyama M, Mochizuki T, Tryggvason K, Reeders ST (1994) Complete primary structure of the human type IV collagen α4(IV) chain. Comparison with structure and expression of the other α(IV) chains. J Biol Chem 269:26172–26177

Leitinger B, Hohenester E (2007) Mammalian collagen receptors. Matrix Biol 26:146–155. doi:10.1016/j.matbio.2006.10.007

Lepescheux L (1988) Spatial organization of collagen in annelid cuticle: order and defects. Biol Cell 62:17–31. doi:10.1111/j.1768-322X.1988.tb00702.x

Lethias C, Exposito JY, Garrone R (1997) Collagen fibrillogenesis during sea urchin development–retention of SURF motifs from the N-propeptide of the 2α chain in mature fibrils. Eur J Biochem 245:434–440. doi:10.1111/j.1432-1033.1997.t01-2-00434.x

Leys SP, Riesgo A (2012) Epithelia, an evolutionary novelty of metazoans. J Exp Zool B Mol Dev Evol. doi:10.1002/jez.b.21442

Li Y, Lacerda DA, Warman ML, Beier DR, Yoshioka H, Ninomiya Y, Oxford JT, Morris NP, Andrikopoulos K, Ramirez F et al (1995) A fibrillar collagen gene, *Col11a1*, is essential for skeletal morphogenesis. Cell 80:423–430. doi:10.1016/0092-8674(95)90492-1

Linsenmayer TF, Gibney E, Igoe F, Gordon MK, Fitch JM, Fessler LI, Birk DE (1993) Type V collagen: molecular structure and fibrillar organization of the chicken α1(V) NH2-terminal domain, a putative regulator of corneal fibrillogenesis. J Cell Biol 121:1181–1189. doi:10.1083/jcb.121.5.1181

Loria PM, Hodgkin J, Hobert O (2004) A conserved postsynaptic transmembrane protein affecting neuromuscular signaling in *Caenorhabditis elegans*. J Neurosci 24:2191–2201. doi:10.1523/JNEUROSCI.5462-03.2004

Luther KB, Hülsmeier AJ, Schegg B, Deuber SA, Raoult D, Hennet T (2011) Mimivirus collagen is modified by bifunctional lysyl hydroxylase and glycosyltransferase enzyme. J Biol Chem 286:43701–43709. doi:10.1074/jbc.M111.309096

Mann K, Mechling DE, Bächinger HP, Eckerskorn C, Gaill F, Timpl R (1996) Glycosylated threonine but not 4-hydroxyproline dominates the triple helix stabilizing positions in the sequence of a hydrothermal vent worm cuticle collagen. J Mol Biol 261:255–266. doi:10.1006/jmbi.1996.0457

Mariyama M, Kalluri R, Hudson BG, Reeders ST (1992) The α4(IV) chain of basement membrane collagen. Isolation of cDNAs encoding bovine α4(IV) and comparison with other type IV collagens. J Biol Chem 267:1253–1258

Marneros AG, Keene DR, Hansen U, Fukai N, Moulton K, Goletz PL, Moiseyev G, Pawlyk BS, Halfter W, Dong S, Shibata M, Li T, Crouch RK, Bruckner P, Olsen BR (2004) Collagen XVIII/endostatin is essential for vision and retinal pigment epithelial function. EMBO J 23:89–99. doi:10.1038/sj.emboj.7600014

Mayne R, Brewton RG, Mayne PM, Baker JR (1993) Isolation and characterization of the chains of type V/type XI collagen present in bovine vitreous. J Biol Chem 268:9381–9386

McAlinden A, Smith TA, Sandell LJ, Ficheux D, Parry DA, Hulmes DJ (2003) α-helical coiled-coil oligomerization domains are almost ubiquitous in the collagen superfamily. J Biol Chem 278:42200–42207. doi:10.1074/jbc.M302429200

McElroy K, Mouton L, Du Pasquier L, Qi W, Ebert D (2011) Characterisation of a large family of polymorphic collagen-like proteins in the endospore-forming bacterium *Pasteuria ramosa*. Res Microbiol 162:701–714. doi:10.1016/j.resmic.2011.06.009

Meier S, Jensen PR, Adamczyk P, Bächinger HP, Holstein TW, Engel J, Ozbek S, Grzesiek S (2007) Sequence-structure and structure-function analysis in cysteine-rich domains forming the ultrastable nematocyst wall. J Mol Biol 368:718–728. doi:10.1016/j.jmb.2007.02.026

Meyer F, Moussian B (2009) Drosophila multiplexin (Dmp) modulates motor axon pathfinding accuracy. Dev Growth Differ 51:483–498. doi:10.1111/j.1440-169X.2009.01111.x

Miura S, Kimura S (1985) Jellyfish mesogloea collagen. Characterization of molecules as α1 α2 α3 heterotrimers. J Biol Chem 260:15352–15356

Mouton L, Traunecker E, McElroy K, Du Pasquier L, Ebert D (2009) Identification of a polymorphic collagen-like protein in the crustacean bacteria *Pasteuria ramosa*. Res Microbiol 160:792–799. doi:10.1016/j.resmic.2009.08.016

Murray LW, Tanzer ML (1983) Characterization of a large fragment from annelid cuticle collagen and its relationship to the intact molecule. Coll Relat Res 3:445–458

Myers JC, Amenta PS, Dion AS, Sciancalepore JP, Nagaswami C, Weisel JW, Yurchenco PD (2007) The molecular structure of human tissue type XV presents a unique conformation among the collagens. Biochem J 404:535–544. doi:10.1042/BJ20070201

Nagy A, Patthy L (2011) Reassessing domain architecture evolution of metazoan proteins: the contribution of different evolutionary mechanisms. Genes 2:578–598. doi:10.3390/genes2030578

Nakano K, Hokamura K, Taniguchi N, Wada K, Kudo C, Nomura R, Kojima A, Naka S, Muranaka Y, Thura M, Nakajima A, Masuda K, Nakagawa I, Speziale P, Shimada N, Amano A, Kamisaki Y, Tanaka T, Umemura K, Ooshima T (2011) The collagen-binding protein of *Streptococcus mutans* is involved in haemorrhagic stroke. Nat Commun 2:485. doi:10.1038/ncomms1491

Nayak BR, Spiro RG (1991) Localization and structure of the asparagine-linked oligosaccharides of type IV collagen from glomerular basement membrane and lens capsule. J Biol Chem 266:13978–13987

Netzer KO, Suzuki K, Itoh Y, Hudson BG, Khalifah RG (1998) Comparative analysis of the noncollagenous NC1 domain of type IV collagen: identification of structural features important for assembly, function, and pathogenesis. Protein Sci 7:1340–1351. doi:10.1002/pro.5560070610

Oh SP, Kamagata Y, Muragaki Y, Timmons S, Ooshima A, Olsen BR (1994) Isolation and sequencing of cDNAs for proteins with multiple domains of Gly-Xaa-Yaa repeats identify a distinct family of collagenous proteins. Proc Natl Acad Sci U S A 91:4229–4233. doi:10.1073/pnas.91.10.4229

Ohkubo H, Vogeli G, Mudryj M, Avvedimento VE, Sullivan M, Pastan I, de Crombrugghe B (1980) Isolation and characterization of overlapping genomic clones covering the chicken α2 (type I) collagen gene. Proc Natl Acad Sci U S A 77:7059–7063. doi:10.1073/pnas.77.12.7059

Ohno S (1970) Evolution by gene duplication. Springer, New York

Ohtani K, Yao T, Kobayashi M, Kusakabe R, Kuratani S, Wada H (2008) Expression of Sox and fibrillar collagen genes in lamprey larval chondrogenesis with implications for the evolution of vertebrate cartilage. J Exp Zool B Mol Dev Evol 310:596–607. doi:10.1002/jez.b.21231

Olinski RP, Lundin LG, Hallböök F (2006) Conserved synteny between the Ciona genome and human paralogons identifies large duplication events in the molecular evolution of the insulin-relaxin gene family. Mol Biol Evol 23:10–22. doi:10.1093/molbev/msj002

Oliver-Kozup HA, Elliott M, Bachert BA, Martin KH, Reid SD, Schwegler-Berry DE, Green BJ, Lukomski S (2011) The streptococcal collagen-like protein-1 (Scl1) is a significant determinant for biofilm formation by group A Streptococcus. BMC Microbiol 11:262. doi:10.1186/1471-2180-11-262

Ozbek S (2011) The cnidarian nematocyst: a miniature extracellular matrix within a secretory vesicle. Protoplasma 248:635–640. doi:10.1007/s00709-010-0219-4

Ozbek S, Engel U, Engel J (2002) A switch in disulfide linkage during minicollagen assembly in hydra nematocysts or how to assemble a 150-bar-resistant structure. J Struct Biol 137:11–14. doi:10.1093/emboj/20.12.3063

Ozbek S, Balasubramanian PG, Chiquet-Ehrismann R, Tucker RP, Adams JC (2010) The evolution of extracellular matrix. Mol Biol Cell 21:4300–4305. doi:10.1083/jcb.201109041

Pace JM, Corrado M, Missero C, Byers PH (2003) Identification, characterization and expression analysis of a new fibrillar collagen gene, *COL27A1*. Matrix Biol 22:3–14. doi:10.1016/S0945-053X(03)00007-6

Pastor-Pareja JC, Xu T (2011) Shaping cells and organs in Drosophila by opposing roles of fat body-secreted Collagen IV and perlecan. Dev Cell 21:245–256. doi:10.1016/j.devcel.2011.06.026

Philippe H, Derelle R, Lopez P, Pick K, Borchiellini C, Boury-Esnault N, Vacelet J, Renard E, Houliston E, Quéinnec E, Da Silva C, Wincker P, Le Guyader H, Leys S, Jackson DJ, Schreiber F, Erpenbeck D, Morgenstern B, Wörheide G, Manuel M (2009) Phylogenomics revives traditional views on deep animal relationships. Curr Biol 19:706–712. doi:10.1016/j.cub.2009.02.052

Plumb DA, Dhir V, Mironov A, Ferrara L, Poulsom R, Kadler KE, Thornton DJ, Briggs MD, Boot-Handford RP (2007) Collagen XXVII is developmentally regulated and forms thin fibrillar structures distinct from those of classical vertebrate fibrillar collagens. J Biol Chem 282:12791–12795. doi:10.1074/jbc.C700021200

Popovici C, Leveugle M, Birnbaum D, Coulier F (2001) Coparalogy: physical and functional clusterings in the human genome. Biochem Biophys Res Commun 288:362–370. doi:10.1006/bbrc.2001.5794

Pöschl E, Schlötzer-Schrehardt U, Brachvogel B, Saito K, Ninomiya Y, Mayer U (2004) Collagen IV is essential for basement membrane stability but dispensable for initiation of its assembly during early development. Development 131:1619–1628. doi:10.1242/dev.01037

Pozzolini M, Bruzzone F, Berilli V, Mussino F, Cerrano C, Benatti U, Giovine M (2012) Molecular characterization of a nonfibrillar collagen from the marine sponge *Chondrosia reniformis* Nardo 1847 and positive effects of soluble silicates on its expression. Mar Biotechnol 14(3):281–93. doi:10.1007/s10126-011-9415-2

Qin XX, Coyne KJ, Waite JH (1997) Tough tendons. Mussel byssus has collagen with silk-like domains. J Biol Chem 272:32623–32627. doi:10.1074/jbc.272.51.32623

Rasmussen M, Edén A, Björck L (2000) SclA, a novel collagen-like surface protein of *Streptococcus pyogenes*. Infect Immun 68:6370–6377. doi:10.1128/IAI.68.11.6370-6377.2000

Rasmussen M, Jacobsson M, Björck L (2003) Genome-based identification and analysis of collagen-related structural motifs in bacterial and viral proteins. J Biol Chem 278:32313–32316. doi:10.1074/jbc.M304709200

Reitzel AM, Sullivan JC, Traylor-Knowles N, Finnerty JR (2008) Genomic survey of candidate stress-response genes in the estuarine anemone *Nematostella vectensis*. Biol Bull 214:233–254

Reuter M, Caswell CC, Lukomski S, Zipfel PF (2010) Binding of the human complement regulators CFHR1 and factor H by streptococcal collagen-like protein 1 (Scl1) *via* their conserved C termini allows control of the complement cascade at multiple levels. J Biol Chem 285:38473–38485. doi:10.1074/jbc.M110.143727

Ricard-Blum S (2011) The collagen family. Cold Spring Harb Perspect Biol 3:a004978. doi:10.1101/cshperspect.a004978

Rigby BJ, Robinson MS (1975) Thermal transitions in collagen and the preferred temperature range of animals. Nature 253:277–279. doi:10.1038/253277a0

Risteli J, Bächinger HP, Engel J, Furthmayr H, Timpl R (1980) 7-S collagen: characterization of an unusual basement membrane structure. Eur J Biochem 108:239–250. doi:10.1111/j.1432-1033.1980.tb04717.x

Roy SW, Gilbert W (2005) Resolution of a deep animal divergence by the pattern of intron conservation. Proc Natl Acad Sci U S A 102:4403–4408. doi:10.1073/pnas.0409891102

Rychel AL, Swalla BJ (2007) Development and evolution of chordate cartilage. J Exp Zool B Mol Dev Evol 308:325–335. doi:10.1002/jez.b.21157

Rychel AL, Smith SE, Shimamoto HT, Swalla BJ (2006) Evolution and development of the chordates: collagen and pharyngeal cartilage. Mol Biol Evol 23:541–549. doi:10.1093/molbev/msj055

Saito M, Takenouchi Y, Kunisaki N, Kimura S (2001) Complete primary structure of rainbow trout type I collagen consisting of α1(I)α2(I)α3(I) heterotrimers. Eur J Biochem 268:2817–2827. doi:10.1046/j.1432-1327.2001.02160.x

Segev F, Héon E, Cole WG, Wenstrup RJ, Young F, Slomovic AR, Rootman DS, Whitaker-Menezes D, Chervoneva I, Birk DE (2006) Structural abnormalities of the cornea and lid resulting from collagen V mutations. Invest Ophthalmol Vis Sci 47:565–573. doi:10.1167/iovs.05-0771

Seppinen L, Pihlajaniemi T (2011) The multiple functions of collagen XVIII in development and disease. Matrix Biol 30:83–92. doi:10.1016/j.matbio.2010.11.001

Sertié AL, Sossi V, Camargo AA, Zatz M, Brahe C, Passos-Bueno MR (2000) Collagen XVIII, containing an endogenous inhibitor of angiogenesis and tumor growth, plays a critical role in the maintenance of retinal structure and in neural tube closure (Knobloch syndrome). Hum Mol Genet 9:2051–2058. doi:10.1093/hmg/9.13.2051

Shah NK, Sharma M, Kirkpatrick A, Ramshaw JA, Brodsky B (1997) Gly-Gly-containing triplets of low stability adjacent to a type III collagen epitope. Biochemistry 36:5878–5883. doi:10.1021/bi963146c

Sicot FX, Exposito JY, Masselot M, Garrone R, Deutsch J, Gaill F (1997) Cloning of an annelid fibrillar-collagen gene and phylogenetic analysis of vertebrate and invertebrate collagens. Eur J Biochem 246:50–58. doi:10.1111/j.1432-1033.1997.00050.x

Sicot FX, Mesnage M, Masselot M, Exposito JY, Garrone R, Deutsch J, Gaill F (2000) Molecular adaptation to an extreme environment: origin of the thermal stability of the pompeii worm collagen. J Mol Biol 302:811–820. doi:10.1006/jmbi.2000.4505

Snellman A, Tuomisto A, Koski A, Latvanlehto A, Pihlajaniemi T (2007) The role of disulfide bonds and alpha-helical coiled-coils in the biosynthesis of type XIII collagen and other collagenous transmembrane proteins. J Biol Chem 282:14898–14905. doi:10.1074/jbc.M609605200

Söder S, Pöschl E (2004) The NC1 domain of human collagen IV is necessary to initiate triple helix formation. Biochem Biophys Res Commun 325:276–280. doi:10.1016/j.bbrc.2004.10.034

Sperling EA, Peterson KJ, Pisani D (2009) Phylogenetic-signal dissection of nuclear housekeeping genes supports the paraphyly of sponges and the monophyly of Eumetazoa. Mol Biol Evol 26:2261–2274. doi:10.1093/molbev/msp148

Srivastava M, Simakov O, Chapman J, Fahey B, Gauthier ME, Mitros T, Richards GS, Conaco C, Dacre M, Hellsten U, Larroux C, Putnam NH, Stanke M, Adamska M, Darling A, Degnan SM, Oakley TH, Plachetzki DC, Zhai Y, Adamski M, Calcino A, Cummins SF, Goodstein DM, Harris C, Jackson DJ, Leys SP, Shu S, Woodcroft BJ, Vervoort M, Kosik KS, Manning G, Degnan BM, Rokhsar DS (2010) The *Amphimedon queenslandica* genome and the evolution of animal complexity. Nature 466:720–726. doi:10.1038/nature09201

Su MW, Suzuki HR, Bieker JJ, Solursh M, Ramirez F (1991) Expression of two nonallelic type II procollagen genes during *Xenopus laevis* embryogenesis is characterized by stage-specific production of alternatively spliced transcripts. J Cell Biol 115:565–575. doi:10.1083/jcb.115.2.565

Sun M, Chen S, Adams SM, Florer JB, Liu H, Kao WW, Wenstrup RJ, Birk DE (2011) Collagen V is a dominant regulator of collagen fibrillogenesis: dysfunctional regulation of structure and function in a corneal-stroma-specific *Col5a1*-null mouse model. J Cell Sci 124:4096–4105. doi:10.1242/jcs.091363

Sundaramoorthy M, Meiyappan M, Todd P, Hudson BG (2002) Crystal structure of NC1 domains. Structural basis for type IV collagen assembly in basement membranes. J Biol Chem 277:31142–31153. doi:10.1074/jbc.M201740200

Suzuki OT, Sertié AL, Der Kaloustian VM, Kok F, Carpenter M, Murray J, Czeizel AE, Kliemann SE, Rosemberg S, Monteiro M, Olsen BR, Passos-Bueno MR (2002) Molecular analysis of collagen XVIII reveals novel mutations, presence of a third isoform, and possible genetic heterogeneity in Knobloch syndrome. Am J Hum Genet 71:1320–1329. doi:10.1086/344695

Sylvestre P, Couture-Tosi E, Mock M (2002) A collagen-like surface glycoprotein is a structural component of the *Bacillus anthracis* exosporium. Mol Microbiol 45:169–178. doi:10.1046/j.1365-2958.2000.03000.x

Tahtouh M, Croq F, Vizioli J, Sautiere PE, Van Camp C, Salzet M, Daha MR, Pestel J, Lefebvre C (2009) Evidence for a novel chemotactic C1q domain-containing factor in the leech nerve cord. Mol Immunol 46:523–531. doi:10.1016/j.molimm.2008.07.026

Takahara K, Hoffman GG, Greenspan DS (1995) Complete structural organization of the human α1(V) collagen gene (*COL5A1*): divergence from the conserved organization of other characterized fibrillar collagen genes. Genomics 29:588–597. doi:10.1006/geno.1995.9961

Than ME, Henrich S, Huber R, Ries A, Mann K, Kühn K, Timpl R, Bourenkov GP, Bartunik HD, Bode W (2002) The 1.9-A crystal structure of the noncollagenous (NC1) domain of human placenta collagen IV shows stabilization via a novel type of covalent Met-Lys cross-link. Proc Natl Acad Sci U S A 99:6607–6612. doi:10.1073/pnas.062183499

Than ME, Bourenkov GP, Henrich S, Mann K, Bode W (2005) The NC1 dimer of human placental basement membrane collagen IV: does a covalent crosslink exist? Biol Chem 386:759–766. doi:10.1515/BC.2005.089

Thom JR, Morris NP (1991) Biosynthesis and proteolytic processing of type XI collagen in embryonic chick sterna. J Biol Chem 266:7262–7269

Tillet E, Franc JM, Franc S, Garrone R (1996) The evolution of fibrillar collagens: a sea-pen collagen shares common features with vertebrate type V collagen. Comp Biochem Physiol B Biochem Mol Biol 113:239–246. doi:10.1016/0305-0491(95)02014-4

Timpl R, Wiedemann H, van Delden V, Furthmayr H, Kühn K (1981) A network model for the organization of type IV collagen molecules in basement membranes. Eur J Biochem 120:203–211. doi:10.1111/j.1432-1033.1981.tb05690.x

Upholt WB, Strom CM, Sandell LJ (1985) Structure of the type II collagen gene. Ann N Y Acad Sci 460:130–140. doi:10.1111/j.1749-6632.1985.tb51161.x

Väisänen MR, Väisänen T, Tu H, Pirilä P, Sormunen R, Pihlajaniemi T (2006) The shed ectodomain of type XIII collagen associates with the fibrillar fibronectin matrix and may interfere with its assembly in vitro. Biochem J 393:43–50. doi:10.1042/BJ20031974

van der Rest M, Mayne R (1988) Type IX collagen proteoglycan from cartilage is covalently cross-linked to type II collagen. J Biol Chem 263:1615–1618

Vanacore RM, Friedman DB, Ham AJ, Sundaramoorthy M, Hudson BG (2005) Identification of S-hydroxylysyl-methionine as the covalent cross-link of the noncollagenous (NC1) hexamer of the α1α1α2 collagen IV network: a role for the post-translational modification of lysine 211 to hydroxylysine 211 in hexamer assembly. J Biol Chem 280:29300–29310. doi:10.1074/jbc.M502752200

Vanacore R, Ham AJ, Voehler M, Sanders CR, Conrads TP, Veenstra TD, Sharpless KB, Dawson PE, Hudson BG (2009) A sulfilimine bond identified in collagen IV. Science 325:1230–1234. doi:10.1126/science.1176811

Veit G, Zimina EP, Franzke CW, Kutsch S, Siebolds U, Gordon MK, Bruckner-Tuderman L, Koch M (2007) Shedding of collagen XXIII is mediated by furin and depends on the plasma membrane microenvironment. J Biol Chem 282:27424–27435. doi:10.1074/jbc.M703425200

Vizzini A, Arizza V, Cervello M, Cammarata M, Gambino R, Parrinello N (2002) Cloning and expression of a type IX-like collagen in tissues of the ascidian *Ciona intestinalis*. Biochim Biophys Acta 1577:38–44. doi:10.1016/S0167-4781(02)00403-7

Vuoristo MM, Pihlajamaa T, Vandenberg P, Prockop DJ, Ala-Kokko L (1995) The human *COL11A2* gene structure indicates that the gene has not evolved with the genes for the major fibrillar collagens. J Biol Chem 270:22873–22881. doi:10.1074/jbc.270.39.22873

Wada H, Okuyama M, Satoh N, Zhang S (2006) Molecular evolution of fibrillar collagen in chordates, with implications for the evolution of vertebrate skeletons and chordate phylogeny. Evol Dev 8:370–377. doi:10.1111/j.1525-142X.2006.00109.x

Waite JH, Qin XX, Coyne KJ (1998) The peculiar collagens of mussel byssus. Matrix Biol 17:93–106. doi:10.1016/S0945-053X(98)90023-3

Waite JH, Vaccaro E, Sun C, Lucas JM (2002) Elastomeric gradients: a hedge against stress concentration in marine holdfasts? Philos Trans R Soc Lond B Biol Sci 357:143–153. doi:10.1098/rstb.2001.1025

Weis MA, Hudson DM, Kim L, Scott M, Wu JJ, Eyre DR (2010) Location of 3-hydroxyproline residues in collagen types I, II, III, and V/XI implies a role in fibril supramolecular assembly. J Biol Chem 285:2580–2590. doi:10.1074/jbc.M109.068726

Wenstrup RJ, Florer JB, Brunskill EW, Bell SM, Chervoneva I, Birk DE (2004) Type V collagen controls the initiation of collagen fibril assembly. J Biol Chem 279:53331–53337. doi:10.1074/jbc.M409622200

Wenstrup RJ, Smith SM, Florer JB, Zhang G, Beason DP, Seegmiller RE, Soslowsky LJ, Birk DE (2011) Regulation of collagen fibril nucleation and initial fibril assembly involves coordinate interactions with collagens V and XI in developing tendon. J Biol Chem 286:20455–20465. doi:10.1074/jbc.M111.223693

Wright GM, Keeley FW, Robson P (2001) The unusual cartilaginous tissues of jawless craniates, cephalochordates and invertebrates. Cell Tissue Res 304:165–174. doi:10.1007/s004410100374

Wu JJ, Weis MA, Kim LS, Carter BG, Eyre DR (2009) Differences in chain usage and cross-linking specificities of cartilage type V/XI collagen isoforms with age and tissue. J Biol Chem 284:5539–5545. doi:10.1074/jbc.M806369200

Wu JJ, Weis MA, Kim LS, Eyre DR (2010) Type III collagen, a fibril network modifier in articular cartilage. J Biol Chem 285:18537–18544. doi:10.1074/jbc.M110.112904

Xiao J, Cheng H, Silva T, Baum J, Brodsky B (2011) Osteogenesis imperfecta missense mutations in collagen: structural consequences of a glycine to alanine replacement at a highly charged site. Biochemistry 50:10771–10780. doi:10.1021/bi201476a

Xu Y, Keene DR, Bujnicki JM, Höök M, Lukomski S (2002) Streptococcal Scl1 and Scl2 proteins form collagen-like triple helices. J Biol Chem 277:27312–27318. doi:10.1074/jbc.M201163200

Yamada Y, Avvedimento VE, Mudryj M, Ohkubo H, Vogeli G, Irani M, Pastan I, de Crombrugghe B (1980) The collagen gene: Evidence for its evolutionary assembly by amplification of a DNA segment containing an exon of 54 bp. Cell 22:887–892. doi:10.1016/0092-8674(80)90565-6

Yandell M, Mungall CJ, Smith C, Prochnik S, Kaminker J, Hartzell G, Lewis S, Rubin GM (2006) Large-scale trends in the evolution of gene structures within 11 animal genomes. PLoS Comput Biol 2:e15. doi:10.1371/journal.pcbi.0020015

Yasothornsrikul S, Davis WJ, Cramer G, Kimbrell DA, Dearolf CR (1997) Viking: identification and characterization of a second type IV collagen in Drosophila. Gene 198:17–25. doi:10.1016/S0378-1119(97)00274-6

Yoneda C, Hirayama Y, Nakaya M, Matsubara Y, Irie S, Hatae K, Watabe S (1999) The occurrence of two types of collagen proα-chain in the abalone *Haliotis discus* muscle. Eur J Biochem 261:714–721. doi:10.1046/j.1432-1327.1999.00313.x

Yurchenco PD (2011) Basement membranes: cell scaffoldings and signaling platforms. Cold Spring Harb Perspect Biol 3(2):pii: a004911

Yurchenco PD, Ruben GC (1987) Basement membrane structure in situ: evidence for lateral associations in the type IV collagen network. J Cell Biol 105:2559–2568. doi:10.1083/jcb.105.6.2559

Zhang G, Cohn MJ (2006) Hagfish and lancelet fibrillar collagens reveal that type II collagen-based cartilage evolved in stem vertebrates. Proc Natl Acad Sci U S A 103:16829–16833. doi:10.1073/pnas.0605630103

Zhang G, Miyamoto MM, Cohn MJ (2006) Lamprey type II collagen and Sox9 reveal an ancient origin of the vertebrate collagenous skeleton. Proc Natl Acad Sci U S A 103:3180–3185. doi:10.1073/pnas.0508313103

Zhang X, Boot-Handford RP, Huxley-Jones J, Forse LN, Mould AP, Robertson DL, Lili AM, Sarras MP Jr (2007) The collagens of hydra provide insight into the evolution of metazoan extracellular matrices. J Biol Chem 282:6792–6802. doi:10.1074/jbc.M607528200

Zhou J, Ding M, Zhao Z, Reeders ST (1994) Complete primary structure of the sixth chain of human basement membrane collagen, α6(IV). Isolation of the cDNAs for α6(IV) and comparison with five other type IV collagen chains. J Biol Chem 269:13193–13199

Chapter 4
The Evolution of Elastin

Fred W. Keeley

Abstract Elastin is the matrix protein imparting the physiologically essential properties of extensibility and elastic recoil to large arteries, lung parenchyma and other vertebrate tissues. Elastin is a polymeric protein formed from tropoelastin monomers and cross-linked, like collagens, through the side chains of lysine residues by the action of lysyl oxidase. Unlike many matrix proteins, elastin appeared relatively late in evolution, present in all species from sharks to humans, but absent at least as a vascular protein from lampreys, hagfish and other lower chordates and invertebrates. As an entropic elastomer, the polypeptide chains of elastin must remain disordered even in the polymeric state. Like many other disordered structural proteins, elastin has a low complexity, highly nonpolar and highly repetitive sequence, with the conservation of an unusual 'style' of sequence but an apparently relaxed requirement for precise sequence conservation, even between closely related species. Here, using both full-length and partial tropoelastin sequences representing the full range of phylogeny over which this protein is found, we describe the sequence elements and domain arrangements that are shared by all tropoelastins and are therefore likely to be required for its fundamental properties as an entropic elastomer. We then examine differences in sequence and domain arrangements between species perhaps representing adaptations to species- or tissue-specific functional requirements that may drive evolutionary changes. Finally, we consider the relationship of tropoelastins to other glycine-rich, nonpolar structural proteins that have been described in lower chordates and invertebrates and the possible evolutionary roots of this unusual protein.

F.W. Keeley (✉)
Molecular Structure and Function Program, Research Institute, The Hospital for Sick Children, 555 University Avenue, Toronto, Canada
e-mail: fwk@sickkids.ca

F.W. Keeley and R.P. Mecham (eds.), *Evolution of Extracellular Matrix*, Biology of Extracellular Matrix, DOI 10.1007/978-3-642-36002-2_4,

4.1 Introduction

Elastin is an extracellular matrix protein that provides the unusual properties of extensibility and elastic recoil to the tissues in which it is found. Although elastin does not have the widespread distribution of collagens and other matrix proteins, it often makes up a significant proportion of the tissues in which it is present. For example, large arterial vessels such as the aorta, parenchyma of the lungs, certain elastic ligaments and cartilages, all of which depend on a resilient extracellular matrix for their physiological function, can contain more than 50 % elastin by dry weight. Functionally the thoracic aorta acts as a damping chamber, expanding with each bolus of blood ejected from the ventricle and then recoiling, smoothing the pressure wave and passing the blood down the arterial tree. Similarly the expansion and recoil of the elastic matrix of the lung is an important factor in normal respiratory activity.

Elastin is a polymeric protein assembled from tropoelastin, its monomeric precursor. The polymer is stabilized by covalent cross-links formed from products of the oxidative deamination of side chains of lysine residues through the action of lysyl oxidase. The major cross-links in elastin are desmosine and isodesmosine, each formed from four lysine residues. Also present in smaller amounts are merodesmosine and lysinonorleucine, formed from three and two lysine residues, respectively. Although the precise mechanism for the assembly and cross-linking of the polymer is still a matter of investigation, both self-organizing properties intrinsic to the monomer itself (Clarke et al. 2006; Yeo et al. 2011) and the participation of other matrix-associated proteins (Wagenseil and Mecham 2007; Cirulis et al. 2008; Cain et al. 2008) appear to be important for determining the final architecture of the polymeric matrix. For general reviews of elastin biochemistry, see Mithieux and Weiss (2005) and Kielty (2006).

Once elastin has been cross-linked into its polymeric form it becomes completely insoluble except in proteolytic reagents. Indeed, classical biochemical methods for the isolation of polymeric elastin rely on extraction of all other tissue components, leaving an insoluble residue of purified elastin. As well as this persistent insolubility, the normal turnover rate for polymeric elastin is very low with, for example, a half-life in human tissues such as the aorta or lung in the order of many decades (Powell et al. 1992; Shapiro et al. 1991; Lillie and Gosline 2002; Ritz-Timme et al. 2003). As a result the matrix must have extraordinary durability. For example, the elastic matrix of the human aorta must be able to undergo literally billions of cycles of extension and elastic recoil over a lifetime, without structural failure.

The earliest insights into the evolution of elastin were provided by the work of Sage and Gray. In a series of papers (Sage and Gray 1979, 1980, 1981), using criteria for elastin based on histological characteristics, persistent insolubility, amino acid composition and desmosine cross-links, they investigated the presence and characteristics of elastins from arterial tissues of a wide range of vertebrate and invertebrate species. They reported that arterial elastin, identifiable by these criteria, was present in all vertebrates with the exception of the agnathans (jawless fish). However, recognizable elastin could not be detected in lower chordates or invertebrates. They concluded that, unlike collagens and many other matrix proteins that are phylogenetically ancient, elastin appeared much later in evolution, probably corresponding to the development of pulsatile, high-pressure closed-circuit circulatory systems in which substantial 'Windkessel' or damping chamber properties were required.

Extensibility and elastic recoil is an uncommon property for a protein. This, together with its impressive durability as a biomaterial and the propensity of the monomer for self-assembly, has prompted attempts to understand how these unusual characteristics can arise from the sequence and domain arrangement of the protein. Comparison of sequences of elastins across as broad a phylogenetic range of species as possible provides one approach both to identify common features responsible for such fundamental properties and to understand the relationship of the evolutionary history of the protein to differing functional requirements among species.

Tropoelastins from all species must share fundamental properties, including the ability to assemble into a polymeric matrix aligning lysine residues for cross-linking. Most importantly, this polymeric matrix must have the unusual properties of elastomeric extension and recoil with low energy loss (i.e. high resilience) and low viscoelasticity. Such fundamental properties are no doubt derived from the sequence and domain arrangements of the soluble monomer as well as the assembly and final architecture of the elastic matrix. We will focus first on the sequence elements that are shared by all tropoelastins and which therefore may be related to these fundamental properties. Subsequently we will address the question of the differences between tropoelastins across species and how these differences may reflect modifications to properties of polymeric elastin matrices in response to species-specific physiological challenges. Finally, since tropoelastin is a relative evolutionary newcomer compared to many or most other matrix structural proteins, we will address the question of the origin of tropoelastin and its possible relationship to earlier matrix proteins.

4.2 Tropoelastin Sequences

4.2.1 Finding Tropoelastins

Searches for tropoelastin sequences in databases can be hindered by the unusually repetitive nature of these sequences both at the DNA and protein levels. Thus, reliance on already processed data or the use of low-complexity filters, often default settings for searches, can result in failure to detect tropoelastin sequences. Furthermore, an apparently relaxed requirement for sequence conservation in tropoelastins can present challenges for identifying exonic regions in genomic sequence. Establishing protein sequence and overall composition is complicated by the common occurrence of several sequence variants arising from species-specific alternate splicing. There are also several species-specific examples of exons which have been lost to the genomic sequence as well as exons that are present in the genomic sequence but are apparently always spliced out of the protein. On the other hand, with experience, exonic sequences of tropoelastin can often be recognized by manual inspection of translated genomic sequence because of the unique style of the sequence, including the characteristic nonpolar tandem repetitive sequences in hydrophobic domains and the spacing and sequence context of the lysine residues in cross-linking domains.

4.2.2 Reference Sequences

Although genomic sequence for tropoelastins from several species, particularly mammalian, can be found in various databases, few of these have been carefully annotated, with intron/exon borders mapped by comparison to cDNA or EST sequences. Therefore, for the purposes of this review, we have provided a set of curated, full-length tropoelastin reference sequences with representation from all phylogenetic groups in which the protein is found. The amniote group includes human, bovine, mouse, chicken and lizard tropoelastins. The amphibian group includes the pair of tropoelastins that have been identified in frog (*Xenopus tropicalis*), designated Xtr eln-1 and Xtr eln-2 (Chung et al. 2006; Miao et al. 2009). The teleost group includes tropoelastins from zebrafish (*Danio rerio*) (Chung et al. 2006; Miao et al. 2007) and fugu (*Takifugu rubripes*), each of which expresses two versions of the protein. Originally these were also called eln-1 and eln-2 but more recently have been officially designated eln-a and eln-b, respectively. In all species in which two genes for tropoelastin have been identified, the general character of the eln-1 (eln-a) version appears to resemble more closely that of the single tropoelastin in amniotes (Table 4.1). A partial sequence of tropoelastin from elephant shark (*Callorhinchus milii*) is also included, representing the most extensive sequence of tropoelastin currently available from elasmobranchs (cartilaginous fishes). All of these sequences and their exon borders have been carefully annotated either by ourselves or others. In all cases exon numbers and borders are indicated. Two hundred bases of the 3′UTR are included (lower case), with the stop codon bolded. For amniote species (human, bovine, mouse, chicken, lizard), sequence similarities allows alignment and cross-identification of most exons, with the exceptions indicated. Some sequence comparisons will also use partial sequences or regions of tropoelastins from other species. In these cases, database references to the sequences are provided.

4.2.2.1 Amniotes: Human (*Homo sapiens*) Tropoelastin, RefSeq NM_000501

Compared to other amniote species, human tropoelastin lacks exons 34 and 35 which have been lost from the genomic sequence during primate evolution (Szabó et al. 1999). Although exon 22 is present in the genomic sequence, this exon never appears in cDNA, EST or protein sequences and is therefore not included in the reference sequence. Similarly exon 26A, originally reported as an extension of exon 26, is seldom if ever found in cDNA or EST sequences and is not included. Compared to the several other RefSeqs for human tropoelastin, each corresponding to different splice variants, NM_000501 represents the most complete version of the protein sequence.

Table 4.1 Characteristics of reference tropoelastins

	Hum	Bov	Mse	Ckn	Liz	Xtr-1	Xtr-2	Zef-a	Zef-b	Fug-a	Fug-b	Esk[a]
ala (A)	22.5	20.9	21.4	15.9	16.9	13.7	10.2	16.6	7.3	9.7	6.5	11.5
arg (R)	0.9	0.8	1.1	0.8	1.4	0.5	1.3	0.9	1.7	1.1	2.2	2.0
asn (N)	0.0	0.0	0.1	0.1	0.3	0.3	4.3	0.1	0.0	0.5	0.8	0.2
asp (D)	0.1	0.4	0.0	0.0	0.0	0.0	0.9	0.0	0.4	0.0	0.6	0.0
cys (C)	0.3	0.3	0.2	0.3	0.0	0.2	0.4	0.2	0.1	0.2	0.1	0.3
gln (Q)	1.1	1.2	0.8	1.2	1.4	2.6	11.0	2.2	1.8	2.6	1.5	7.0
glu (E)	0.3	0.1	0.0	0.0	0.0	0.0	1.6	0.0	0.4	0.0	0.1	0.2
gly (G)	29.8	31.7	37.1	34.2	37.5	38.8	19.4	35.6	45.5	41.7	47.5	32.4
his (H)	0.0	0.0	0.0	0.0	0.0	0.0	0.0	0.1	0.0	0.4	0.1	0.0
ile (I)	2.3	2.6	2.3	1.8	3.1	2.4	2.3	2.1	3.2	1.5	2.4	3.0
leu (L)	5.7	6.0	5.6	5.7	5.8	6.2	4.8	6.4	7.5	7.4	2.8	4.0
lys (K)	5.0	5.2	4.6	4.4	4.7	4.6	6.5	4.7	2.6	4.3	2.1	3.8
met (M)	0.0	0.0	0.0	0.0	0.0	0.0	0.0	0.1	0.0	0.1	0.1	0.0
phe (F)	2.3	3.1	1.8	1.9	1.1	1.0	2.3	1.1	0.9	2.3	3.9	0.7
pro (P)	12.3	12.1	10.8	14.5	11.4	15.4	13.6	12.7	10.7	9.5	12.6	13.1
ser (S)	0.7	1.1	1.3	0.5	1.1	0.6	7.4	2.0	2.6	1.8	2.6	0.3
thr (T)	1.3	1.1	1.6	0.9	1.5	2.1	2.2	3.1	5.4	4.7	7.0	1.2
trp (W)	0.0	0.0	0.0	0.0	0.0	0.1	0.0	0.0	0.0	0.0	0.0	0.0
tyr (Y)	2.2	1.0	2.6	1.4	3.1	4.4	8.3	4.8	2.9	4.6	2.9	5.5
val (V)	13.1	12.5	8.5	16.4	10.7	7.3	3.6	7.3	6.7	7.4	4.3	15.0
# Exons[b]	33	35	36	33	32	54	39	55	57	38	34	38[a]
% Coding[c]	5.8	7.2	5.9	12.7	–	6.1	8.0	6.8	18.0	8.4	36.2	–
Total aa	697	736	832	773	718	1,156	557	1,139	2,029	903	1,542	602[a]
Predicted MW(kD)	59.9	62.8	69.1	65.7	60.8	99.0	56.5	98.5	170.0	77.9	129.5	54.9[a]
% Nonpolar[d]	85.7	85.8	85.7	88.5	85.4	83.8	53.9	80.7	80.9	77.2	76.1	79.0
D + E[e]	0.4	0.5	0.0	0.0	0.0	0.0	2.5	0.0	0.8	0.0	0.7	0.2
K + R[e]	5.9	6.0	5.7	5.2	6.1	5.1	7.8	5.6	4.3	5.4	4.3	5.8
GRAVY[f]	0.697	0.691	0.479	0.686	0.447	0.147	−0.709	0.225	0.144	0.140	−0.161	0.231

[a]Partial sequence only
[b]Excluding signal peptide (exon 1)
[c]# Bases in exons/total bases between start and stop codons
[d]Sum of A, G, I, L, P and V
[e]Residues/100
[f]Grand average of hydropathy (http://web.expasy.org/protparam/)

^{1}MAGLTAAAPRPGVLLLLLSILHPSRPG|^{2}GVPGAIPGGVPGGVFYP|^{3}GAGLG
ALGGG|^{4}ALGPGGKPLKP|^{5}VPGGLAGAGLGA|^{6}GLGAFPAVTFPGALVPGGVADA
AAAYKAAKA|^{7}GAGLGGVPGVGGLGVSA|^{8}GAVVPQPGAGVKPGKVP|^{9}GVGLPGVY
PGGVLP|^{10}GARFPGVGVLPGVPTGAGVKPKAP|^{11}GVGGAFAGIP|^{12}GVGPF
GGPQPGVPLGYPIKAPKLP|^{13}GGYGLPYTTGKLPY|^{14}GYGPGGVAGAAGKAGYP
TGT|^{15}GVGPQAAAAAAAKAAAKF|^{16}GAGAAGVLPGVGGAGVPGVPGAIPGIG
GIA|^{17}GVGTPAAAAAAAAAAKAAKY|^{18}GAAAGLVPGGPGFGPGVVGVPGAGV
PGVGVPGAGIPVVPGAGIPGAAVP|^{19}GVVSPEAAAKAAAKAAKY|^{20}GARPGV
GVGGIPTYGVGAGGFPGFGVGVGGIPGVAGVPGVGGVPGVGGVPGVGIS|^{21}PEA
QAAAAAKAAKY|^{23}GVGTPAAAAAKAAAKAAQF|^{24}GLVPGVGVAPGVGVAPGVGV
APGVGLAPGVGVAPGVGVAPGVGVAPGIGPGGVA|^{25}AAAKSAAKVAAKAQL|^{26}RAA
AGLGAGIPGLGVGVGVPGLGVGAGVPGLGVGAGVPGFGA|^{27}VPGALAAAKAA
KY|^{28}GAAVPGVLGGLGALGGVGIPGGVV|^{29}GAGPAAAAAAKAAAKAAQF|^{30}GLVG
AAGLGGLGVGGLGVPGVGGLG|^{31}GIPPAAAAKAAKY|^{32}GAAGLGGVLGGAGQ
FPLG|^{33}GVAARPGFGLSPIFP|^{36}GGACLGKACGRKRK**uga**gcttcctaggacccct
gactcacgacctcatcaacgttggtgctactgcttggtggagaatgtaaaccctttt
gtaaccccatcccatgcccctccgactccccaccccaggagggaacgggcaggc
cgggcggccttgcagatccacagggcaaggaaacaagaggggagcggccaagtg
ccccgaccaggaggcccccct

4.2.2.2 Amniotes: Bovine (*Bos taurus*) Tropoelastin, Based on RefSeq NM_175772.1

RefSeq NM_175772.1 does not include exon 33. However, exon 33 is present in the genomic sequence and is reported in a number of ESTs, notably those derived from foetal spinal column (e.g. EE231886) and foetal lung (e.g. DV852340). This exon is therefore included in the protein sequence.

^{1}MRSLTAAARRPEVLLLLLCILQPSQPG|^{2}GVPGAVPGGVPGGVFFP|^{3}GAGLGGL
GVG|^{4}GLGPGVKPAKP|^{5}GVGGLVGPGLGA|^{6}EGSALPGAFPGGFFGAGGGAAGA
AAAYKAAAKA|^{7}GAAGLGVGGIGGVGGLGVST|^{8}GAVVPQLGAGVGAGVKPGK
VP|^{9}GVGLPGVYPGGVLPGA|^{10}GARFPGIGVLPGVPTGAGVKPKAQ|^{11}VGAGAF
AGIP|^{12}GVGPFGGQQPGLPLGYPIKAPKLP|^{13}AGYGLPYKTGKLPY|^{14}GFGPG
GVAGSAGKAGYPTGT|^{15}GVGPQAAAAAAKAAAKL|^{16}GAGGAGVLPGVGVGGPGI
PGAPGAIPGIGGIA|^{17}GVGAPDAAAAAAAAAKAAKF|^{18}GAAGGLPGVGVPGVGV
PGVGVPGVGVPGVGVPGVGVPGVGVPGVGVPGVGVPGVGVPGVGVP|^{19}GALSPA
ATAKAAAKAAKF|^{20}GARGAVGIGGIPTFGLGPGGFPGIG|^{21}DAAAAPAAAAAKA
AKI|^{22}GAGGVGALGGVVPGAPGAIPGLPGVGGVP|^{23}GVGIPAAAAAKAAAKAA
QF|^{24}GLGPGVGVAPGVGVVPGVGVVPGVGVAPGIGLGPGGVIGAGVP|^{25}AAA
KSAAKAAAKAQF|^{26}RAAAGLPAGVPGLGVGAGVPGLGVGAGVPGLGVGAGVPG
PGA|^{27}VPGTLAAAKAAKF|^{28}GPGGVGALGGVGDLGGAGIPGGVA|^{29}GVVPAA
AAAAKAAAKAAQF|^{30}GLGGVGGLGVGGLGAVPGAVGLG|^{31}GVSPAAAAKAA
KF|^{32}GAAGLGGVLGAGQPFPIG|^{33}GVAARPGFGLSPIFP|^{34}GGAGGLGVG|
^{35}GKPPKPFGGALGALGFP|^{36}GGACLGKSCGRKRK**uga**gcttcccgggacc

cctgactcacgacctcatcaacgttggtgctactgcttggtggagaatgtaaacc
ctttgtgaccccacccccctttcccatgccccctcctaagttcccaccccgggtg
ggaacagggcaggccgggcggccttggaaatccacagggcaaggaggcaagagag
tggtggccaagtgggccccggg

4.2.2.3 Amniotes: Mouse (*Mus musculus*) Tropoelastin, RefSeq NM_007925

Mouse and rat tropoelastins contain an additional exon (designated exon 4a) located between amniote exons 4 and 5. This additional exon is highly represented in mouse ESTs (e.g. CJ092255) and is included in the protein sequence. There is no indication of a nascent or vestigial exon 4a in other amniote species.

^{1}MAGLTAVVPQPGVLLILLLNLLHPAQPG|^{2}GVPGAVPGGLPGGVPGGVYYP|^{3}GAG
IGGLGGGGG|^{4}ALGPGGKPPKP|^{4a}GAGLLGTFGA|^{5}GPGGLGGAGPGA|^{6}GLGAFPA
GTFPGAGALVPGGAAGAAAAYKAAAKA|^{7}GAGLGGVGGVPGGVGVGGVPGGVGVG
GVPGGVGVGGVPGGVGGIGGIGGLGVST|^{8}GAVVPQVGAGIGAGGKPGKVP|^{9}GVGL
PGVYPGGVLPGT|^{10}GARFPGVGVLPGVPTGTGVKAKAP|^{11}GGGGAFAGIP|^{12}GVG
PFGGQQPGVPLGYPIKAPKLP|^{13}GGYGLPYTNGKLPY|^{14}GVAGAGGKAGYPT
GT|^{15}GVGPQAAAAAAKAAAKL|^{16}GAGGAGVLPGVGGGGIPGGAGAIPGIGG
IA|^{17}GAGTPAAAAAAKAAAKAAKY|^{18}GAAGGLVPGGPGVRLPGAGIPGVGGIP
GVGGIPGVGGPGIGGPGIVGGP|^{19}GAVSPAAAAKAAAKAAKY|^{20}GARGGVGIPT
YGVGAGGFPGYGVGAGAGLG|^{21}GASPAAAAAAAKAAKY|^{22}GAGGAGALGGLVPG
AVPGALPGAVPAVPGAGGVP|^{23}GAGTPAAAAAAAAAKAAAKA|^{24}GLGPGVGGV
PGGVGVGGIPGGVGVGGVPGGVGPGGVTGIGAGPGGLG|^{25}GAGSPAAAKSAAKA
AAKAQY|^{26}RAAAGLGAGVPGFGAGAGVPGFGAGAGVPGFGAGAGVPGFGAGAGV
PGFGAGA|^{27}VPGSLAASKAAKY|^{28}GAAGGLGGPGGLGGPGGLGGPGGLGGAGVP
GRVA|^{29}GAAPPAAAAAAAKAAAKAAQY|^{30}GLGGAGGLGAGGLGAGGLGAGGLGA
GGLGAGGLGAGGLGAGGLGAGG|^{31}GVSPAAAAKAAKY|^{32}GAAGLGGVLGARPFP
GG|^{33}GVAARPGFGLSPIYP|^{34}GGGAGGLGVG|^{35}GKPPKPYGGALGALGYQ|^{36}GGG
CFGKSCGRKRK**uga**tcttctggggacccctgactcgcgacctcatcaacgttggt
gctactgcttggtggagaatgtaaaccttctatgaccacccccttccatcccccct
gacccccacctgggaggggacaacaggccagtggccttggaaacccacaggacaa
ggaaatcagacagcagcagccatgcagccctaaccagaaactcccccca

4.2.2.4 Amniotes: Chicken (*Gallus gallus*) Tropoelastin

This sequence has been personally curated based on the genomic sequence (WAG-71G10), cDNA and EST sequences and RefSeq XM_003642386.1, but differs slightly from RefSeq XM_003642386.1. Exons corresponding to 21, 22 and 32 in other amniote species are missing from the chicken genomic sequence, with no trace of vestigial exonic sequence in the adjacent introns. Exon 29a was placed between exons 29 and 30, in order to allow for the clear cross-species alignment of exon 30 in amniotes.

[1]MARQAAAPLLPGVLLLFSILPASQQG|[2]GVPGAIPGGGVPGGGFFP|[3]GAGVGGL
GAGLGA|[4]GLGAGGKPLKP|[5]GVSGLGGLGPLGLQP|[6]GAGVGGLGAGLGAFPGAA
FPGAASAAALKAAAKA|[7]GAGLGGVGGIGGLGGVGGVGVPG|[8]GLGVPGVVQPGVGAA
GKPPKVP|[9]GAGIPGAFPGGGVLPGA|[10]GIRFPGVGVLPGVPTGTGIKAKGP|[11]GAG
AFAGIP|[12]GLGGFGGQQPGVPLGYPIKAPKLP|[13]GGYRLPFVN|[14]GLGPGGIGA
GVLAGKAGYPTGT|[15]GVGAQAAAAKAAAKY|[16]GAGVLPGAGGIPGVGGVVPGVGV
VPGAG|[17]VGGPAAAAAAAKAAAKAGAY|[18]GAGVLPGAGGVPGVVPGVGVVPGLV
PGVGGIPGVA|[19]GVGTPAGAAAAAAKAAKY|[20]GAGVPGVGVPGVGIGGVPGVPG
VPGVPGVPGVPGVPGVPGVPGVPGVPGVVPGVG|[23]VGGPAAAAAAKAAAKA
AAF|[24]GAGRVPGVGVPGAVPGVGVPGVGVPGVGVPGVGVPGVGVPGVGVPGV
GVPGVGVPGVGVPGVGVPGVGVPGVGVPGLVP|[25]GAGPAAAAKAAAKAAKY|[26]GAG
GLAPGVGGLAPAVGGLAPGVGGLVPGVGGLVPGVGGLAPGVGGLAPGVGA
VP|[27]GVGGPAAAAKAAAKAAKY|[28]GAGVGGVPGAVPGAVPGVPGVPGVTPGVGG
VPSLVPGVGVPGVGVLPGA|[29]GIPQVGVQPGAKPPKF|[29a]GVPGAGVPGVG
GIP|[30]GGLGVGGLGVGGLGAGGLGA|[31]GLLYPGAAGKPPKP|[33]GVGVPGFGVSPI
FP|[34]GGVGGQLGFG|[35]GKPPKTYGGALGALGFR|[36]GGVGCAQGKYCGRKRK**uaa**c
cgcccaccatcaccgatgacctcatgaactttggtgctactgcatggtccaga
atgtaaatccccaccaagctgagacccccccggccccacgctcccggcagg
gaccccacggcagcgtcaggggccgccccccacccctggaaccccccagcccc
agcagggagcgcggaataaactgcggggagcagccgtcc

4.2.2.5 Amniotes: Lizard (*Anolis carolinensis*) Tropoelastin, Based on NW_003339591.1

This sequence has been personally curated based on the genomic sequence NW_003339591.1, using EST data and splice site prediction software and by analogy to chicken tropoelastin sequence. Exon 1 was not present in the genomic sequence. As in chicken, exons 21 and 22 are absent from the genomic sequence and exon 29a is similar to that in chicken. Protein sequence between exons 32 and 36 is recognizable but prediction may be less reliable, and exons 32 and 33 may also be absent from the genome.

[2]GVPGAGGGIPGRGLYP|[3]GIGGGGLGAGGLGTGLGGA|[4]GYGAGAKPPKP|[5]GVG
GAGSL|[6]GGLYPGGAYPGGVFQGAGTPAQAAAAYKAAAKA|[7]GLGVPGGGIGGGVG
VPGGGVGVGVPGGGVGVPGGGVGVGVPGGGIGVPG|[8]GGVIPQTGAQPGAGVGAK
PGKIP|[9]GVGIPGVFPGGVLPGT|[10]GIRYPGVGVLPGVPTGNGLKAKVP|[11]GAGGG
AFAGIP|[12]GFGGFGGGQQPGLPLGYPIKAPKLP|[13]GSYYGKLSF|[14]GAGPGGVF
GGAGGKAGYPTGT|[15]GVGAQAAAAKAAAKY|[16]GVGAGRLPGAGALGVGGLLPGA
GGVPGAI|[17]GAGGPAAAAAAAAKAAAKARAY|[18]GAGGLPGAGVGGVGVPGGVG
IPGVGGLPGVGPGVGGVP|[19]GLSPAAAAAAAKAAKY|[20]GLGPGGVGAIPGGVG
VGGIPGVGGVPGVGVGGVPGAGIGGVPGLPAAP|[23]GAGASAAAAAKAAAKAA
AY|[24]GVAPGYGGIGGIGTGGIGTGGLVPGGVGVG|[25]APGTQAAAVAAKAAAKAA
KY|[26]GAGGARGLVPGAGVGGLVPGVGGVGGVGGIGGVGGVGGGLVPGAG|[27]GAG
SISAAVAAKAAAKAAKY|[28]GAGRGGVLPGGGVLPGGGVLPGGGVLPGGGVLPGGG

VLPGGGVVPGGGVLPG|[29]GIPQIGVQPGAKPPKY|[29a]GVPGAGLGGLGVP|[30]GGG
LVPGAGGAGYGGLGAGGLSP|[31]GLTYPGAGGKPPKP|[34]GYGGGLP|[35]AKAAKY
GVGVAVPKRNFP|[36]GALGRRRK**uag**ccccgaactgaccgccgatctttcacctc
atgaactttggtgctactgcctggtctcgaatgtacccccacccacccacccac
cctccttcagccacggaagccctctcccttggcaccgtccttgcctctgccccac
acagccgcccaaagacggcccgccgagcccctttgccctgtctttggtgctactt
cgacca

4.2.2.6 Amphibians: Frog (*Xenopus tropicalis*) Tropoelastin-1, Provisional RefSeq NM_001078709.1

This sequence was curated using genomic sequence in overlapping scaffolds 32 (containing exon 1) and 534, positions 77111–125079 (containing the remainder of the gene). Exon borders and expressed sequence were confirmed using cDNAs and ESTs (Chung et al. 2006). *Xenopus tropicalis* possesses two elastin genes, designated eln-1 and eln-2 (see below), with differential tissue expression patterns (Miao et al. 2009).

[1]MESIRPAAPLRSIILFLAFCQISLQG|[2]GVPGAGVGGQFYPGAGAYPGAGAYPG
AGAYPGAGAYPGAGAYP|[3]GAGVGTGGIAPGIGA|[4]GVYPGGKPPKP|[5]GVGGAG
AGGLGAGGLGGLGVGGLGAGGKPGKA|[6]GYLPGGPGAGAGV|[7]GGVPGAYPGAG
LP|[8]GAGVVPGVST|[9]GAVVPQAGVQPGAAGKPGKLP|[10]GAGIPGVFPGGLVP
GT|[11]GGRFPGVGVLPGVATGNGVKQKAP|[12]GGGAFAGIP|[13]GFNGFGGQQPGV
PLGYPIKSPKLP|[14]GGYGLPYTAGKPGY|[15]GYGAFGAGAAGKAGKAGYPTGT|[16]GVG
ALTAAQQAAAKAAKY|[17]GIGGIPGAVPGTGLGGVPGLTPGGGAVPGVVPGLVPG
AGGVPGA|[18]GIPQLGVQPGAKASKY|[19]GLPGVGGVPGVGGVPGVGGVPGVGGV
PGVGGVPGVGGVPGVGGVPGP|[20]TISGLGAKPPKIG|[21]GVGAGGFPAG|[22]VGVG
GVPGA|[23]GAYPAGGKPPKP|[24]GYGAGAGLIPGGGITPGAGLGAGLVPGAGT
GA|[25]GLYPGGKPPKP|[26]GYGLPGGALTPGAGLQPGAGLQPGAGLQPGAGLQPG
AGLQPGAGLQPGAATGA|[27]GTYPGGKPPKP|[28]GLGIGGLTPGVAGLGPGAGLVP
GAGGLAPGAAG|[29]GAYPGGKPPKP|[30]GAGLVPGAGGLAPGAGGLAPGAGGLVPG
AGGLAPGAAA|[31]GAYPGGKPPKP|[32]GYGPGAGLFPGAGVQPGGIGAGVAPG
AG|[33]GLYPGAGGKPPKP|[34]GYGLGAGLVPGGGLVQPGGIATGVAPGGAG|[35]GLY
PGAGGKPPKP|[36]GYGPGAGFVPGAGIGAGAGAGLQPGGIGAGVAPGGVGAG|[37]GLY
PGAGGKPPKP|[38]GYGPGAGWVPGAGVGLGTGVQPGGTGTGVAPGGIGAG|[39]GLYP
GAGGKPPKP|[40]GYGLGAGLVPGAGVGPGAGLQPGGIGAGGVGAG|[41]GLYPGGKPP
KP|[42]GYGPGAGYVPGAGYVPGAGVGPGAGLQPGGIGAGVAPGGVGAG|[43]GLYPGG
KPPKP|[44]GYGPRAGLGTGAGIGAGLQPGGIGAGVAPGGVGAG|[45]GLYPGGKPP
KP|[46]GYGPGAGFVPGVSNGAGQQPGGIGTGGVPGGIGAG|[47]GLYPGGKPPKP|[48]GYG
PGAGLGTGAGVGAGLQPGGIGAGVAPGGIGAG|[49]GLYPGAGGKPPKP|[50]GYGA
GVGPGAGVQPGAIGGGAGLGGVGAA|[51]GLYPGAGGKPPKP|[52]GYGGGIGAGLGL
GVGGLG|[53]GAGQLGYPLGG|[54]AQKPSKSYYR|[55]GGGLCGQGKYCGRRRK**uaa**g
cagccaaagagctacgagtaacctcatcaactatggtgctactgcatggtgtgaa
tgtaaaaggattcaagaatcccctctgtttttccatccccgcattccttccccag

cccttatttgtccatttagcatccatctctctgaacacggagtcaggcagctac
acataatatagggaagcctgcaatctgacactca

4.2.2.7 Amphibians: Frog (*Xenopus tropicalis*) Tropoelastin-2, Provisional RefSeq NP_001136298

This sequence was curated using genomic sequence in scaffold 638, positions 715450–755450. Exon borders and expressed sequences were confirmed using cDNAs and ESTs (Miao et al. 2009).

^{1}MRGSILLQVSLLLCLVGCSLQG | ^{2}GTGAVQSPVSSPLAVKAG | ^{3}GRYPQFGPQG
YQQGYQQGPT | ^{4}GIRNGYEAGYGVKAGKP | ^{5}AAGSLGVLPAGKQKP | ^{6}GYGNGRYP
SNFQQQ | ^{7}GIGAKPSKA | ^{8}GYGQPVGGVRGIGAQP | ^{9}GYGNGLYPSNLQQQ | ^{10}GIG
GKPPKA | ^{11}GYGAPAGGLPNVGVQPGYGNGLYPSNVQQP | ^{12}GIGGKPPKA | ^{13} GYGSP
AGVLPGAGVQP | ^{14}GFLNGAGRFPSNIQQP | ^{15}GFGSKPSKA | ^{16}GYGPGNYPRFG
AQA | ^{17}GYGAKASKA | ^{18}GLGSKPSKA | ^{19}GAFQQPYPS | ^{20}GLGSKPSKA | ^{21}GYPQG
VGNYPSNIPQQ | ^{22}GAFQQPYPG | ^{23}GPQEYIQGAAGYPNTYPQQ | ^{24}GLGAKPS
KA | ^{25}GAYQQPYPN | ^{26}VAGGYPQPIGNYPSSIQQQ | ^{27}GLGAKPSKA | ^{28}GAFQ
QPYPN | ^{29}GAYQQQPYPN | ^{30}SAYQQQPYPN | ^{31}GLNNFLGNGKGQ | ^{32}GVKSPYGS
LGALGKSSKQ | ^{33}GALGKFPYKSQPLPADALGYDSKSLKN | ^{34}SGAQLPFASQAGYP
DPASVKY | ^{35}GGGPVPYGPQGAYPDPSAIKY | ^{36}GGVPQYPETASSPISTLE | ^{37}EDL
SQNVPLE | ^{38}GTQLIAAGPTTPPTTQASPSTLKQKAYK | ^{39}VPEEQAAYRFYGTG
YQ | ^{40}GCAEC**uaa**gctctgaagtccaacatttgtccttggattggatgttttgt
gggttggggtgtatataggggagggtttgcagaggattaatcaataaaaataagta
aaatgtcaacaggttgccctgtttgtgtctttctttcagggtaaggactaagtag
gatgtgttcctattgggcgtttgctgccctgggcacatgtgtacagt

4.2.2.8 Teleosts: Zebrafish (*Danio rerio*) Tropoelastin-a, RefSeq NP_001073532.2

This sequence was curated using genomic sequence in gi:34366544 (Zv5 scaffold 824), which contains the entire gene. Exon borders and expressed sequences were confirmed using cDNAs and ESTs (Chung et al. 2006). Note that teleosts possess two elastin genes, designated eln-a and eln-b (see below), with differential tissue expression patterns (Miao et al. 2009).

^{1}MANRSVALLLLGFFLSVQLSPSLQG | ^{2}GVYVPQAGGGGAGPGGAGVLPGAGL
YP | ^{3}GAGSSPYKAAKAA | ^{4}AGGYGGAAGAGALGAGGYGA | ^{5}GGYGGYGGYGGAGR
FYPMAGGLKPAKS | ^{6}GAGLPGGVLPGGVLPGGVLPGGVLPGGVLPGGGLP | ^{7}GGA
GGKPPKG | ^{8}PGVGGAGTGLP | ^{9}GLGIPQTGVQGLGPGGVGPGAKAAKAGKAPVP | ^{10}GVG
VPGPYQGGLPGQ | ^{11}GFNGRGVLPGVATGTGLKPKS | ^{12}GGGQGLVPGA | ^{13}GGR
GPFGGPLQQGVFHGYPLKSPKAP | ^{14}GAGLPYTGGKLPY | ^{15}GYGGFGGGAGLPGGKA
GAGSKPGYPTGT | ^{16}GVGSLGVSAAQAKAAKY | ^{17}GAGAGLGGAGAFPGGVGAGLG
GAGAFPGGAGGFYPGAV | ^{18}GTGGLTPAQAKAAKY | ^{19}GAVPGATGIGGLPGAGG
LFPGPGVG | ^{20}GAGALSPAQAKAAKY | ^{21}GAGAVPGGAGVLPGGAGVLPGGAGGI

YPAPG|^{22}GAGALSPAQAKAAKY|^{23}GAVPGGAGVLPGGAGVLPGGAGVLPGGAG
IVPGAGGVYPAPG|^{24}GAGALSPAQAKAAKY|^{25}GAGAVPGGAGILPGGAGILPGG
AGILPGGAGILPGGAGIVPGAGGVYPGAG|^{26}GAGALSPAQAKAAKY|^{27}GAGAVP
GGAGILPGGAGILPGGAGIGPGGVYPGTG|^{28}GAGALSPAQAKAAKY|^{29}GAVPGG
AGILPGGAGILPGAGGVYPGAG|^{30}GAGALSPAQAKAAKY|^{31}GAGAVPGGAGVLP
GGAGIVPGRTGIVPAGPGIVPGAGGVYPATG|^{32}GLTPAQAKAAKY|^{33}GLGAAGG
AGAVPGVGGLYPGAG|^{34}GAGVAPGYGSVAGLGGQLGAG|^{35}GLAAGAKPPKY|^{36}GVP
GGTGFGVPGGVPGGIPITGPG|^{37}GYPAGAKALKY|^{38}GLGSGGLPAGTGVVPGLQ
RP|^{39}GLGGALGGAGTGTGYPFG|^{40}GYGAGAKPPKY|^{41}GVPGGVPGGVPGGVPG
GVPG|^{42}GYPAGVKPPKY|^{43}GVAGGAGTVPGAGGVPLAGTGTGLAPGAGVG|^{44}GYP
GGVKPPKT|^{45}GAGIAGTGIAAPGATPGGTGIVVPGVPGGTGIAVPSGTGVPILP
SGKP|^{46}AKLPTLAPAGTAVP|^{47}GGAQRVPGA|^{48}VPGGAYPYGS|^{49}GQYYGSAGP
SGVGTGVLPGAKPLKAP|^{50}GVGGAGGVGGVGGLYPGA|^{51}GVGLGTGGVGTGAKP
PKP|^{52}GYGSVGTGYAQP|^{53}GGVAAGGYGAYP|^{54}GGYGAGLSPQQ|^{55}AKAAKYG
GGLTGFLGGGYR|^{56}GGAGCQGKYCGRRRK**uag**aggagctcttgagatgtgaagt
gacctcaagaaacatggtgctactgcatgctctagaatgtacttccgatactcaa
accgaccgccccatcatgcacaggtgttgttttgtactgatagtatatatttgta
acagatccagatccctgaaagctggaataaccccagtcctgttcgtttatggcac
atcccgatcatgt

4.2.2.9 Teleosts: Zebrafish (*Danio rerio*) Tropoelastin-b, RefSeq NP_001041529.1.2

This sequence was curated using genomic sequence gi:50978520 (scaffold CH211-257O19), which contains the entire gene. Exon borders and expressed sequences were confirmed using cDNAs and ESTs (Chung et al. 2006).

^{1}MARRKVPLLLHGFLLLVLCRSSLQG|^{2}GIYIPAGAGGGPGAGLGAGVGTGAGLG
IGTGPGGGYGAGIGGAGPGVGPGGVGTGLGLGTGLGGLGTGTGIRPGT|^{3}GAGGLGGIG
PGGGGLGTGGLGTGGLGTAGLGT|^{4}GPFVKPPKT|^{5}GGYGGLPGIGAGGTGVGL
GGLGLRPAGPGVGGLVPGGFGVGTGVGGTGTGVP|^{6}AIGVGTGVGGLGIGGVGPGGA
GI|^{7}GVGGKPPKT|^{8} GVSTGGIGPGGVVPGGIRPGGIVPGGVGPGGIVPGGVGP
GGIFPGGVGPGGIVPGGVGPGGIVPGGVGPGGIVPGGVGPGGTVPGAGRKPG
KP|^{9} GYVGPGGGAGPFGTGTLGAGDPRFGAIPLTPGGTGTGYPSG|^{10}GGYGGYG
GPGGIYPGAAGQKPPKP|^{11}GIGPGGTGTGLGGAGFGLGGAGVGPGGAGIGPGG
TGIGSGGTGLGPGGTGVGTGGVGIGPGGTGTGGVGIAPGG|^{12}SPFLPQTGLPG
IGPGTGGKVGKPGKAPVP|^{13}GVGIPGLYQGGIVPGQ|^{14}GFGGRGVLPGVATGT
QT|^{15}GQLGQGALGAN|^{16}GRGGFGQQIPGVFGGYPLISPKSG|^{17}GAAKSQAKAA
KY|^{18}GAAGVPGLGGGLGVGGVPGAGGITGTSGVPGLGGVPGAGTVPGIGSVPGA
AGIPGVGGIPGAGGATPGGLYPGAGGIYPG|^{19}GAGAKARKY|^{20}ASGLGTGGLP
GSGVGLGGLGTGGLPGSGVGVGGLGTGGLPGSGVGVGGLGTGGLPGSGVGVGGLGT
GGLPGSGLGGGGLGAGGLPGSGVGGGGLGAGGLPGSGISP|^{21}GYAAAKAR
KY|^{22}AQLGRGGTLGTGGLPGGAAGVGLPGGGLGAGGQLGTGGLGTYGLPGGGLGG
LPGGGAGGLPVGGIGGLPGSGT|^{23}GYADAKARKY|^{24}AQLGRGGTLGTGGLPGG

TAGVGLPGGGLGAGGQLGGGLGTGGLPGGGLGTGGQLGGGLGTGGLPRGGLGTG
GQLGGGLGTGLPGGGIGTGGLGTGGLPTGGLGGLPGGGVGGLPGGGVGGLPG
SGI|^{25}GYADAKARKY|^{26}AQLGRGGTLGSGGLPGGTAGVGLGGLPGGGLGTGGQ
LGGGLGTGGLPGGGLGGLPGGGAGGLPGGGAGGLPGSGI|^{27}GYAEAKARKY|^{28}AQL
GRGGTLGTGGLPGGAAGVGLGGLPGGGLGTGGQLGGGLGTGGLGTGLGGLPGGG
AGGLPSGGVGGLP|^{29}GYAEAKARKY|^{30}GLLGRGTGLGPGTGLAAGIPGGLPSTG
IAPGGAGGLPGSGVGQGGIPGVGYGPGGVPSGGFGPGSVPGGGYGPG|^{31}GISAA
KARKY|^{32}GIIAGAGGIGLPGGQLGSGTVPGVGAGGGTPVAGIGPGGVPVGGYR
PGGVPAGGYGPGGVPAGGYGPGGVPAGGYGPGGVPAGGYGPGGVPAGGYGPGGV
PAGGYGPGGVPAGGFGPGSG|^{33}AYPGGAKALKY|^{34}GPGGSGGIPGLGLQGQVG
TGPAGGL|^{35}GYGPGSKAAKY|^{36}GLPGFGGALGTGALPGAGTGAG|^{37}GYGGAQ
KPPKY|^{38}GQPVGGVGSVPGGSVPATGGTGVTGTGIST|^{39}GGTAIPLATGTDRGGLR
PG|^{40}VIGSPDGGTRGTGEAQ|^{41}GGTKAPKT|^{42}ESGGPVEGS|^{43}GVTGGPIDTDG
LYGVAGTGIPILAGERP|^{44}GGTGVPRPE|^{45}IGDSGTLPGAKPLKPP|^{46}GTGG
GAIAGRGDTPGTIEGGPGSIGSGIGGVGGGPGGVGATPGGVGGVPGGVGGVPGG
VGGVPGGVGGVPGGVGGVPGGVGGVPGGVGGVPGGVGGVPGGVGGVPGGVGGVP
GGVGGVPGGVGGVPGGVGGVPGGVGSGLTGTGGVKPPK|^{47}VYGGAGGTGALGV
GGVGGLGPAGGGGAGLLPGGGGLLPGGGAAGTGFGLQKPGK|^{48}SYGGAGSLGAG
GILPGT|^{49}GIRFPSGAAVGSKPGK|^{50}VYGAGTLGGLGGPGGYGAGPGGYGGGP
GGYGSGPGGGYGSGPGSAGGLGGPGFG|^{51}GLGYGPGGVKPPK|^{52}SYGGAGALGG
AGQGGIGGGPGRLGVGPGGAGGIGGGLGVGPGGVGGLGGGQGVGPGGVGGGPGGL
GG|^{53}GFGGYGGVGGGPGGTGG|^{54}GLRYPGGAGAGKPGKA|^{55}GGSVPGYGTGAG
PGAGYQQQYP|^{56}GFGGPGAGGPGSAPLTPQQ|^{57}SKAAKYAALQGFLGAGG
YR|^{58}GGAGCQGKYCGRRRK**uga**cagtccagccgatagaagtgacctcataaataa
gtggtgctactgcctgatctcaaatgcactttttacagtctaaaagtgcctcatt
gcaatatgtgagagcgagagaaatttggacctgcattacaggaattctgcagcc
gtagtattagatgtgaacctctcaaggcacttgttgcactgcctcgactccctt
tact

4.2.2.10 Teleosts: Fugu (*Takifugu rubripes*) Tropoelastin-a

This sequence was curated using genomic sequence from IMCB scaffold 9 (positions 2251787 to 2285131). Exon borders were determined using splice site prediction software and by analogy to zebrafish tropoelastin sequences.

^{1}MASRHSLLLFCGLFVLALVQPALQG|^{2}GVYVPAAAGIRPGTTGGGTGFFP|^{3}GSA
GGVSGGYKPGKAA|^{4}GGYGGGRGTGPGGLVPGGVGQGGLGGGI|^{5}GQGGKGP
KP|^{6}AFYLFSQAKAPKSHT|^{7}SYGSLGGGGLGGGGLGGGGLGGGGLGGGGLGGGGQ
GFGGFGGYGG|^{8}GGHGGYGGGAGVGYQLGVGQKAAKR|^{9}GAGGSLLPYG|^{10}GT GG
TVLGAGVGPGGIGTGAAGKTQT|^{11}LPVIPQTGLPGGTVGGAGTKASKVP|^{12}GVG
VPGLFHGLVPGQ|^{13}GFGGRGVLPGVAIGSNLNPKS|^{14}VPGVGQGTQG|^{15}GGY
GGQMQPGVFHGYPLTSPK|^{16}GYGTAAKAAKY|^{17}GVPGGVPGGIPGGVPGGFPG
GVPGGGFPGGVPGGGFPGGVPGGGFPGGVPGGFPGGVPGGFPGGVPGGFPGGAP
LGGISPAKAAKY|^{18}GGAGGYGGLGGTGGLGGGGPGGYSAAAKAAKY|^{19}GVGGL

GGAGGLGGAGGLGGAGGLGGAGGLGGLGGTGGYNAAAKAQKY|^{20}GAGGLGGTG
LGGTGLGGTGLSGLGLGGTGGYNAAAKAQKY|^{21}AQALGGAGALGGAGLGGTGLG
GTGLGGTGGLGGTGGLGATGLGGLGGTGGYNAAAKAQKY|^{22}GAGALGGAGLGGT
GLGGRGLGGTGLGGTGGVGGGTYSAAAKAAKY|^{23}GGGAGIVPGGGPGGVIPGRV
PGYLPGYG|^{24}LGVPGGGAVPGGPGQFFPGGGVYG|^{25}GYGGKPPKY|^{26}GVGGGT
GIVPGGTGVVPGGTGLVPGGTGLVPGGTGVIPGGVGVIPGGVGGYSAAKAA
KY|^{27}GLTGGVGGALPVGGVGG|^{28}GYPGGVKPPKP|^{29}AGLVVPGATP|^{30}GGVG
VPAPSG|^{31}GAGLGAGGKAPKPLLP|^{32}GAGGVGLVPGAGGVGLVPGAVGTGLVPGA
GGYPY|^{33}VPYGVGTGTLPGAKPLKPP|^{34}GGGGVGVGRGVGIPLTACGFP|^{35}GYG
VLGGGGIYQP|^{36}GVVPGYGGLYPQQYPQ|^{37}GGYVPAPLTPQQ|^{38}AKAAKYGALQGFLG
GAGGAFR|^{39}GGAAGCQGKFCGRRK**uaa** aaggatctcttgagattcggagt
gacctcaagaaactatggtgctactgcatgatctagaatgtacatttt atatgc
aatcaagccccattagaagccatatagagacatttgttacttgacttcattgtt
ttttattaaaaagc caaaacccaccgacccagaatttgtacacgttctgaag
tgatggaggctgttt

4.2.2.11 Teleosts: Fugu (*Takifugu rubripes*) Tropoelastin-b

This sequence was curated using genomic sequence from IMCB scaffold 17 (positions 2251787 to 2285131). Exon borders were determined using splice site prediction software and by analogy to zebrafish tropoelastin sequences.

^{1}MARGTVVTYLHGVLLLALWKPSLQG|^{2}GVYVPAGGATGTGAGTGGAGAGFGP
GGAGTGLQPGGGAGPGTGGL|^{3}GGYGGQGPHGFGPGGIRPGGVGPGGFGPGGVG
PGGVGPGGFGPGGVGPGGVAPGGFGPGGVGPGGVGPGGFGPGGTGPGGLRPGTF
GP|^{4}GGKPPKT|^{5}GGYGNLGGVLGGFGTGPGGTGGGIGPGGFGPGGAGNYGTGPG
GVGPGGQGT|^{6} GGYGGYGPGGAGQYPGSGLKPPKT|^{7}GFGPGGSGVRPGGTGFGP
GGAGIGPGGTGFGPGGAGIGPGGTGFGPGGAGIGPGGTGFGPGGTGFGPGGQGF
GPGGAGVGPGGTGFRPGGAGVGPGGTGFRPGGAGVGPGGTGFGPGGAGVGPGGT
GFGPGGAGVGPGGTGFGPGGAGFGPGGTGLGPGG|^{8}VLPQTGGGTGGTTGGKAS
KLP|^{9}GIVMPGPYQGGYVPGQ|^{10}AFGGRGILPGVAIGSGLGTQN|^{11}GAGVLGQ
GGIGPGGTS|^{12}NGRGQPLPGVIRGYPLISPKL|^{13}GVGKANKSPAKVAAKY|^{14}GIA
AAGGLGGVLGIGAGPGSVGGIGPGGAGIGPGGTGYGPGLGVGPNAGGVGFGPG|^{15}GIP
AGGVLPGRTGFGPGGYGPGVGYGPGAGYGP|^{16}GYGSGLKPAKY|^{17}GYGSTGSVG
AAGVVPGR|^{18}DSSGGVVPGVVPGSGGTGTGGASTGPGTRTSPGGAGTSPGTSVN
GTGSRGTTGQS|^{19}GGLGVTGRPGGAGGDGDGLGGTGIPIIALKP|^{20}GDGGTLPGAR
PLRPP|^{21}GVPRGDTPGEGDGDGGPDGDRDRTGSRGPTGAAGGTGGTPTAGTG
VRP|^{22}GSGPGGSTIGGVGPGGGVAVGGSGGRGPALGAGGLPLGAGGVGGLKP
GK|^{23}NYGAGAFGILPTG|^{24}GIRYPTGAGVGPGALKPGK|^{25}VYGALGALGQGRVGP
GTYGTGPGGYGTGPGGYGGRGVGPGGALGIGGVGPGGVGGGLGGLGRGPGGYGVGV
GPG|^{26}GTRYGPGQGQGNGKPPK|^{27}NYGAGPAGTGIGPGVYGTGPGGAGSRPGG
FGPGGFGPGGFGPGGVGPGGFGPGGVGPGGFGPGGVGPGGFGPGGIGPGTTSGG
SYGPGGTRPGGYGPGTGGIGTGPAGTGPGGFGPGGIGPGATGGGSYGPGGIGPGG
YGPSTGPGGYGPGGAGTGPGIYTPGGQGLGKNKPSKS|^{28}GYGGAGFGQ|^{29} GTAV

GPGATGTGPGTYGPGGFGPGGVRPGGTGTGTGGFGPGGAGIGTGGFGPGGAGT
GPGGFGPGGAGTGPGGFGPGGAGTGPGGFGPGGAGTGPGGFGPGGAGTGPGGFGP
GGAGTGPGGFGPGGAGIATGGFGPG GVGKIKSKS | [30]GYGAGYGT | [31]GVLGGGT
GGGPAGVGPGGFGPGGAGTGGFGPGGAGTGGFGPGGAGTGGFGPGGAGTGGFGPG
GAGTGGFGPGGAGTGGFGPGGAGTGGFGPGGAGTGGFGPGGAGTGGFGPGGAGT
GGFGPGGAGTGGFGPGGAGTGGYGPGGAGTGGFGPGGAGTGGFGPGGAGTGGFG
PGSTGIGSGGFGPGGQALGSGKPPKPGYGSSLGGTGYRP | [32]GGGIGGASGTILT
GPGVGGGSQ | [33]GGVGPGYGGGQYP | [34]AIAQKAAKYAALQAILGPGGNR | [35]GVGC
QGKYCGRRRK**uaa** gagagaccaacagaaaggaagcgacctcattgaagaaaaaa
aaaaagcggtgctattgcatta acccaaatggtaaatgtcttgtaatcattgca
gttaaaccttcagcacattccagatcaattttatacatgaa acttttgaagatcaa
caactcccagcctgcaatcaaagactacttttaaaacatggccacgta

4.2.2.12 Elasmobranchs: Elephant Shark (*Callorhinchus milii*) Tropoelastin

This is a partial sequence only. Exons are numbered backwards, beginning at the 3′-terminal exon. Initial sequence was obtained from blast searches of the whole-genome shotgun sequence of *Callorhinchus milii*. Two search sequences were used. A consensus sequence from the highly conserved 3′UTR of tropoelastins yielded AAVX0115889 (2,408 bp), containing the final two exons of tropoelastin and including 600 bp of 3′UTR. Similarly, AAVX01251361 (1,628 bp), containing four internal tropoelastin exons was retrieved by searching with a common replicated cross-linking sequence from fugu tropoelastin-a (AKAAKY). Using putative exonic regions of the retrieved sequences for primers, PCR was then performed on RACE-ready *Callorhinchus milii* cDNA obtained from the Elephant Shark Genome Project, Institute of Molecular and Cell Biology, Singapore (http://esharkgenome.imcb.a-star.edu.sg/). PCR fragments were cloned, sequenced and used for WGS blast searches. In this way a total of 38 exons upstream from the 3′-terminal were identified.

[38]GGYAGAKAAKYG | [37]GPIGYQV | [36]GVPGGVQPGGVA | [35]GGVQPVPGAYP | [34]GFPG
AVPGPYQ | [33]GLYAGAKAAQYR | [32]GVAGGVQPGA | [31]GVPGAVPGLYP | [30]GQYAG
AKAAKYG | [29]AGVPGGVPGGIGT | [28]GVAGGVQPGA | [27]GVPGAVPGLYP | [26]GQY
AGAKAAKYG | [25]AGVPGGVPGGIGT | [24] GVAGGVQPGA | [23]GVPGAVPGLYP | [22] GQY
AGAKAAKYG | [21]AGVPGGVPGGIGT | [20]GVAGGVQPGA | [19]GVPGAVPGLYP | [18]GQYA
GAKAAKYG | [17]AGVPGGVPGGIGT | [16]GVAGGVQPGA | [15] GVPGAVPGLYP | [14]GQY
AGAKAAKYG | [13]AGVPGGVPGGIGT | [12]GVAGGVQPGA | [11]GFPGGVAGGYYP | [10]GLF
SRAKAAKY | [9]GGGVQPGAI | [8]GGLRPRLGAYP | [7]GLVPGLYP | [6]GVVPRAYP | [5]GLAA
AQKAAKYR | [4]GAGVGVQPGIGVGVQPGIGVGVQPGIGVGVQPGLGVGVQPG
LGVGVQPGLGVGVQPGIGVGVQPGIGVGVQPGIGVGVQPGLGVGVKPGVG
VG VQPGLGVGVQPGIGVGVQPGLGVGVQPGLGVGVQPGLGVGVKPGVGVGIQP
GLGVGIQPGLGVGVQPGVGVGIQPGLGVGVQP | [3]GITEVRPSPAPLTPQP | [2]GKPP
KPYGAVGQYGQRYQR | [1]GFGCANGKYCGRRRK**uga** acaaataatttcaagtga

cctcaagaactttggtgctactacctgattgggaatgtaaaggatggtgtgc aa
cggaactctctccctctcagctaagaaaacatttgtctgtattcattatcgata
tgtaatggaccttatca caggaacctctgttgtaggatagtgctgtgtgcttca
aacactcagggaccatctta

4.3 Characteristics Shared by all Tropoelastins

4.3.1 Composition and Elastomeric Properties

Table 4.1 compares characteristics of the reference tropoelastins detailed in Sect. 4.2.2. Note that amino acid compositions derived from genomic or cDNA sequences will differ in lysine content from those based on purification of insoluble elastin, since most lysines are consumed in the process of covalent cross-linking. In addition, post-translational modifications of some proline residues to form hydroxyproline will not be represented. Despite these limitations, since isolation of insoluble elastin from tissue sources involves purification by the extraction of contaminants, compositions of elastins purified in this way will always be suspect, due to either incomplete removal of all contaminating proteins or the use of overly harsh extracting reagents that damage and partially solubilize the elastin.

This table highlights the unusually high proportion of small side-chain, nonpolar amino acids shared by all tropoelastins, with glycine, alanine, proline, isoleucine, leucine and valine accounting for close to 80 % of all amino acids. The only exception to this is eln-2 from frog, which is also unusual in several other respects. With few exceptions, tryptophan, histidine and internal methionine residues are absent from tropoelastins. Note also the consistent low proportion of negatively charged amino acids. Indeed, with the exception of human and bovine and the eln-2/eln-b tropoelastins in frog and fish, negatively charged amino acids are entirely absent from most tropoelastins. As reported previously (Sage 1983; Chalmers et al. 1999), the overall hydrophobicity of tropoelastins appears to be decreased in frog, fish and shark species, with a particularly striking decrease in eln-2 of frog and eln-b of fish.

The consistently high proline and glycine content of all tropoelastins has recently been shown to be of fundamental importance for the elastomeric properties of elastins (Rauscher et al. 2006; Muiznieks et al. 2010; Rauscher and Pomès 2012; Miao et al. 2003; Cheng et al. 2010). As primarily entropic elastomers, elastins depend on the maintenance of a dynamically disordered state, with multiple, rapidly interchanging polypeptide chain conformations. However, while many intrinsically disordered proteins are highly charged and rely on charge repulsion effects to maintain their disordered state, elastins are mostly uncharged and highly hydrophobic in nature, risking collapse into a hydrophobic core incompatible with elastomeric properties. Both molecular dynamic simulations and experimental investigations of entropic elastomers such as elastins indicate that the content and distribution of proline and glycine residues are fundamental to the maintenance of the disordered state, proline because of its fixed bond angle preventing the formation of secondary structures such as β-sheet and α-helix and glycine because of its propensity for inducing chain hydration and mobility (Rauscher et al. 2006; Muiznieks et al. 2010; Rauscher and Pomès 2012; Miao et al. 2003; Cheng et al. 2010; Muiznieks and Keeley 2010). Thus, the conservation of glycine and proline contents in elastins across all species appears to be driven by their common requirements to perform as entropic elastomers.

4.3.2 Short Sequence Elements Contributing to Elastomeric Behaviour

While entropic elastomers such as elastins require maintenance of a disordered state, this does not necessarily imply a completely random structure. Both molecular dynamic simulations and circular dichroism spectroscopy and other structural analysis tools identify both polyproline II structure and β-turns in short sequence elements comprising only a few residues of tropoelastins. In a recent cross-species sequence analysis of tropoelastins, He and his colleagues (He et al. 2007) have identified VPG as a dominant motif that is highly conserved in amniotes, amphibians and teleosts. Domain 20 of chicken tropoelastin provides a particularly striking example of an extensive tandem VPG repeat. Together with similar but less well-conserved elements such as IPG, LPG and perhaps others, these are likely to act as sites for β-turn formation. Recognition of the possibility of repeated β-turns prompted several early models of tropoelastin structure (Gray et al. 1973), including the highly structured β-spiral proposed by Urry (Venkatachalam and Urry 1981). However, given a preponderance of more recent data pointing to a disordered state for elastins (Rauscher et al. 2006; Muiznieks et al. 2010; Rauscher and Pomès 2012), it appears more likely that β-turns at any one site are transient, more consistent with the 'sliding β-turn' model of Tamburro and his colleagues (Tamburro et al. 2003). Conservation of the rich distribution of such dynamic and flexible hinges throughout tropoelastins (He et al. 2007) (Fig. 4.1) would clearly be important for preserving the constitutive conformational disorder required for an entropic elastomer.

4.3.3 Alternating Domain Structure

The presence of an alternating domain arrangement in elastin was apparent even from the earliest partial sequence information (Gray et al. 1973). Subsequent full-length sequences of several mammalian tropoelastins (Yoon et al. 1984; Indik et al. 1987, 1989) as well as chicken tropoelastin (Bressan et al. 1987) confirmed that much of the protein consisted of alternating hydrophobic and

```
    2                    3                      4           4a         5               6
Hum GVPGAIPGGVPGGV----FYP|--------GAGLGALGGG----|ALGPGGKPLKP|----------|VPGGLAGAGLGA---|---------GLGAFPAVTFPGA--
Bov GVPGAVPGGVPGGV----FFP|--------GAGLGGLGVG----|GLGPGVKPAKP|----------|GVGGLVGPGLGA---|---EGSALP--GAFPGG-FFGA--
Mse GVPGAVPGGLPGGVPGGVYYP|--------GAGIGGLGGGGG--|ALGPGGKPPKP|GAGLLGTFGA|GPGGLGGAGPGA---|---------GLGAFPAGTFPGAGA
Ckn GVPGAIPGG---GVPGGGFFP|--------GAGVGGLGAGLG-A|GLGAGGKPLKP|----------|GVSGLGGLGPLGLQP|GAGVGGLGAGLGAFPGAAFPGA-A
Liz GVPGAGGGIPGRG-----LYP|GIGGGGL-GAG--GLGTGLGGA|GYGAGAKPPKP|----------|GVGGAGSL-------|----GGLYPG-GAYPGGVFQGA--
                        7                                                             8                         9
Hum LVPGGVADAAAAYK-AAKA|GA-GL---GGVPG----------------------------------------VGGLGVSA|GAVVPQP------GAGVKPGKVP|GV
Bov --GGGAAGAAAAYKAAAKA|GAAGLG-V------------------------------------GGIGGVGGLGVST|GAVVPQLGA--GVGAGVKPGKVP|GV
Mse LVPGGAAGAAAAYKAAAKA|GA-GLGGVGGVPG--GVGVGGVPGG--VGVGGVPGGVG-VGGVPG-GVGGIGGIGGLGVST|GAVVPQVGA--GIGAGGKPGKVP|GV
Ckn LVPGGAAGAAAAYKAAAKA|GA-GLGGV-----GGIGG--L------------GGVGGVG-VPG---------------|GLGVPGV-VQPGVGAAGKPPKVP|GA
Liz GTP---AQAAAAYKAAAKA|---GLG----VPGGGIGGVGVPGGG-VGV-GVPGGG--VG-VPGGGVGVGVPGGGIGVPG|GGVIPQTGAQPGAGVGAKPGKIP|GV
                    10                       11          12                        13             14
Hum GLPGVYPGG-VLP--|GARFPGVGVLPGVPTGAGVKPKAP|GVG-GAFAGIP|GVGPFGG-PQPGVPLGYPIKAPKLP|GGYGLPYTTGKLPY|GYGPGGVAGAA---
Bov GLPGVYPGG-VLPGA|GARFPGIGVLPGVPTGAGVKPKAQ|-VGAGAFAGIP|GVGPFGG-QQPGLPLGYPIKAPKLP|AGYGLPYKTGKLPY|GFGPGGVAGSA---
Mse GLPGVYPGG-VLPGT|GARFPGVGVLPGVPTGTGVKAKAP|G-GGGAFAGIP|GVGPFGG-QQPGVPLGYPIKAPKLP|GGYGLPYTNGKLPY|-----GVAGAG---
Ckn GIPGAFPGGGVLPGA|GIRFPGVGVLPGVPTGTGIKAKGP|GA--GAFAGIP|GLGGFGG-QQPGVPLGYPIKAPKLP|GGYRLPF----VN-|GLGPGGI-GAGVLA
Liz GIPGVFPGG-VLPGT|GIRYPGVGVLPGVPTGNGLKAKVP|GAGGGAFAGIP|GFGGFGGGQQPGLPLGYPIKAPKLP|GSY---Y--GKLSF|GAGPGGVFGGA--G
               15                  16                                      17                      18
Hum GKACYPTCT|GVGPQAAAAAAAKAAAKF|GAGAAGVLPGVGGAGVPG-VPGA----IPG---IGGIA|GVGTP--AAAAAAAAAKAAKY|GAAAGLVPGGPGFGPGV
Bov GKAGYPTGT|GVGPQ-AAAAAAKAAAKL|GAGGAGVLPGVGVGG-PG-IPGA-PGAIPG---IGGIA|GVGAPD--AAAAAAAAKAAKF|GAAGG-LPG--------
Mse GKAGYPTGT|GVGSQ-AAAAAAKAA-KY|GAGGAGVLPGVG-GG--G-IPGGA-GAIPG---IGGIA|GAGTP--AAAAAAKAAAKAAKY|GAAGGLVPGGPGV----
Ckn GKAGYPTGT|GVGAQ---AAAAKAAAKY|---GAGVLPGAG-----G-IPGVGG-VVPGVGVVPGAG|-VGGPA-AAAAAAKAAAKAGAY|GA--GVLPGAGGVPG-V
Liz GKAGYPTGT|GVGAQ---AAAAKAAAKY|GV-GAGRLPGAGALGVGGLLPGAGG--VPG--AI----|GAGGPAAAAAAAAKAAAKARAY|GA-GG-LPGA-GV-GGV
                                                            19                  20
Hum VGVPGAG-VPGVG-VPGAG-IPV--VPGAG-IPGAAVP-------------------|GVVSPEAAAKAAAKAAKY|GARPGVGVGGIPTYGVGAGGFPGFGVGVGG
Bov VGVPGVG-VPGVG-VPGVG-VPGVGVPGVG-VPGVGVPGVGVPGVGVPGVGVPGVGVP|GALSPAATAKAAAKAAKF|GARGAVGIGGIPTFGLGPGGFP--------
Mse -RLPG-AGIPGVGGIPGVGGIPGVGGPGIGG-PGI-VGGP-----------------|GAVSPAAAAKAAAKAAKY|GARGGV---GIPTYGVGAGGFPGYGVGAGA
Ckn --VPGVGVVPGL--VPGVGGIPGVA--------------------------------|GVGTPAGAAAAAKAAKY|GAGVPGVGVPGV--------------GIGG
Liz -GVPGGVGIPGVGGLPGV-------GPGVGGVP------------------------|GL-SPA-AAAAAAKAAKY|-----GLG-PG---GVGA-----------
                                            21                22                                  23
Hum IPGVAGVPGVGGVPGVGGVPGV-----------------GIS|--PEAQAAAAAKAAKY|GAAGAGVLGGLVPGPQAA-----VPGVPGTGGVP|GVGTPAAAAA-KA
Bov ---------------------------------------GIG|DAAAAPAAAAAKAAKI|GAGGVGALGGVVPG-----APGAIPGLPGVGGVP|GVGIPAAAAA-KA
Mse ---------------------------------------GLG|GASPAAAAAAAKAAKY|GAGGAGALGGLVPGAVPGALPGAVPAVPGAGGVP|GAGTPAAAAAAAA
Ckn VPGVPGVPGVPGVPGVPGVPGVPGVPGVPGVPGVVPGVG|----------------|----------------------------------|-VGGPAAAAAAKA
Liz IPGGVGVGGIPGVGGVPGV-GVGGVPGA-GIGGVPGL-PAAP|----------------|----------------------------------|GAGASAAAAA-KA
         24                                                                                              25
Hum AAKAAQF|GLVPGVGV-APGVGVAPGVGVAPG-VGL--APGVGVAPGVGVAPGVGVAPGIG--PGGVA-------------------------------|-------
Bov AAKAAQF|GLGPGVGV-APGVGVVPGVGVVPG-VGVA-------------------PGIGLGPGGVIGAG--------------------------VP|-------
Mse AKAAAKA|GLGPGVG-GVPG-GV--GVGGIPGGVGVGGVPG-GVGPGGVT--------GIGAGPGGL---G-----------------------------|GAGSP--
Ckn AAKAAAF|GAGRVPGVGVVPGA--VPGVG-VPG-VG---VPGVG-VPGVG-VPGVG-VPGVG-VPGVG-VPG-VGVPGVGVPGVGVPGVGVPGVGVPGLVP|GAG-PAA
Liz AAKAAAY|GVAPGYG---------------------------------GIGGIGT-GGIGTGGL---VPGGVGV-G-----------------------|APGTQAA
                  26                                                             27                    28
Hum -AAAKSAAKVAAKAQL|RAAAGLGAGIPGLGVGVG--VPGLGVGAGVPGLGVGAGVPGFGA------------------|VPGAL-------AAAKAAKY|GAAVPGV
Bov -AAAKSAAKAAAKAQF|RAAAGLPAGVPGLGVGAG--VPGLGVGAGVPGLGVGAGVPGPGA------------------|VPGTL-------AAAKAAKF|------G
Mse -AAAKSAAKAAAKAQY|RAAAGLGAGVPGFGAGAG--VPGFGAGAGVPGFGAGAGVPGFGAGAGVPGFGAGA-------|VPGSL-------AASKAAKY|GAAGGLG
Ckn A--AKAAAKAA-K--Y|-GAGGLAPGVGGLAPAVGGLAPGVG-GL-VPGVG-GL-VPGVG-GL-APGVGGLAPGVGAVP|GVGGP--AAAAKAAAKAAKY|GAGVGG-
Liz AVAAKAAAKA-AK--Y|-GAGGARGLVPGAG--VGGLVPGVGGVGGVGGIG-GVG--GVGGGL-VPGAG----------|GAGSISAAVAAKAAAKAAKY|GAGRGGV
                                                      29                      29a           30
Hum LGGLGALGGV-------------GIPGGVV----------------------|GAGP-AAAAAAAKAAAKAAQF|-------------|GLVGAAGLGGLGVGGLG-
Bov PGGVGALGGVGDLGGA-------GIPGGVA----------------------|GVVP--AAAAAAKAAAKAAQF|-------------|GLGGV---GGLGVGGLGA
Mse GPGGLGGPGGLGGPGGLGGAG---VPGRVA----------------------|GAAPPAAAAAAAKAAAKAAQY|-------------|GLGGA---GGLGAGGLGA
Ckn VPG--AVPGA--VPGVPGVPGV--TPGVGGVPS--LVPGVG-VPGVGVLPGA|GI-PQ----VGVQPGAKPPKF|GVPGAGVPGVGGIP|--------GGLGVGGLGV
Liz LPGGGVLPGGGVLPGGGVLPGGGVLPGGGVLPGGGVVPGGGVLPG-------|GI-PQ----IGVQPGAKPPKY|GVPGAGLGGL-GVP|--------GGGLVPGA--
                                       31            32                  33              34         35
Hum ----------------------VPG-VGGLG|G-IPPAAAAKAAKY|GAAGLGGVLGGAGQFPLG|GVAARPGFGLSPIFP|----------|-------------
Bov ----------------------VPGAVG-LG|G-VSPAAAAKAAKF|GAAGLGGVLGAGQPFPIG|GVAARPGFGLSPIFP|GG-AGGLGVG|GKPPKPFGGALGAL
Mse GGLGAGGLGAGGLGAGGLGAGGLGAGGLGAGG|G-VSPAAAAKAAKY|GAAGLGGVLGA-RPFPGG|GVAARPGFGLSPIYP|GGGAGGLGVG|GKPPKPYGGALGAL
Ckn GGLGAGGLGA---------------------|GLLYPGAAGKPPKP|------------------|GVGV-PGFGVSPIFP|GGVGGQLGFG|GKPPKTYGGALGAL
Liz GGAGYGGLGAGGL----------SP-------|GLTYPGAGGKPPKP|------------------|---------------|-GYGG--GLP|AKAAK-YGVAVPKR
     36
Hum ---|GGA-CLG-KACGRKRK
Bov GFP|GGA-CLG-KSCGRKRK
Mse GYQ|GGG-CFG-KSCGRKRK
Ckn GFR|GGVGCAQGKYCGRKRK
Liz NFP|-GAL-------GRRRK
```

Fig. 4.1 Alignments of full-length tropoelastins from amniote species. All amniote tropoelastins have approximately 36 exons with phase 1 exon borders. Within this group it is possible to identify exon-to-exon relationships in spite of the relaxed requirement for sequence conservation, variable tandem repeats in corresponding exons and the gain or loss of some exons from the genomic sequence. The alternate domain structure, cross-linking exons alternating with hydrophobic exons, throughout most of the sequence is also evident. VPG and other putative β-turn-forming motifs are shaded

cross-linking domains. Hydrophobic domains are generally rich in nonpolar amino acids, often present in tandem repeats and pseudo-repeats. Cross-linking domains contain the lysines destined to take part in covalent cross-linking, predominantly in KAAK or KAAAK motifs and often preceded by polyalanine sequence (KA-type cross-linking domains). Alternatively, cross-linking domains may be of the KP-type, consisting of KxxK motifs in which one or both of the x residues are proline.

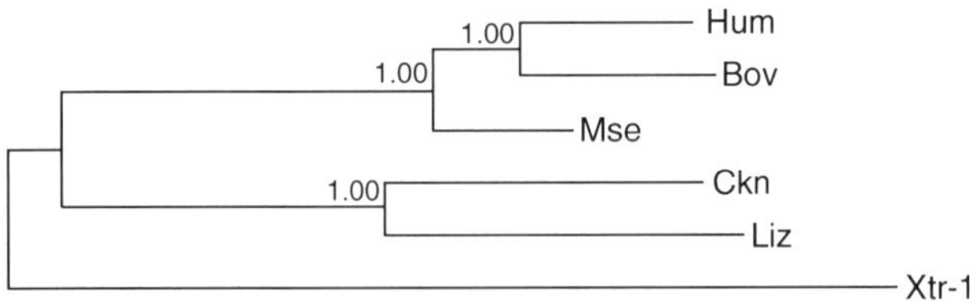

Fig. 4.2 Phylogenetic tree of full-length tropoelastin sequences of amniotes given in Fig. 4.1, using *Xenopus tropicalis* eln-1 as an outgroup. This Bayesian tree was constructed with MrBayes 3.1.2 (Ronquist and Huelsenbeck 2003) using one million generations, with the first 25,000 trees discarded as burn-in. Numbers at nodes are posterior probabilities, with the maximum probability represented as 1.00. Relationships between species are in agreement with generally accepted phylogenetic relationships among amniotes and are consistent with a common ancestral sequence for all amniote tropoelastins

Sequence alignment of full-length tropoelastins from amniotes (Fig. 4.1) provides a good illustration of the alternating domain arrangement of these elastins, the distribution of VPG and similar putative β-turn-inducing sequence elements and the relaxed sequence conservation, particularly of hydrophobic domains, even between relatively closely related species. In all of these amniote tropoelastin sequences domain borders correspond to exon divisions, and all splice sites are phase 1. Furthermore, all of these sequences contain approximately 36 domains (exons).

While sequences are sufficiently similar to allow recognition and alignment of most domains in amniotes, many differences are apparent. In hydrophobic domains these include modifications to the nature and number of tandem repeats in the domain (e.g. domains 18, 20, 24, 26 and 30). Variations also include loss or gain of domains. For example, human tropoelastin has lost domains (exons) 34 and 35 from the genomic sequence (Szabó et al. 1999), and exon 22, although present in the genomic sequence, is apparently never expressed. Similarly, exons 21, 22 and 32 are missing from the genomic sequence of both chicken and lizard tropoelastins. In contrast, mouse and rat tropoelastins contain an additional exon (designated 4a), and chicken and lizard tropoelastins contain exon 29a, which is not present in other members of the amniote group.

Cross-linking domains appear to be comparatively better conserved. In general there is a predominance of KA-type cross-linking domains with preceding polyalanine sequences in amniotes, although the less common KP-type domains are also well conserved (e.g. domains 4, 8 and 12). Exceptions to this are domains 29 in the chicken and lizard, which are KP rather than KA, and exon 35 in the lizard, which is KA rather than KP. For a possible explanation of these exceptions, see Sect. 4.3.5.4.

With the availability of full-length sequences for amphibian and teleost tropoelastins (Chung et al. 2006; Miao et al. 2007, 2009), it became apparent that the general ability for domain-to-domain recognition and alignment across amniote tropoelastins could not be extended to amphibians and teleosts, suggesting that amniote tropoelastins could be considered a clade, with a common ancestral sequence. This relationship is supported by a phylogenetic tree based on full-length protein sequences of reference tropoelastins in the amniote group (human, bovine, mouse, chicken and lizard) using the full-length reference Xtr eln-1 sequence as an outgroup (Fig. 4.2).

4.3.4 Multiple Tropoelastin Genes in Some Species

The presence of two tropoelastin genes appears to be a feature shared by all teleosts and is consistent with a genome duplication event that has been reported to have taken place in ray-finned fish after their divergence from the line leading to the tetrapods (Taylor et al. 2001; Christoffels et al. 2004). More unexpected was the presence of the two tropoelastin genes in frog. This gene duplication event has evidently taken place in both *tropicalis* and *laevis* species (Miao et al. 2009), but as yet there is no evidence for tropoelastin gene duplication in other amphibians such as salamanders, which underwent early divergence (~300 million years ago) from the line leading to frogs. Tropoelastin gene duplication in teleosts and frogs has been followed by considerable mutation particularly of one of the genes, designated *eln*-b or *eln*-2. Nevertheless, it has been established that both genes are expressed and their protein products are deposited in tissues, although with differential tissue distribution (Miao et al. 2007, 2009). In spite of the increased variability of sequence when amphibians, teleosts and elasmobranch sequences are considered, all of these are recognizable as tropoelastins by their alternating hydrophobic and cross-linking domain arrangements, their high glycine and proline contents and the abundance of VPG-like sequence motifs. Exon numbers are generally increased in non-amniotes, up to 58 in the case of zebrafish *eln*-b, and expressed proteins are generally much larger, with putative molecular weights up to ~170 kD, compared to 60–70 kD in amniotes (Table 4.1). Particularly in zebrafish eln-b and fugu eln-b, increased molecular weight is due to increases in the size of hydrophobic domains, with the cross-linking domains remaining relatively small (Chung et al. 2006).

The predominant cross-linking domains in both teleost tropoelastins and in the partial sequence so far available for shark elastin are of the KA-type, although in both cases they lack the preceding polyalanine sequences seen in KA cross-linking domains of amniotes. However, KP-type cross-linking domains predominate in both tropoelastins of frogs. In contrast to amniote tropoelastins, in which strict, exon-specific delineation of hydrophobic and cross-linking domains is the rule, teleost elastins contain several examples of what appears to be fusion of hydrophobic and cross-linking exons (e.g. exon 46 of zebrafish *eln*-b; exons 21, 22 and 26 of fugu *eln*-a; and exons 21, 27 and 31 of fugu *eln*-b). In addition, zebrafish *eln*-a contains an example of a KP-type cross-linking motif bridging between exons 45 and 46. It is interesting to note that the density of coding sequence in fugu *eln*-a is ~8 %, consistent with most of the other reference tropoelastin genes (Table 4.1), while that of fugu *eln*-b is ~36 %, suggesting that decreased intron size as a factor in the compaction of the fugu genome (Venkatesh et al. 2000) has only taken place in *eln*-b.

Overall, with the exception of the regions to be discussed below, it is not possible to relate tropoelastin domains in amphibians, teleosts or elasmobranchs by sequence alignment either to the amniote group or to each other. Similarly, full-length alignments between eln-a and eln-b of zebrafish or fugu or between eln-1 and eln-2 of frog are not possible. However, eln-1 sequences are highly similar between *tropicalis* and *laevis* species of frogs, as are eln-2 sequences (Miao et al. 2009). Indeed, there is evidence for further duplication of *eln*-1 and *eln*-2 genes in the tetraploidization of the *laevis* species, which is suggested to have taken place ~40 million years ago (Hellsten et al. 2007).

4.3.5 Regions of Sequence Similarity Across All Tropoelastins

4.3.5.1 Central Conserved Region

In spite of the remarkable sequence diversity across elastins from humans to sharks, there are nevertheless three regions that can be related and aligned with reasonable assurance across all tropoelastins. The first of these, referred to as the central conserved region, corresponds to domains 8–14 in amniote tropoelastins, although domain numbers vary for the non-amniote groups (Fig. 4.3). Available tropoelastin sequences of this region from several other species are also included, together with their database references. Interestingly, hydrophobic domains in this region are not rich in the tandem repeats that characterize such domains in other regions, and cross-linking motifs are either of the KP-type or lack preceding polyalanine sequences, even in the amniote group. In addition, this region also includes conserved KxK motifs and single K residues, which are otherwise either absent or uncommon in tropoelastins. As such, this region would not have been considered a 'classical' elastin-like sequence. The considerable areas of sequence similarity in this region are highlighted by shading in Fig. 4.3. Note that sequence alignment is particularly limited in eln-2 of frog.

In spite of the relatively short sequences available for alignment of this central conserved region, phylogenetic trees constructed from these sequences (Fig. 4.4) are in good agreement with generally accepted phylogenetic relationships, including the grouping of medaka, stickleback and fugu tropoelastins closer to each other than to zebrafish (Wittbrodt et al. 2002) and the placement of the coelacanth in the lineage leading to the tetrapods. Note also that teleost eln-a tropoelastins are grouped separately from their corresponding eln-b tropoelastins, consistent with the wholesale genomic duplication in the ray-finned fish line preceding the separation of these teleost species in evolution. Eln-2 of *Xenopus tropicalis* appears as an outgroup in the tree, reflecting the particularly unusual character of this sequence.

```
Hum    GAVVPQP-------------GAGVKPGKVP--GVGLPGVYPGG-VLP--GARFPGVGVLPGVPT
Bov    GAVVPQLGA---------GVGAGVKPGKVP--GVGLPGVYPGG-VLPGAGARFPGIGVLPGVPT
Mse    GAVVPQVGA---------GIGAGGKPGKVP--GVGLPGVYPGG-VLPGTGARFPGVGVLPGVPT
Rat    GAVVPQLGA---------GVGAGGKPGKVP--GVGLPGVYPGG-VLPGTGARFPGVGVLPGVPT
Ckn    GLGVPGV-VQP-------GVGAAGKPPKVP--GAGIPGAFPGGGVLPGAGIRFPGVGVLPGVPT
Liz    GGVIPQTGAQP-------GAGVGAKPGKIP--GVGIPGVFPGG-VLPGTGIRYPGVGVLPGVPT
Xtr-1  GAVVPQAGVQP---------GAAGKPGKLP--GAGIPGVFPGG-LVPGTGGRFPGVGVLPGVAT
Xtr-2  ------------GIRNGYEAGYGVKAGK-P---------------------------AAGSLGV
Coe    GVVLPQTGVAP-------GLGAGGKPPKIP--GVGIPGVYPGG-VVPGTGLRYLGIGVLPGVPT
Zef-a  GLGIPQTGVQGLGPGGVGPGAKAAKAGKAPVPGVGVPGPYQGG-L-PGQG--FNGRGVLPGVAT
Fug-a  LPVIPQTGLPG-----GTVGGAGTKASKVP--GVGVPGLFHG--LVPGQG--FGGRGVLPGVAI
Sbk-a  VPVIPQTGLQG-----GGGAGAGKKAAKVP--GVGVPGLYQGG-LVPGQG--FGGRGVLPGVAI
Med-a  APVIPQTGLQG-----GGGGAAGKKAAKVP--GVGVPGLYQGG-LMQGQG--FNGRGVLPGVAT
Zef-b  SPFLPQTGLPGIGP--GTGGKVG-KPGKAPVPGVGIPGLYQGG-IVPGQG--FGGRGVLPGVAT
Fug-b  --VLPQTG-------GGTGGTTGGKASKLP--GIVMPGPYQGG-YVPGQA--FGGRGILPGVAI
Sbk-b  VPMLPGSGTGE----VGPGGNSGGKASK-QVPGVGVPGLYQGG-YVPGQG--FGGRGVLPGVAT
Med-b  ---VPQTGSTGV---VGPGGKSGSKASK-QLPGIGVPGLYQGG-FAPSQG--FGARGILPGVAT

Hum    GAGVKPKAP-GVG-GAFAGIPGVGPFGG-PQ-PGVPLGYPIKAPKLPGGYGLPYTTGKLPYGYG
Bov    GAGVKPKAQ--VGAGAFAGIPGVGPFGG-QQ-PGLPLGYPIKAPKLPAGYGLPYKTGKLPYGFG
Mse    GTGVKAKAP-G-GGGAFAGIPGVGPFGG-QQ-PGVPLGYPIKAPKLPGGYGLPYTNGKLPYG--
Rat    GTGVKAKVP-GGGGGAFSGIPGVGPFGG-QQ-PGVPLGYPIKAPKLPGGYGLPYTNGKLPYG--
Ckn    GTGIKAKGPGA---GAFAGIPGLGGFGG-QQ-PGVPLGYPIKAPKLPGGYRLPFVNG-----LG
Liz    GNGLKAKVPGA-GGGAFAGIPGFGGFGGGQQ-PGLPLGYPIKAPKLPGSY---Y--GKLSFGAG
Xtr-1  GNGVKQKAP---GGGAFAGIPGFNGFGG-QQ-PGVPLGYPIKSPKLPGGYGLPYTAGK-PGYGY
Xtr-2  LPAGKQK-P--------GYGNGRYPSNFQQQ------GIGAKPSKA--GYGQPVGGVRGIGAQP
Coe    GTGVKPKA--GVGGGGLAGFPGVGLTFGGQQPGVPLGYPIKSPKLPGGYGLPYSTGRLPYGLG
Zef-a  GTGLKPKS-GGGQGLVPGAGGR-GPFGGPLQQ-GVFHGYPLKSPKAPGAG-LPYTGGKLPYGYG
Fug-a  GSNLNPKS-----VPGVGQGTQGGGYGGQMQ-PGVFHGYPLTSPK
Sbk-a  GTNLNPKS-GGGQG--GPVQGGRG-YGGPMQ-PGVFHGYPLKTPKVQGGYGAKPGGGKLPFGYG
Med-a  GTDLNPKS-----GSGALGQSGLGGVHGQMQ-PGVFHGYPLKSPTAQGGYGAKPGGSKLPFGYG
Zef-b  GT----QT---GQLGQGALGA-NGRGGFGQQIPGVFGGYPLISPKSG
Fug-b  GSGLGTQN-GAGVLGQGGIGPGGTSNGRGQPLPGVIRGYPLISPKLG
Sbk-b  GSGLGPQR-GAGVLGQGGAGQGGA--GHGQQLPGVFRGYPLISPKTG
Med-b  GSGFGPQT-GSGALGQSGLGPGGA--GRGQQLPGVFRGFPLVTPKSG

Hum    PGG--VAGAA------G-KAGYPTGT
Bov    PGG--VAGSA------G-KAGYPTGT
Mse    -----VAGAG------G-KAGYPTGT
Rat    -----VAGAG------G-KAGYPTGT
Ckn    PGG--I-GAGVL---AG-KAGYPTGT
Liz    PGGV-FGGAG------G-KAGYPTGT
Xtr-1  G---AFGAGAAGK--AG-KAGYPTGT
Xtr-2  GYGNGLYPSNLQQQ
Coe    NGGVG-----------A-KLGYPTGT
Zef-a  GFGG-GAGLPGGKAGAGSKPGYPTGT
Fug-a
Sbk-a  -FRNVGPRLTAGVGGPASRPGYPAGT
Med-a  PYGNTGAGLP
Zef-b
Fug-b
Sbk-b
Med-b
```

Fig. 4.3 Alignments of central conserved regions of tropoelastins. Regions of particular similarity are shaded. KxxK or KxK motifs and singlet K residues are underlined. Exons included in these regions are as follows: Hum/Bov/Mse/Rat/Ckn/Liz eln (8–14), Xtr eln-1/Zef eln-a (9–15),

4.3.5.2 C-Terminal Region

The second region in which reasonable sequence alignment across all tropoelastins is possible consists of the last two domains of the protein sequence (Fig. 4.5). Sequence similarity is particularly strong in the C-terminal domain, which has been suggested to be important for the assembly of tropoelastins into the proper extracellular architecture (Brown-Augsburger et al. 1996; Hsiao et al. 1999; Broekelmann et al. 2005, 2008; Yeo et al. 2012). With the exceptions noted below, most of these C-terminal domains include the only two cysteine residues found in tropoelastin as well as a tetra-basic (or tri-basic) terminal sequence consisting of lysines and arginines. The C-terminal exons of lizard tropoelastin and frog eln-2 are unusual in this respect, with this domain in lizard lacking the region containing the cysteine pair and that in frog eln-2 containing the cysteine pair but lacking the terminal lysines and arginines. In spite of this unusual sequence, frog eln-2 is expressed and incorporated into extracellular matrix (Miao et al. 2009), suggesting that this tetra-basic C-terminal sequence is not absolutely essential for polymeric assembly.

The penultimate domains are less well conserved but contain a few common features worth noting. For example, this domain contains a cross-linking sequence (either KA or KP) essentially at the beginning of the domain, the only examples of such a placement of a cross-linking motif across all tropoelastin domains in all species. Similarly, the last two/three residues of the penultimate domain have a similar character across all species. Note that human tropoelastin lacks the cross-linking sequence in the penultimate exon, since exons 34 and 35 have been lost from the genomic sequence (Szabó et al. 1999). However, the penultimate domain in human tropoelastin (corresponding to domain 33 of other amniotes) does end with an FP sequence, consistent with at least some other species. Consistent with the unusual character of frog eln-2, the penultimate domain for this tropoelastin bears little resemblance to this domain in other species, with the exception of the last three amino acids. It is clear, however, that this anomaly is not the result of sequencing errors, since the aberrant eln-2 sequences in the final exons of *tropicalis* and *laevis* species of frogs are identical (Miao et al. 2009).

Fig. 4.3 (continued) Xtr eln-2 (4–9), Zef eln-b (12–16); Fug eln-a (11–15), Fug eln-b (8–13), Sbk eln-a (8–15), Sbk eln-b (8–12), Med eln-a (9–14) and Med eln-b (9–13). Coelacanth exon numbers have not been verified. Database references for tropoelastin sequences not included as reference sequences are as follows: rat eln NP_036854 coelacanth (Coe) eln AFYH01004509, AFYH01004508, AFYH01004507; stickleback (Sbk) eln-a AANH010030; medaka (Med) eln-a BAAF04035070; stickleback (Sbk) eln-b AANH01000630; and medaka (Med) eln-b BAAF04066300

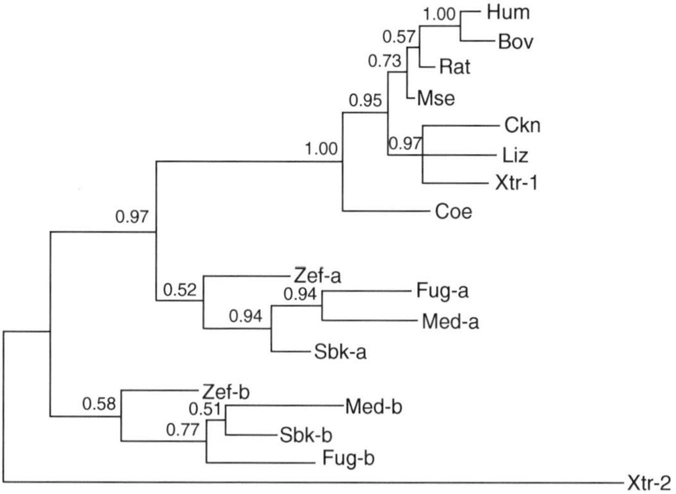

Fig. 4.4 Phylogenetic tree generated using sequences of the central conserved regions of tropoelastins given in Fig. 4.3. This Bayesian tree was constructed with MrBayes 3.1.2 (Ronquist and Huelsenbeck 2003) using one million generations, with the first 25,000 trees discarded as burn-in. Numbers at nodes are posterior probabilities, with the maximum probability represented as 1.00. Relationships between species are in good agreement with generally accepted phylogenetic relationships. Xtr-2 is an exception, treeing as an outgroup, reflecting its particularly unusual character relative to other tropoelastins

4.3.5.3 3′ Untranslated Region

The third region of sequence conservation, perhaps the most impressive and the most unexpected, occurs in the first ~80 bases of the 3′UTR of tropoelastins (Fig. 4.6), with a cluster of sequence motifs conserved over all species, from amniotes to elasmobranchs, including the eln-b sequences of teleosts. These include GACCUCA and AAUG sequences whose functions are unknown as well as a UGGUGCUA sequence (underlined in Fig. 4.6) corresponding to the binding site of a microRNA (miR-29). miR-29 has not only been implicated generally in regulation of production of extracellular matrix components (van Rooij et al. 2008; Du et al. 2010; Villarreal et al. 2011; Cushing et al. 2011) but more specifically has been suggested to control developmental downregulation of tropoelastin production (Zhang et al. 2012; Ott et al. 2011). Again, the notable exceptions to this sequence conservation are restricted to the eln-2 tropoelastins of both *tropicalis* and *laevis* species of frogs, whose 3′UTRs show no sequence resemblance to the others. Two other areas of the tropoelastin sequence have previously been proposed to be important for post-transcriptional regulation of production through modulation of mRNA stability, including a coding sequence in domain 30 of the rat tropoelastin (Zhang et al. 1999) and a sequence motif further downstream in the 3′UTR of chicken tropoelastin (Hew et al. 1999, 2000). However, currently there is no evidence that such a role extends beyond the amniote group.

```
Hum     GVAARPGFGLSP--------------------IFP|-GG-ACL-GKACGRKRK
Bov     ---GKPPKPF------GGALGAL---------GFP|-GG-ACL-GKSCGRKRK
Mse     ---GKPPKPY------GGALGAL---------GYQ|-GG-GCF-GKSCGRKRK
Rat     ---GKPPKPY------GGALGAL---------GYQ|-GG-GCF-GKSCGRKRK
Ckn     ---GKPPKTY------GGALGAL---------GFR|-GGVGCAQGKYCGRKRK
Liz     ---AKAAK-Y----GVGVAVPKR---------NFP|-GAL--------GRRRK
xtr-1   --AQKPSKSY-----------------------YR|-GGGLCGQGKYCGRRRK
Xlv-1   --AQKPSKSY-----------------------YR|-GGAQCGQGKYCGRRRK
Sal     ---VKPAKSY-----GGGALGAL---------GYR|-GGAACGQGKYCGRRRK
Xtr-2   VPEEQAAYRFYGT-------------------GYQ|----GCAE---C-----
Xlv-2a  VPEEQAAYRFYGT-------------------GYQ|----GCAE---C-----
Coe     ---GKAPKPY------AGALGLLG--------GYR|-GGVLCGPGKYCGRRRK
Zef-a   ---AKAAK-Y--GGGLTGFLGG----------GYR|-GGAGC-QGKYCGRRRK
Fug-a   ---AKAAK-Y---GALQGFLGGAGG-------AFR|GGAAGC-QGKFCGRR-K
Sbk-a   ---AKAAK-Y---GPLQSFLGGAGGGGGGGVGVYR|GGVAGC-QGKYCGRR-K
Med-a   ---AKAAK-Y---APLQGFLGGAGGGGGG----PYR|----GC-QGKYCGRR-K
Zef-b   ---SKAAK-Y---AALQGFLGAG---------GYR|-GGAGC-QGKYCGRRRK
Fug-b   AIAQKAAK-Y---AALQAILGPG---------GNR|--GVGC-QGKYCGRRRK
Sbk-b   SKAQKAAK-Y---AAMQAFLGSG---------GYR|--GAGC-QGRYCGQRRK
Med-b   AKQQKAAK-Y---AAMQALLRGG---------GYR|--GVG--------HRRK
Esk     ---GKPPKPY--GAVGQYGQR-----------YQR|--GFGCANGKYCGRRRK
Skt     ............GAVGQYGQI-----------GQR|--GSPCAMGKYCGRRRK
```

Fig. 4.5 Alignment of final two domains (exons) of tropoelastins. Exon border is indicated as |. Regions of particular similarity are shaded. KxxK motifs are underlined. *Dotted line* indicates missing sequence. Exons included in these regions are as follows: Hum eln (33/36), Bov/Mse/Rat/Ckn/Liz eln (35/36), Xtr eln-1 (54/55), Xtr eln-2/Xlv eln-2a (39/40), Zef eln-a (55/56), Zef eln-b (57/58), Fug eln-a (38/39) and Fug eln-b (34/35). Exon numbers for Xlv eln-1, Sal eln, Coe, Sbk eln-a, Med eln-a, Sbk eln-b, Med eln-b, Esk and Skt have not been verified. Database references for tropoelastin sequences not included as reference sequences are as follows: rat eln NM_012722.1, *Xenopus laevis* (Xlv) eln-1 EG568683, salamander (Sal) eln CN040369, *Xenopus laevis* (Xlv) eln-2a BK006571.1, coelacanth (Coe) eln AFYH01003045, stickleback (Sbk) eln -a AANH01003045, medaka (Med) eln-a BJ732935, stickleback (Sbk) eln-b AANH01000629, medaka (Med) eln-b BAAF04066300 and skate (Skt) eln AESE010099687/AESE010039636 (only partial sequence of next-to-last exon)

4.3.5.4 Region of Gene Expansion by Exon Replication

The earliest partial sequence data (Gray et al. 1973) as well as the initial full-length sequences of mammalian and avian tropoelastins (Yoon et al. 1984; Indik et al. 1987, 1989; Bressan et al. 1987) prompted suggestions that sequence motifs were not only replicated within exons, leading to the tandem repeats and pseudo-repeats common within exons/domains, but also that substantial portions of the overall tropoelastin sequences arose through exon replication. Subsequent availability of non-amniote tropoelastin sequences (Chung et al. 2006; Miao et al. 2009; He et al. 2007) provided evidence that such replication was present in all tropoelastins, as well as additional details on the site and nature of these replication events. It is now clear for all tropoelastins that a number of the exons/domains downstream of the central conserved region and preceding the conserved C-terminal region are the

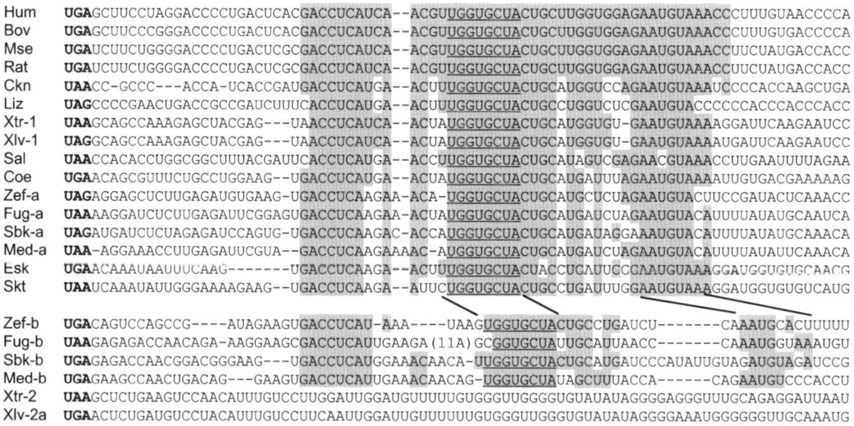

Fig. 4.6 Alignment of first ~80 bases of 3′UTR of tropoelastins. Regions of exact sequence homology to human eln are shaded. The stop codon is bolded. The binding sequence for miR-29 is underlined. Database references for tropoelastin sequences not included as reference sequences are as follows: rat eln NM_012722.1, *Xenopus laevis* (Xlv) eln-1 EG568683, salamander (Sal) eln CN040369, coelacanth (Coe) eln AFYH01004502, stickleback (Sbk) eln-a AANH01003045, medaka (med) eln-a BJ732935, skate (Skt) eln AESE010039636, stickleback (Sbk) eln-b AANH01000629, medaka (med) eln-b BAAF04066300 and *Xenopus laevis* (Xlv) eln-2a BK006571.1

result of replication of pairs of cross-linking and hydrophobic domains. Using mouse and zebrafish eln-a tropoelastins as examples, Fig. 4.7 clearly demonstrates the location of the replicated regions in relation to the central conserved region as well as the abundance of tandem sequence repeats within exons (Chung et al. 2006). Such characteristics are common to all tropoelastin sequences. The detailed nature of the replication events are particularly apparent in dotplots of genomic sequence of the *eln*-a gene of zebrafish, showing an area of replication between exons 16 and 34 (Fig. 4.8a) (Chung et al. 2006). Moreover, expansion of the genomic sequence dotplots, focusing on the exon 19–31 region (Fig. 4.8b), demonstrates not only replication of exonic sequences but also remnants of sequence similarity in the flanking intronic sequence (Chung et al. 2006). The sequence conservation within the amniote group as well as the evident domain-to-domain correspondences would suggest that the replication event or events that expanded these tropoelastins took place in the common ancestor of the amniotes.

Replication events in non-amniotes were more variable. Indeed, it appears that the increased numbers of exons and sizes of proteins in many non-amniote tropoelastins (Table 4.1) can be attributed to greater extents of exon replication (Chung et al. 2006; He et al. 2007). While replication has clearly taken place in all of these non-amniote tropoelastins, sequence similarities even between replicated exon pairs in zebrafish eln-a and fugu eln-a are difficult to recognize.

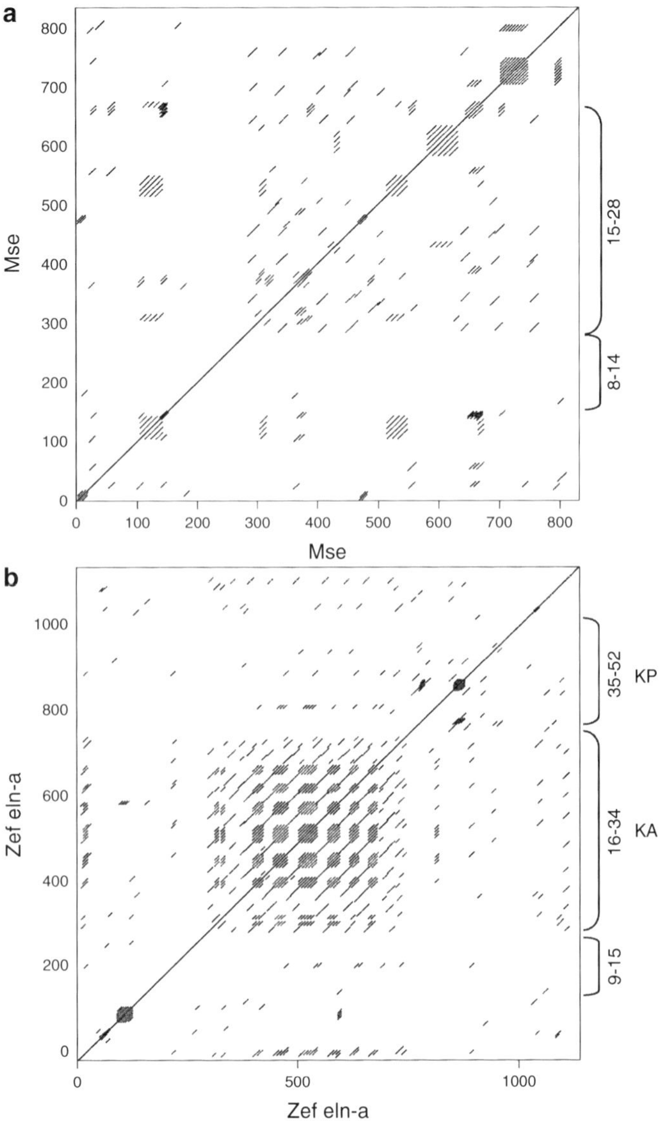

Fig. 4.7 Dotplots of full-length protein sequences, excluding signal peptide (exon 1), of mouse eln (**a**) and zebrafish eln-a (**b**) tropoelastins showing central conserved region (exons 8–14 for Mse eln and 9–15 for Zef eln-a) followed by regions of exon pair replication. Mse eln has a single KA replication region (exons 15–28). Zef eln-a includes a KA replication region (exons 16–34) followed by a KP replication region (exons 35–52). Dotplots were prepared using dotmatcher (http://emboss.bioinformatics.nl/cgi-bin/emboss/dotmatcher) using window size = 10 and threshold = 40 (Figure is modified from Chung et al. (2006) with permission)

A closer analysis of this replication region reveals sequence relationships that provide further insights into the replication events in different species (Fig. 4.9). This simplified model assumes a prototypical tropoelastin containing the central

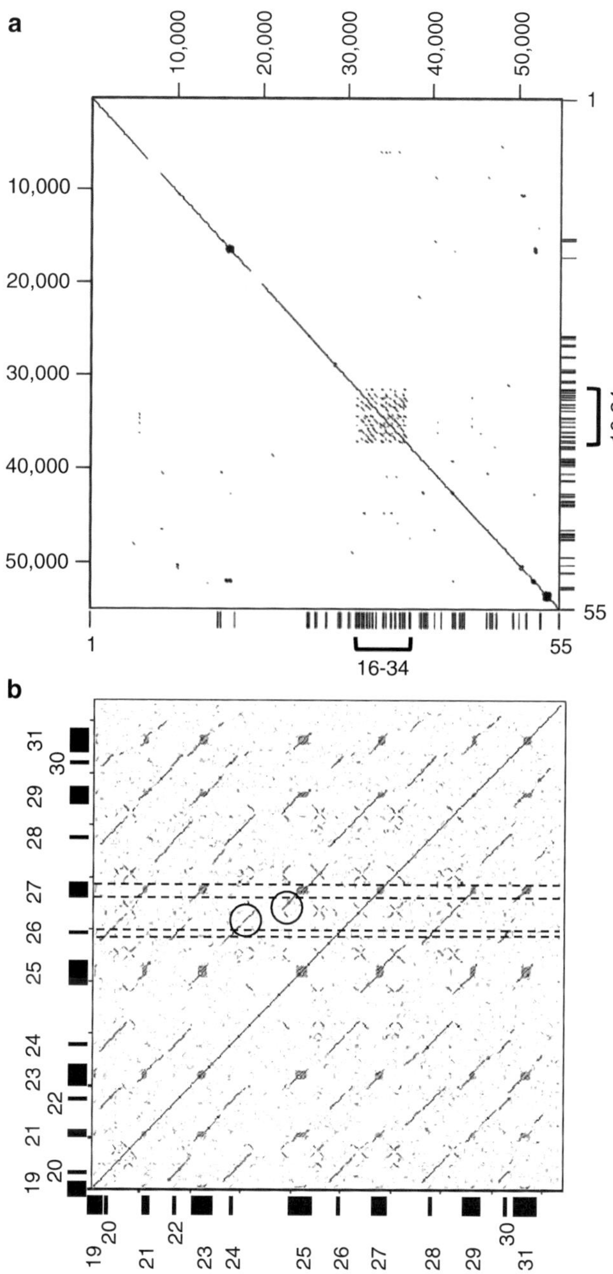

Fig. 4.8 Dotplots of genomic sequence of Zef *eln*-a. (**a**) Dotplot of full genomic sequence. Exons are indicated by bars, with exon 1 at *top right* and *bottom left* of the dotplot. The position of the KA exon pair replication (exons 16–34) is evident. The KP replication region is not visible at this sensitivity. (**b**) Expansion of the genomic dotplot for the region from exons 19 to 31. Examples of remnants of sequence similarities in flanking introns are circled. Dotplots were prepared using Dotmatcher (http://emboss.bioinformatics.nl/cgi-bin/emboss/dotmatcher) using window size = 10 and threshold = 40 (Figure is modified from Chung et al. (2006) with permission)

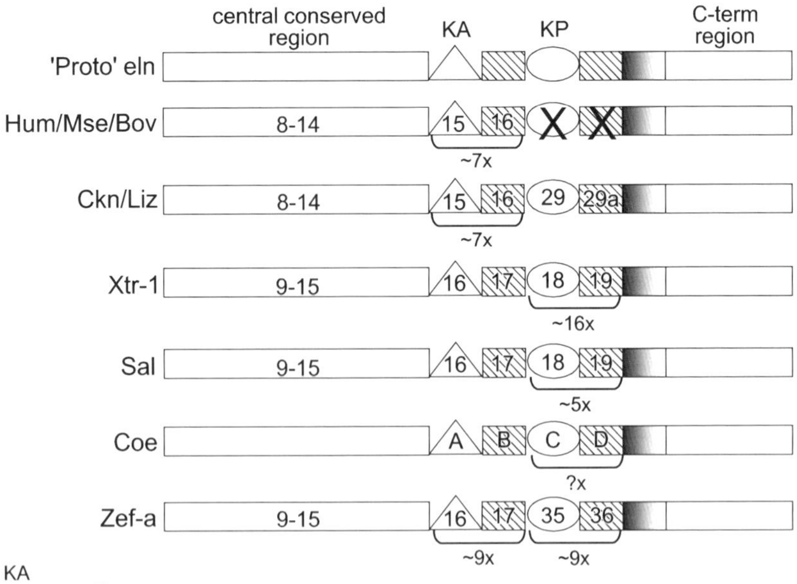

KA

```
Hum    [15]GVGP------QAAAA-----AAAKAAAKF
Bov    [15]GVGP------QAAAA------AAKAAAKL
Mse    [15]GVGP------QAAAA------AAKAAAKL
Ckn    [15]GVGA------QAAAA--------KAAAKY
Liz    [15]GVGA------QAAAA--------KAAAKY
Xtr-1  [16]GVGAL--TAAQQAAA--------KAA-KY
Sal    [16]GVGA------QAAAL----KAAYKAA-KY
Coe    [A]GVGA------QVAAA(QAAAA)_6KAAAKS
Zef-a  [16]GVGSLGVSAAQA-----------KAA-KY

Hum    [16]GAGAAG----VLP------GVGGAG-VPGV------PGAIPGIGGIA
Bov    [16]GAGGAG----VLPGV----GVGG----PGIPGA---PGAIPGIGGIA
Mse    [16]GAGGAG----VLPGVGG----GG---IPG--GA----GAIPGIGGIA
Ckn    [16]---GAG----VLPGA------GG---IPGV--G----GVVP--GVGVVP-GAG
Liz    [16]GV-GAG----RLPGAGAL-GVGGL--LPGA---GGVPGAI
Xtr-1  [17]GIGG-------IPGAVPGTGLGG---VPGLTPGG---GAVPGV--VPGLVPGAGGVPGA
Sal    [17]GAGA(GGVPGA)_3-------- GG---VPGLYPGV--(GGVPGA)_2-GGVPGV(GGVPGA)_3
Coe    [B]GAGA-------FPG-------GG---IPSYYPG--V-GLYPGAGGIPGLIPGA
Zef-a  [17]GAGAGLGGAGAFPGGV-GAGLGGAGAFPGGA-GGFYPGAV
```

KP

```
Ckn    [29]GIPQVGVQ-PGAKPPKF|  [29a]GVPGAG-VPGVGGIP
Liz    [29]GIPQIGV-QPGAKPPKY|  [29a]GVPGAG-LGGL-GVP
Xtr-1  [18]GIPQLGV-QPGAKASKY|  [19]GLPGVGGVPGVGGVP
Sal    [18]GIPQVGA-QPGAKPPKY|  [19]GVPGVGGVPGVGGVPGAGGAGGVGRVPGV
Coe    [C]GIPQLGLGQAGAKALKY|  [D]GIPGVGGIPGVGGVPGIGGVPGVGGIPGVGGVPGVGGVPGQ
Zef-a  [35]----GLA-AGAKPPKY|  [36]GVP--GGT-GF-GVPG--GVP
```

Fig. 4.9 Pattern of exon pair replication in tropoelastin genes. A 'prototypical' tropoelastin including a central conserved region, a KA/hydrophobic exon pair, a KP/hydrophobic exon pair and a C-terminal region is assumed as a starting point. Approximate replications of the KA exon pair, KP exon pair or both are indicated for each species. Sequences of KA and KP exon pairs across different species are given below. Sequences for Coelacanth eln are from AFYH 01004507. Sequences for salamander eln are from EU686405

conserved region and the C-terminal region and includes a single copy of a cross-linking/hydrophobic domain pair of the KA type (i.e. a proto-KA exon pair), followed by a single copy of a cross-linking/hydrophobic domain pair of the KP type (a proto-KP exon pair). In human, bovine and mouse *eln*, the proto-KA exon pair (e.g. exons 15/16) was replicated ~7 times. Subsequently the proto-KP exon pair was lost, resulting in the alternating domain arrangement present in human, bovine and mouse tropoelastins. Consistent with the other amniotes, chicken and lizard *eln* also underwent a sevenfold replication of the proto-KA exon pair. However, in these species the proto-KP exon pair remained in the sequence, accounting for the presence of exons 29 and 29a. In the case of frog (*Xenopus tropicalis*) *eln*-1 and salamander *eln*, the replication event took place in the proto-KP exon pair, resulting in a pattern of alternating domains that includes a single KA exon pair followed by multiple KP exon pairs. Similarly, in the coelacanth, a lobe-finned fish that is evolutionarily closer to tetrapods than teleosts (Zardoya and Meyer 1997), replication appears to have taken place in the proto-KP exon pair, although more sequence needs to be identified in order to establish the extent of that replication. Finally, in the case of zebrafish *eln*-a, replication events appear to have included both proto-KA and proto-KP exon pairs.

In spite of mutational events taking place after replication, particularly in hydrophobic exons, sequence similarities supporting the proposed pattern of replication in Fig. 4.9 are apparent in KA exons 15 in amniotes, 16 in amphibians and teleosts and A in coelacanths (exon numbers in coelacanths have not yet been established) and to a much lesser extent in their paired hydrophobic exons (16, 17 and B) in these species. With respect to the KP exon pair, there are clear similarities among exons 29/29a of chicken and lizard *eln*, exons 18/19 of frog (*Xenopus tropicalis*) *eln*-1 and salamander *eln*, exons C/D of coelacanth *eln* and exons 35/36 of zebrafish *eln*-a, as would be predicted from this simplified scheme. Note that a similar analysis including *eln*-2 of frog and *eln*-b of zebrafish has not been possible.

4.4 Species- and Tissue-Specific Functional Differences

The fundamental requirement of an elastomeric material is the ability to undergo repetitive extension and recoil without distortion or structural failure. We have already discussed the commonalities of 'style' of sequence and domain arrangements that allow a highly nonpolar protein such as elastin to perform as an entropic elastomer. These principles are, of necessity, shared by all elastins across phylogeny. Nevertheless, in spite of the relaxed requirement for sequence conservation to achieve these basic properties, it is reasonable to assume that at least some of the differences among tropoelastins evolved to meet species-specific functional requirements.

Direct correlation of sequence/domain differences among tropoelastins with altered functional properties is not simple, in part because these monomers must be incorporated into a cross-linked, polymeric matrix in order to behave as an elastomer. Polymeric assembly of the extracellular elastic matrix is currently thought to involve both an inherent ability of tropoelastin monomers for self-organization into networks (Kozel et al. 2006; Clarke et al. 2006; Cirulis et al. 2008; Yeo et al. 2011) as well as interactions with other matrix-associated components (Wagenseil and Mecham 2007; Kozel et al. 2003; Cain et al. 2008), including cells, which together determine the final architecture of the elastic matrix. In this way, mutations or modifications to tropoelastins affecting either self-organization or interactions with other matrix components may alter assembly, matrix architecture and consequent mechanical properties, including elastic modulus (stiffness), extensibility, resilience (efficiency of recoil) or viscoelasticity. While direct linkage of mutations to functional properties is problematic, it is certainly possible to identify several types of modifications with plausible influences on functional properties of the elastic matrix.

4.4.1 Alternate Splicing and Interactions with Other Matrix Components

Although virtually all of the current information relates to mammalian and chicken tropoelastins, it is evident that tropoelastin variants due to alternate splicing are both common and species specific (Pollock et al. 1990; Heim et al. 1991; Parks et al. 1992) and may also vary with tissue and developmental stage. Excision of hydrophobic and/or cross-linking domains through such a process would be expected to affect alignment for formation of the lysine-dependent intra- and intermolecular cross-links which stabilize polymeric elastin.

Loss or gain of specific interaction sites with other matrix components through alternate splicing or mutation also provide examples of situations in which matrix properties may be affected. For example, the unusual sequence of the C-terminal domain of tropoelastins has been implicated in the assembly of the elastic matrix (Brown-Augsburger et al. 1996; Hsiao et al. 1999; Broekelmann et al. 2005, 2008; Yeo et al. 2012). This sequence has been significantly mutated in the case of lizard and medaka eln-b, with the loss of an otherwise well-conserved cysteine pair and, in the case of eln-2 of *Xenopus tropicalis* and *Xenopus laevis*, with the loss of the C-terminal tetra-/tri-basic sequence (Fig. 4.5). Nevertheless, in the case of Xtr eln-b it is clear that loss of this apparently important sequence does not preclude assembly of the monomer into an elastic matrix, although the tissue distribution of that elastin is altered (Miao et al. 2009), perhaps reflecting differential functional properties. In the case of the lizard, the presence of only a single version of tropoelastin implies that the tropoelastin product can also be incorporated into matrix.

4.4.2 *Cross-Linking Domains and Polymer Assembly*

The nature of the predominant cross-linking domains may also influence properties of the polymer. For example, although KP-type cross-linking domains are present in all tropoelastins, KA-type cross-linking domains dominate in most species. KA cross-linking domains, particularly those with proceeding polyalanine sequences, have traditionally been associated with α-helical structure (Tamburro et al. 2006), at least in part because a KAAK or KAAAK motif in an α-helix would display the pair of lysines on the same side of the helix, facilitating the alignment of four lysine residues (two from each polypeptide chain) for the formation of desmosine or isodesmosine, the major covalent cross-links of elastin. In contrast, KP-type cross-linking domains are not embedded in a polyalanine sequence and, in any case, the presence of even a single proline residue in this motif would preclude α-helical structure. Such considerations have led to suggestions that cross-links formed from KP-type domains would be limited to lysinonorleucine, bridging between two lysine residues (Brown-Augsburger et al. 1995). If that is the case, it might be expected that frog and salamander elastins, in which KP-type cross-linking domains are dominant, would be relatively deficient in desmosine and isodesmosine cross-links. In fact, contents of these cross-links in frog and salamander elastins are reported to be approximately the same as in teleost elastins (Sage and Gray 1979), in which KA domains predominate. In the absence of data on native forms of these elastins, it is not clear whether the differing character of KP cross-linking domains would result in changes in the physical properties of these elastic matrices.

4.4.3 *Functional Requirements Driving Evolutionary Changes*

An alternative approach to establishing a connection between tropoelastin sequence diversity and differential functional properties is to consider species-specific functional or environmental requirements that could drive alterations in the character of tropoelastins.

4.4.3.1 Blood Pressure

The appearance of elastin in phylogeny coincides with the development of higher pressure, pulsatile, closed circulatory systems. However, blood pressures can vary substantially across phylogeny. For example, the mean arterial pressure in the ventral artery of teleosts is approximately 35 mmHg (Stevens and Randall 1967), substantially lower than that of the amniotes. Such differences suggest that varying physical demands on the elastic matrix may result in altered properties that would, in turn, be reflected in differential sequence and domain arrangements of tropoelastins. While most cross-linking domains in teleosts remain short (~15 amino acids), hydrophobic domains are generally substantially larger in teleosts compared to amniotes (Chung et al. 2006). A consequence of this would be an increased inter-cross-link distance in the polypeptide chain and a decreased cross-link density in the polymer, which should result in a decreased elastic modulus (stiffness). Indeed, the elastic modulus of purified elastin from ventral aorta of teleosts is significantly lower than that of amniotes (unpublished data).

The presence of two tropoelastin genes with distinctive characters and distributions in teleosts provides what is probably the clearest example of differential evolution of tropoelastins to fill specific functional roles (Miao et al. 2007, 2009). The general consequence of a gene duplication event is that one of the genes is free to mutate at a greater rate, either degrading to an unexpressed pseudogene or significantly altering the functional properties of its protein product. In the case of teleost tropoelastins, eln-a, primarily expressed in ventral and dorsal aorta, appears to be providing the common function of elastin in vascular tissues. In contrast, eln-b is the exclusive form of tropoelastin present in the bulbus arteriosus (Miao et al. 2007). The bulbus arteriosus of teleosts is a post-ventricular organ that acts as a damping chamber, modulating pulsatile pressure in the ventral aorta and protecting the delicate gill vessels that are immediately downstream. The significant sequence and domain modifications present in eln-b are likely adaptations to this specific physiological role. Indeed, polymeric elastin isolated from teleost bulbus arteriosus has both a lower elastic modulus and decreased resilience (unpublished data), properties which are particularly suited for the role of the organ as a pressure damping chamber.

4.4.3.2 Homeothermia, Poikilothermia and Ambient Environmental Temperature

Poikilothermic species such as teleosts and amphibians living in environments with lower ambient temperatures must adjust their metabolic processes to adapt to these conditions. Differential evolution of elastic matrix components, including tropoelastin sequences and domain arrangements, may therefore have taken place to adjust for the ambient temperature at which the elastomer must function. Interestingly, Gosline and his colleagues have demonstrated that elastomeric properties of isolated polymeric elastin are essentially invariant over such a temperature range, due to offsetting influences of temperature on hydrophobic effects and solvent interactions (Chalmers et al. 1999). On the other hand, environmental temperature may nevertheless affect the assembly of the polymeric matrix. Self-organization of tropoelastin monomers through a process known as coacervation is thought to be an early step in the assembly of the elastic matrix (Kozel et al. 2006; Wagenseil and Mecham 2007). Coacervation consists of a phase separation resulting in the formation of tropoelastin-rich droplets and is highly dependent on temperature (Clarke et al. 2006; Yeo et al. 2011). For example, human tropoelastin, under what is thought to be in vivo conditions, undergoes phase separation at approximately 37 °C but remains in solution at 20 °C. Experimental data on recombinant elastin-like polypeptides show a clear relationship between increased molecular weight, essentially a greater number of domains in the polypeptide, and lower coacervation temperatures (Miao et al. 2003). Thus, increased molecular weights of frog and teleost elastins, generally due to additional exon pair replications, may have the effect of allowing coacervation to take place at lower ambient temperatures. Unfortunately, in the absence of recombinant production of full-length frog and teleost tropoelastins, this hypothesis has not been tested experimentally. The molecular weight of frog eln-2 is inconsistent with this hypothesis, although this tropoelastin is also unusual in many other respects. In the case of the poikilothermic lizard, this adaptation may not be required because the ambient temperature of its environment is generally high.

4.4.3.3 Cycles of Extension and Recoil, Heart Rate and Lifespan

One of the remarkable features of elastins, especially those present in lung and aortic tissues, is their lack of turnover (Powell et al. 1992; Shapiro et al. 1991; Ritz-Timme et al. 2003). As a consequence, this polymeric biomaterial must undergo millions if not billions of cycles of extension and elastic recoil over a normal lifetime. However, the frequency of cycling of extension and recoil in the aorta is dependent on heart rate, which varies widely. For example, within the mammalian group alone heart rates can vary between ~30 beats per minute in the whale and elephant, to ~70 beats per minute in the human, dog and cow, to ~500 beats per minute in the mouse. Since frequency of cycling of extension and recoil is known to affect performance of elastomers (Gosline 1980), it is entirely likely that some evolutionary adaptations to tropoelastin sequence and domain arrangements would have been made in response to the range of frequencies over which the elastic matrix had to operate.

Domain 30 has been identified as particularly important for the assembly process of the elastic matrix in mice (Kozel et al. 2003), perhaps through the ability of this domain to form β-sheet structures. In general, at least in mammalian elastins, formation of extended β-sheet structures is avoided by regular spacing of proline residues, for example, at an average periodicity of every 5–6 residues in human tropoelastin (Muiznieks and Keeley 2010) (Fig. 4.10a). Such close spacing is thought to ensure independence and free mobility of polypeptide chains and therefore elastomeric character. However, domain 30 is exceptional in this respect, with proline spacing increased to >15 residues in the human and >50 residues in the mouse. Using model elastin-like polypeptides, such increased proline spacing has been shown to have significant effects on elastomeric assembly and properties, and has been correlated with increased formation of β-sheet structures (Muiznieks and Keeley 2010). Indeed, wholesale replacement of the regularly spaced proline residues of human domain 24 with glycine residues results in formation of amyloid-like aggregates (Miao et al. 2003). Moreover, Tamburro and his colleagues have

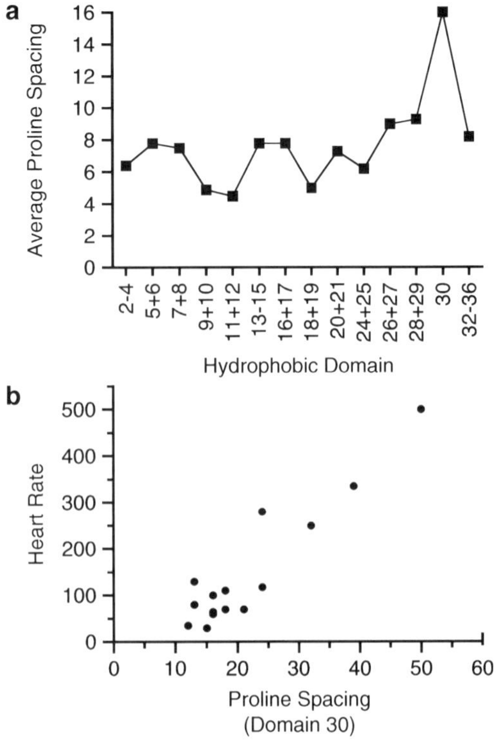

Fig. 4.10 Average spacing between proline residues in hydrophobic regions of human tropoelastin and its relationship to heart rate. (**a**) Spacing between proline residues in human elastin averages about 6 amino acid residues, with the exception of the hydrophobic region of exon 30 (Reproduced from Muiznieks et al. (2010) with permission). (**b**) Relationship between heart rate and spacing of proline residues in the domain 30 region of several amniote species suggests a correlation between increased heart rate and increased stretches of proline-free sequence, with a potential for β-sheet formation. Species included are rat, mouse, chicken, guinea pig, rock hyrax, rabbit, cat, dog, panda, chimpanzee, human, cow, pig, sheep, horse and elephant

reported that, in isolation from the rest of the molecule, domain 30 of human tropoelastin will undergo amyloid-like aggregation (Tamburro et al. 2005).

This unusual contribution of domain 30 to elastin assembly and the wide variability of proline spacing in this region, even within mammalian species, have led to speculations about a functional role for these differences. Interestingly, there is a strong positive correlation between heart rate and proline spacing in the region of domain 30 across mammalian species (Fig. 4.10b), prompting speculation that increased capacity for β-sheet formation in this region may be an adaptation for the requirement of some elastins (e.g. in rodents) to perform at higher frequencies of extension and recoil.

As well as differences in frequency, it is also interesting to consider whether a capacity to withstand an increased total number of cycles of extension and recoil without significant deterioration in elastomeric properties or mechanical failure might act as an adaptive influence for species-specific evolution of elastins, especially in longer-lived animals. However, in spite of large variations in heart rates, it is estimated that the total number of beats/lifetime is approximately equal in all mammalian species (Azbel 1994; Levine 1997), although the explanation for this observation usually relies on considerations of metabolic rates rather than mechanical properties of arterial tissues. For example, a rat with a heart rate of 350 beats per minute and a lifespan of 5 years would accumulate ~900 million cycles of extension and recoil over a lifetime. Similarly, an elephant with a heart rate of 30 beats per minute and a lifetime of 70 years would accumulate 1.1 billion cycles (http://www.sjsu.edu/faculty/watkins/longevity.htm). Humans are exceptional in this respect, accumulating more that 2.9 billion cycles over an 80 year lifespan. While it might be tempting to speculate that this exceptional durability in humans may be related to specific sequence and domain arrangements or other factors affecting matrix assembly and architecture, it is difficult to argue for significant evolutionary pressure to achieve such long-term durability.

4.5 The Origin of Elastin

Whereas many other extracellular matrix structural proteins such as collagens and fibrillins have deep evolutionary roots, elastin is a relative newcomer on the scene, present in the vascular tissues of elasmobranchs (sharks and rays) but absent, at least as a vascular protein, in agnathans (lamprey and hagfish) as well as other lower chordates and invertebrates (Sage and Gray 1979, 1980, 1981). Previous to the appearance of elastin, the functional property of elastomeric damping in lower pressure pulsatile circulations appears to have been provided solely by microfibrillar proteins similar to those present in vertebrate vascular and other tissues (Davison et al. 1995; Faury 2001).

At one time it was postulated that elastin might be evolutionarily related to earlier collagens, based on their similarly high contents of glycine and proline (Rudall 1968). However, even before sequences became available for these proteins others had argued that this was unlikely (Partridge 1962). Although the origin of tropoelastin remains unclear, it is improbable that this protein appeared de novo but rather that ancestral elastin was assembled and modified from earlier proteins with different functional properties through evolutionary processes likely involving exon shuffling (Patthy 1996; Kawashima et al. 2009). Unfortunately, at the moment identification of plausible ancestral proteins is a matter of speculation. Based on the growing availability of detailed information on the sequences of elastins across the entire spectrum of species in which this protein is found, a reasonable approach might be to identify potential fundamental elements of elastin sequences and separately consider their possible origins.

4.5.1 Glycine-Rich Sequences

Glycine-rich regions are characteristic of the sequences of many structural proteins, both intracellular and extracellular. Examples of non-collagenous extracellular proteins include keratins (Wu et al. 2008) (NM_006121.3, NP_006112.3, NP570940.1), intermediate filament-like proteins of hagfish slime (Koch et al. 1995; Fudge et al. 2003), tick cement proteins (Maruyama et al. 2010) (NP_608973.1XM_002415100.1), mussel byssus proteins (Waite et al. 2004), egg envelope proteins of invertebrates (Pau 1984) (NP_608973.1), insect silk proteins including spider silks (Colgin and Lewis 1998; Hayashi and Lewis 1998; Gosline et al. 1999; Knight and Vollrath 2002; Xu and Lewis 1990; Goldsmith and Kafatos 1984) as well as a large variety of non-collagen cartilaginous proteins that are particularly abundant in lower chordates and invertebrates (Robson et al. 1993, 1997, 2000a, b; Wright et al. 1998, 2001). Self-aggregation into fibrillar structures is a common characteristic of most of these proteins.

Among these structural proteins, sequences containing GGhGh or similar motifs (where ‘h’ is a nonpolar amino acid), often as tandem repeats, are remarkably common. These include the GGLGY motifs of lamprin, insect eggshell proteins and some spider silks (Robson et al. 1993; Pau 1984; Xu and Lewis 1990; Goldsmith and Kafatos 1984). Indeed, a recent survey of the bed bug transcriptome has identified the lamprin motif, GGLGY, as the most abundant protein domain in this species (Bai et al. 2011). Similar GGhGh tandem repeats are present in exon 30 of all amniote tropoelastins, although the number of repeats varies from two in the human to eight in the mouse. As discussed previously, these types of sequences have been associated with formation of β-sheet structures and have been postulated to be important for interactions leading to self-aggregation (Robson et al. 1993), including a particular role in the self-assembly of tropoelastin (Kozel et al. 2003; Muiznieks and Keeley 2010). The similarity of sequence ‘style’ in such self-aggregating structural proteins is remarkable, considering that the last common ancestor of species in which these proteins are found existed more than 500 million years ago.

On this basis it is tempting to speculate that glycine-rich regions of tropoelastin may share an evolutionary origin with other glycine-rich structural proteins, particularly those of lower chordate and invertebrate cartilage. Indeed, tropoelastin remains the major matrix protein of elastic cartilages of the external ear, larynx and epiglottis in humans. However, since extensive β-sheet stacking or amyloid-like aggregation is inconsistent with the chain flexibility required for an entropic elastomer such as elastin, modification to the sequences of these glycine-rich regions must have taken place to inhibit formation of extended β-sheets and facilitate chain disorder. Based on current knowledge of the sequence of elastin-like proteins, that modification would likely have included the substitution of prolines for glycines in first position of the GGhGh motif, resulting in the PGhGh and similar motifs that are extremely common in tropoelastins. Indeed, as pointed out earlier, VPG or more generally hPG sequences are the most common sequence elements of tropoelastins across all species (He et al.

2007), introducing the 'sliding β-turns' that have been suggested to play a crucial role in maintaining chain disorder (He et al. 2007; Tamburro et al. 2003). The effect of such G to P mutations would be amplified by both exon replication and tandem sequence replication within exons, both evidently common events in the evolution of tropoelastins. Supporting this role for proline are several recent reports correlating increased proline content in types of spider silks with enhanced elastic properties (Liu et al. 2008a, b; Savage and Gosline 2008a, b). Furthermore, either substitution of glycine for proline in PGhGh-type repeats (Rauscher et al. 2006; Miao et al. 2003) or extending the spacing between proline residues in elastin-like polypeptides (Muiznieks and Keeley 2010) promotes the formation of amyloid-like aggregates.

4.5.2 Cross-Linking Domains

The evolutionary origin of the KxxK cross-linking domains of tropoelastin is more problematical. While lysine-derived cross-links, produced through the oxidative deamination activity of lysyl oxidase, are also present in collagens, the desmosine/isodesmosine cross-links involving four lysine residues are characteristic of elastins and not present in collagens. Although there are several reports of the presence of desmosines in egg membrane proteins, including those of reptiles (Starcher and King 1980; Leach et al. 1981; Crombie et al. 1981; Cox et al. 1982), no KxxK motifs can be identified in available protein sequences (Hyllner et al. 2001; Litscher and Wassarman 2007). This may suggest that desmosine-like cross-links can, in fact, form in the absence of paired lysine residues and that the particular KxxK motif developed later in the evolutionary history of tropoelastin and was subsequently amplified by the exon pair replication process. KP-type domains as well as KxK and singlet lysines are present in the central conserved region of all tropoelastins, and it is worth noting that interconversion of KP and KA motifs through mutation of proline to alanine residues requires only a single-base substitution.

The similarities in sequence and domain arrangements between spider silks and tropoelastins is particularly striking. For example, at least some spider silks consist of repetitive, glycine-rich sequences interspersed with polyalanine sequences (Gosline et al. 1999), which form the β-sheet 'liquid crystal' regions responsible for tensile strength (Gosline et al. 1999; Knight and Vollrath 2002). Although very reminiscent of the glycine-rich hydrophobic domains interspersed with polyalanine-based cross-linking domains seen in amniote tropoelastins, this is perhaps more likely to be an interesting example of sequence convergence for a particular functional purpose.

Searches of sequence databases using short KxxK types of sequences have yielded no convincing hits for proteins other than tropoelastins. The only exception to this is calymmin, a structural protein found in the notochord of zebrafish embryos (Cerdà et al. 2002). The sequence of calymmin consists of glycine-rich domains alternating with 'cross-linking' domains of the KP-type (Fig. 4.11). Furthermore,

[1]MLRWTILRMMLLLWMTRVTLQGGT|[2]GYQPQKSNGNAAPISPG|[3]GYGSNNNAANLKGY|[4]
GVYGGAQNKPANRGHLPH|[5]GYGNGNTNGNNGYKAPSYG|[6]GYSNLLGTHPRKGAGL|[7]GF
GAGPAATKGQGSNPG|[8]GSSQYMGNTNGGTKGPKP|[9]GYGAQAGQTNGQIAKPN| GYGG
YPNGGAANQPKT|[11]GFSQYMGYPNGGTKGPKP|[12]GYGGYPNGGAAVQPGT|[13]GSSQYKG
FPSGGIKGPNP|[14]GFAANAGASNEKGVKPN|[15]GYGGYPNGGAANQPNG|[16]GSSHYMGYP
NGGTKGPKP|[17]GYGANAGPSAGQVAKPN|[18]GYGGYPNAGATNQPNG|[19]GPFQNMGYPNG
GTKGPKP|[20]GYGAKAGPSAGHVAKPN|[21]GNGGYPNGGATSQHNG|[22]GSSQFMGYPNGGT
KGPKS|[23]GYGANAGPSAGQVAKPN|[24]GNGRYPIGGVANQPNR|[25]GSSQNMGYPNGRTKG
PKQ|[26]GYGGYPNGGAANQPNG|[27]GSTQNMGYSNGGTKGPKQ|[28]GYGANAGPSAGQVAKP
N|[29]GYGGYPNGGATNQPNG|[30]GSSQYMGYPNGGAKGPKA|[31]GYGAKSGPSSGQVSKPN|[32]
GSSQYMGYPNGGTNGLKP|[33]GYGAKARTTYGQGAKPQN|[34]GYGVPGYTSKGPKAANS|[35]
GYMPVITGNKGASIGKGPKGEILSAEAPSQAPLSNTKGVLPVAQQEPVTLLPNQGTKGLP
VLQQTKGQILQASLPQSKTQNSMVPQFASNPFMQGPSSYKPSKMYKPPTSTIPIEKPEIP
VIPQSVPAPESAFIPQTSQTTPLEQEQVLSLEQKSQVTAQPVTQQGSISSQAAQSTPVLQ
QAEIQSNPALPQLKNPTTTNP|[36]ELTGLEQPNGQVQGAVPAKP|[37]DCGPEGRPNGQWVK
LTNP|[38]GFGYPNGKGEGLSQP|[39]GFGAGIGGYPTQGMNNGHKA|[40]GYGGYGNQLKNHA|[41]
GLQGGPPARLGTKGKSQTKY|[42]GIGGLPFGGSPSGYQSNPYGQY|[43]GNDGQRYGAKPYN
PKAPGKY|[44]GYGSLPYNSQAFPESVARSP|[45]DFGGFPYNGQPLGYGSGKSTGKY|[46]GQK
GPQYGGQSLGHGPDAISGKY| GSPESLYTPETVNLGGNAKSTKY|[48]GNPAVSYELLGP
VTDGQSVE|[49]ENRSIETLHQGPEIDGQKAVDNF|[50]GDGEIPPHPDVPGAGEANAQSTEK
Y|[51]GTEDYNAGKPQP|[52]EAVSFPAVPTVGPAVTSLDTTLSLDGSSPVPVSVPDAVHEVP
VSPILPQPDLPTLMLQPAPPQQIQIQQHLKLHFQPHSNSQTG|[53]KEGKYKLSGFFGNKYQ
G**uga**acagcguuucugccuggaagugaccucaugaacuauggugcuacugcaugauuuag
aauguaaaauugugacgaaaaag

Fig. 4.11 Sequence of full-length calymmin, an elastin-like protein found in the notochord sheath of zebrafish embryo (NM_131014). Exon borders are indicated by |. Exon numbers are in superscript. KxxK/KxK motifs are shaded. Stop codon is bolded. First ~80 bases of the 3′UTR are indicated in lowercase

exon pair replication similar to that seen in tropoelastins clearly has taken place in this protein. In addition, lysyl oxidase activity has been reported to be essential for notochord development in zebrafish (Gansner et al. 2007), although cross-linking of notochordal collagens may account for this dependence. In spite of these similarities, neither the central conserved region nor the C-terminal region nor the conserved 3′UTR region of tropoelastin is recognizable in the calymmin sequence. Glycine-rich structural proteins are also found in the notochordal sheath of the lamprey (Robson et al. 1997) and likely also in other lower chordates. However, searches of databases using the sequence of calymmin have not identified similar proteins in other chordates, including lamprey and amphioxus. At this point the evolutionary relationship between tropoelastin and calymmin is unknown.

4.5.3 C-Terminal Domains and 3′UTR

Just as was the case for the cross-linking motifs, searches of sequence databases for the well-conserved C-terminal region, including the first 60–80 bases of the 3′UTR, produce no convincing hits for proteins other than tropoelastins. In particular, we have been unable to identify similar sequences in lower chordate or invertebrate databases. Indeed, it has been our experience that this C-terminal/3′UTR sequence is the most selective search tool for tropoelastins. As such, at this point it is not possible to even speculate on the origins of this domain.

4.5.4 Follow the Synteny

A possible alternate approach to identifying proteins with an early evolutionary relationship to elastin could take advantage of the remarkable conservation of synteny between tropoelastin and its nearest downstream neighbour gene, LIM kinase (*LIMK*). Two distinct genes for LIMK, *LIMK*1 and *LIMK*2, are present in most species, excluding insects (Scott and Olson 2007). In all species from humans to teleosts, *LIMK*1 is found downstream from tropoelastin, as part of a group of genes in the Williams-Beuren cluster (Pober 2010). Indeed, in the case of teleosts, where two tropoelastins have been generated by a genome duplication event in the line leading to the ray-finned fish, there are also two *LIMK*1 genes, designated *LIMK*1a and *LIMK*1b. Each of these genes is downstream from its respective tropoelastin gene. However, while this synteny is also maintained for the *eln*-1 gene of frog, we have been unable to locate a *LIMK*1 gene downstream of *eln*-2 in the frog. Although both tropoelastin and *LIMK*1 genes have been located in skate, the spatial relationship between these genes has not yet been established.

The conserved relationship between tropoelastin and *LIMK*1 genes is reflected in a phylogenetic tree constructed from LIMK sequences across phylogeny (Fig. 4.12), which shows the expected evolutionary relationships between species, and is essentially identical to that generated using sequences from the central conserved region of tropoelastins (Fig. 4.4). If the synteny between *LIMK*1 and tropoelastin genes can also be shown to be conserved in elasmobranchs (shark and skate), it will be of interest to search upstream from *LIMK*1 genes in lower chordates (e.g. lamprey, hagfish, amphioxus, ciona) for a glycine-rich structural protein. The presence and character of such a protein might provide important insights into the early evolutionary history of tropoelastin.

4.6 Summary and Conclusions

Independent of species, the two most important properties required of tropoelastins are the ability to assemble and cross-link into a stable polymeric network and for that network to have the unusual properties of extensibility and elastic recoil. Although the number of well-curated full-length sequences of tropoelastins is still limited, the past 10 years have seen a rapid increase in available, if sometimes partial, sequence information, especially from non-mammalian species. At the same time there has been an increased understanding of the process of assembly of elastic matrices, both in terms of the intrinsic properties of tropoelastins for self-organization and the influence of a variety of matrix-associated proteins on network assembly and architecture. Simultaneously, computational and other methods have

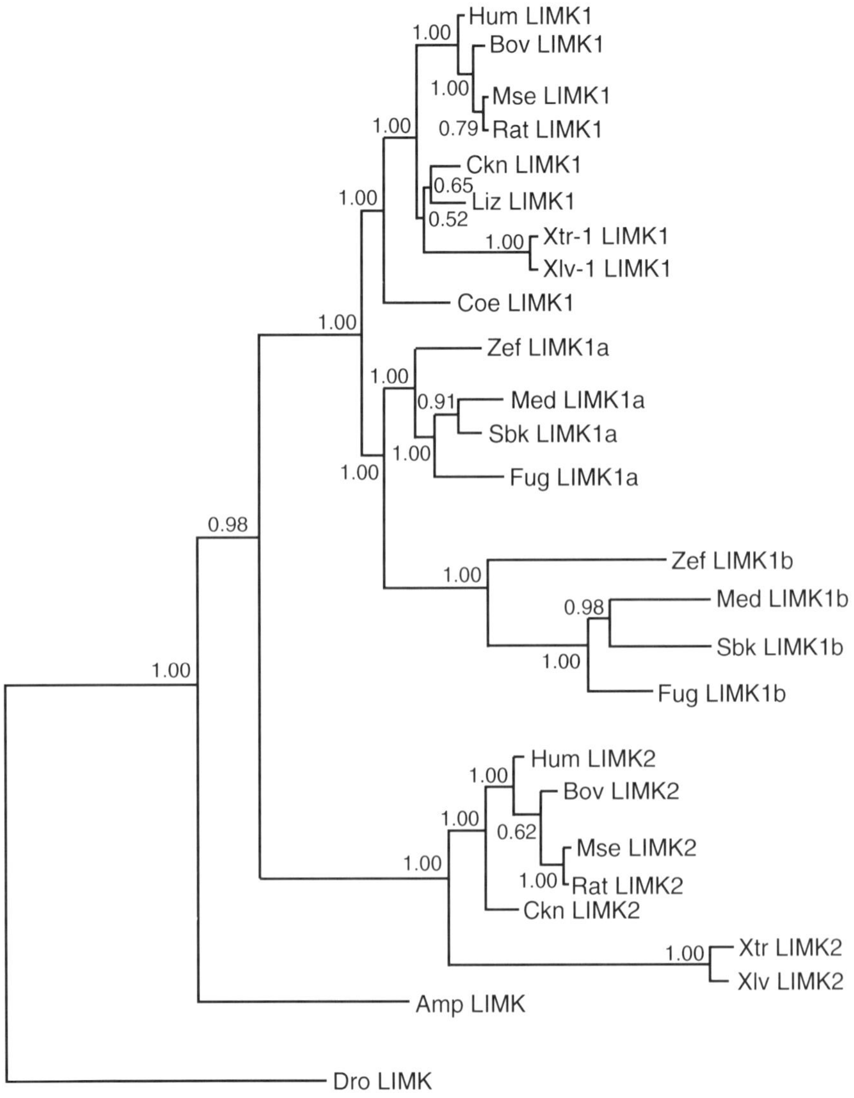

Fig. 4.12 Phylogenetic tree generated using LIM kinase-1 sequences downstream from tropoelastin genes in several species, LIM kinase-2 sequences from amniotes and amphibians and LIM kinase sequence from amphioxus (Amp), a lower chordate. The sequence of LIM kinase from *Drosophila melanogaster* (Dro) was used as an outgroup. This Bayesian tree was constructed with MrBayes 3.1.2 (Ronquist and Huelsenbeck 2003) using one million generations, with the first 25,000 trees discarded as burn-in. Numbers at nodes are posterior probabilities, with the maximum probability represented as 1.00. Relationships between species are in good agreement with generally accepted phylogenetic relationships, including the placement of the coelacanth in the lineage leading to the tetrapods, and the genomic duplication in the lineage of the ray-finned fish occurring before the separation of teleost species. This phylogenetic tree for KIM kinase-1 is essentially identical to that generated from the central conserved regions of tropoelastins (Fig. 4.4)

defined the 'structural' requirements for maintaining the conformational disorder required for the physical properties of entropic elastomers such as elastin. All of this information, taken together, has provided significant insights into the relationship between sequences of elastins and their elastomeric properties.

High glycine and proline contents, and the distribution of proline residues along the polypeptide chain in transient β-turn-forming sequences exemplified by VPG, appear to be essential for maintaining conformational flexibility in hydrophobic domains, even in the polymeric state. However, within such constraints for maintaining a dynamically disordered structure, it is clear that tropoelastins can preserve this essential 'style' of sequence while varying significantly in precise sequence and domain arrangements. Differences in sequence of tropoelastins can evidently arise in a number of ways, including tandem replication of short sequences within exons, replication of entire exons or exon pairs, loss or gain of exons from the genomic sequence, complete gene duplication and species-specific alternate splicing. This is consistent with suggestions that, in general, such low-complexity disordered proteins evolve differently and perhaps more rapidly than ordered proteins, tolerating more insertions and deletions as well as different types of point mutations (Brown et al. 2011).

A comparison of full-length tropoelastin sequences of amniotes provides examples of many of these methods for generating sequence diversity within a consistent style of protein sequence and domain arrangement. At the same time, the fact that exon-to-exon relationships can be recognized among all tropoelastin sequences from amniotes, including both synapsid (mammals) and sauropsid (reptiles and birds) subdivisions of this group, supports the view that all amniotes share a common ancestral sequence of tropoelastin and that the expansion of the replication region through one or more exon pair replication events took place in the ancestral tropoelastin before divergence into these two subgroups approximately 300 million years ago.

Teleosts provide an example of a further method for generating sequence diversity of tropoelastin, specifically the whole-genome duplication event that took place in the lineage of the ray-finned fish. Both tropoelastin genes appear to have remained functional in teleosts, although in all cases one of these genes has clearly mutated more extensively. At least in the case of the zebrafish, this modified gene appears to have developed a specific functional role in the bulbus arteriosus (Miao et al. 2007). Expansion of protein size through repeated exon pair duplication in the replication region appears to be particularly extensive for both genes of the zebrafish. The presence of remnants of sequence homology in flanking introns suggests that this replication took place relatively recently. However, the lack of recognizable sequence similarity between the replicated regions of zebrafish *eln*-a and *eln*-b would argue that a common exon pair sequence was not replicated in these two genes. It is possible that teleost elastins have undergone more than one independent cycle of expansion followed by exon loss, accounting for these observations. Note that, as would be predicted from its evolutionary position on the tetrapod rather than the ray-finned fish lineage (Meyer and Van de Peer 2005), it appears that the coelacanth possesses only a single version of the tropoelastin gene.

Both *Xenopus tropicalis* and *Xenopus laevis* species of frogs also have two tropoelastin genes, although this is likely the result of a duplication event specific to the elastin gene. Again one of these genes (*eln*-2) has undergone extensive mutation. However, in spite of its highly unusual nature, *eln*-2 is expressed and its protein product incorporated into extracellular matrix (Miao et al. 2009). Clearly this gene duplication event took place before the divergence of *tropicalis* and *laevis* species, approximately 65 million years ago. On the other hand, the salamander appears to possess only a single tropoelastin gene, suggesting that the duplication event took place after the divergence of the salamander and frog lineages, approximately 300 million years ago.

Within all this diversity, it is remarkable that there are recognizable regions of sequence similarity retained across all tropoelastins so far identified. These include the central conserved region, the final two protein domains and the striking region of sequence conservation in the first 80 bases of the 3′UTR, flanking and including a fully conserved miR-29 binding sequence. While the C-terminal domain has been suggested to be important for elastin assembly, it is clear that this region is not an absolute requirement for matrix deposition (Miao et al. 2009). On the other hand, the 3′UTR and particularly the miR-29 binding site almost certainly plays an essential role in regulation of expression of all tropoelastins and may be responsible for the distinct temporal pattern of tropoelastin production during development (Miao et al. 2007). Because of its strong sequence conservation, the central conserved region has been speculated to form the core of the 'primordial' tropoelastin sequence (Chung et al. 2006), although a specific contribution of this region to tropoelastin assembly or in determining the physical properties of the elastin network has not been identified. The fact that the replication region is immediately downstream of this central conserved region in all tropoelastins suggests that tropoelastin genes also share a susceptible site in this position for replication events, although the nature of that site has not been identified.

Finally, although it is possible to speculate that some sequence differences between tropoelastins must represent adaptations driven by species-specific functional requirements (e.g. heart rate, poikilothermy), the background sequence diversity that appears to be permitted in tropoelastins while retaining the style of sequence essential for elastomeric properties makes it difficult to identify such functional adaptations with any certainty, at least at the present time. However, as more sequence data emerges and with the ability to recombinantly express and compare assembly characteristics of these tropoelastins and network properties of the polymeric forms of these proteins, using both in vitro and in vivo model systems, a greater appreciation of subtle effects of sequence and domain arrangements will no doubt be possible in the future. In addition, the emergence of additional genomic sequence information, particularly of elasmobranchs as well as of agnathans and other non-vertebrate chordates, should allow a better understanding of the relationship of tropoelastin to other glycine-rich, nonpolar structural proteins and ultimately of the evolutionary roots of this unusual protein.

Acknowledgements The author acknowledges the essential contributions to this work by present and former members of his research laboratory, including Richard Stahl, Lisa Muiznieks, Ming Miao, Eva Sitarz, Sean Reichheld and Martin Chung. Noeleen Loughran provided valuable assistance in the production of phylogenetic trees. This work was supported by operational grants from the Heart and Stroke Foundation of Ontario, the Canadian Institutes of Health Research and the Natural Sciences and Engineering Research Council of Canada. The author holds the Heart and Stroke Foundation of Ontario/Robert M. Freedom Chair in Cardiovascular Science at the Hospital for Sick Children.

References

Azbel MY (1994) Universal biological scaling and mortality. Proc Natl Acad Sci USA 91:12453–12457

Bai X, Mamidala P, Rajarapu SP, Jones SC, Mittapalli O (2011) Transcriptomics of the bed bug (*Cimex lectularius*). PLoS One 6:e16336. doi:10.1371/journal.pone.0016336

Bressan GM, Argos P, Stanley KK (1987) Repeating structure of chick tropoelastin revealed by complementary DNA cloning. Biochemistry 26:1497–1503

Broekelmann TJ, Kozel BA, Ishibashi H, Werneck CC, Keeley FW, Zhang L, Mecham RP (2005) Tropoelastin interacts with cell-surface glycosaminoglycans via its COOH-terminal domain. J Biol Chem 280:40939–40947

Broekelmann TJ, Ciliberto CH, Shifren A, Mecham RP (2008) Modification and functional inactivation of the tropoelastin carboxy-terminal domain in cross-linked elastin. Matrix Biol 27:631–639

Brown CJ, Johnson AK, Dunker AK, Daughdrill GW (2011) Evolution and disorder. Curr Opin Struct Biol 21:441–446

Brown-Augsburger P, Tisdale C, Broekelmann T, Sloan C, Mecham RP (1995) Identification of an elastin cross-linking domain that joins three peptide chains. J Biol Chem 270:17778–17783

Brown-Augsburger P, Broekelmann T, Rosenbloom J, Mecham RP (1996) Functional domains on elastin and microfibril-associated glycoprotein involved in elastic fibre assembly. Biochem J 318:149–155

Cain SA, Raynal B, Hodson N, Shuttleworth A, Kielty CM (2008) Biomolecular analysis of elastic fibre molecules. Methods 45:42–52

Cerdà J, Gründ C, Franke WW, Brand M (2002) Molecular characterization of calymmin, a novel notochord sheath-associated extracellular matrix protein in the zebrafish embryo. Dev Dyn 224:200–209

Chalmers GWG, Gosline JM, Lillie MA (1999) The hydrophobicity of vertebrate elastins. J Exp Biol 202:301–314

Cheng S, Cetinkaya M, Gräter F (2010) How sequence determines elasticity of disordered proteins. Biophys J 99:3863–3869

Christoffels A, Koh EGL, J-m C, Brenner S, Aparicio S, Venkatesh B (2004) Fugu genome analysis provides evidence for a whole-genome duplication early during the evolution of ray-finned fishes. Mol Biol Evol 21:1146–1151

Chung MIS, Miao M, Stahl RJ, Chan E, Parkinson J, Keeley FW (2006) Sequences and domain structures of mammalian, avian, amphibian and teleost tropoelastins: clues to the evolutionary history of elastins. Matrix Biol 25:492–504

Cirulis JT, Bellingham CM, Davis EC, Hubmacher D, Reinhardt DP, Mecham RP, Keeley FW (2008) Fibrillins, fibulins, and matrix-associated glycoprotein modulate the kinetics and morphology of *in vitro* self-assembly of a recombinant elastin-like polypeptide. Biochemistry 47:12601–12613

Clarke AW, Arnspang EC, Mithieux SM, Korkmaz E, Braet F, Weiss AS (2006) Tropoelastin massively associates during coacervation to form quantized protein spheres. Biochemistry 45:9989–9996
Colgin MA, Lewis RV (1998) Spider minor ampullate silk proteins contain new repetitive sequences and highly conserved non-silk-like "spacer regions". Protein Sci 7:667–672
Cox DL, Mecham RP, Sexton OJ (1982) Lysine derived cross-links are present in a non-elastin, proline-rich protein fraction of Iguana iguana eggshell. Comp Biochem Physiol B 72:619–623
Crombie G, Snider R, Faris B, Franzblau C (1981) Lysine-derived cross-links in the egg shell membrane. Biochim Biophys Acta 640:365–367
Cushing L, Kuang PP, Qian J, Shao F, Wu J, Little F, Thannickal VJ, Cardoso WV, Lü J (2011) miR-29 is a major regulator of genes associated with pulmonary fibrosis. Am J Respir Cell Mol Biol 45:287–294
Davison IG, Wright GM, DeMont ME (1995) The structure and physical properties of invertebrate and primitive vertebrate arteries. J Exp Biol 198:2185–2196
Du B, Ma L-M, Huang M-B, Zhou H, Huang H-L, Shao P, Chen Y-Q, Qu L-H (2010) High glucose down-regulates miR-29a to increase collagen IV production in HK-2 cells. FEBS Lett 584:811–816
Faury G (2001) Function-structure relationship of elastic arteries in evolution: from microfibrils to elastin and elastic fibres. Pathol Biol (Paris) 49:310–325
Fudge DS, Gardner KH, Forsyth VT, Riekel C, Gosline JM (2003) The mechanical properties of hydrated intermediate filaments: insights from hagfish slime threads. Biophys J 85:2015–2027
Gansner JM, Mendelsohn BA, Hultman KA, Johnson SL, Gitlin JD (2007) Essential role of lysyl oxidases in notochord development. Dev Biol 307:202–213
Goldsmith MR, Kafatos FC (1984) Developmentally regulated genes in silkmoths. Annu Rev Genet 18:443–487
Gosline JM (1980) The elastic properties of rubber-like proteins and highly extensible tissues. In: Vincent JFV, Currey JD (eds) The mechanical properties of biological materials, Symposia of the Society for Experimental Biology, Number 34. Cambridge University Press, Cambridge
Gosline JM, Guerette PA, Ortlepp CS, Savage KN (1999) The mechanical design of spider silks: from fibroin sequence to mechanical function. J Exp Biol 202:3295–3303
Gray WR, Sandberg LB, Foster JA (1973) Molecular model for elastin structure and function. Nature 246:461–466
Hayashi CY, Lewis RV (1998) Evidence from flagelliform silk cDNA for the structural basis of elasticity and modular nature of spider silks. J Mol Biol 275:773–784
He D, Chung M, Chan E, Alleyne T, Ha KCH, Miao M, Stahl RJ, Keeley FW, Parkinson J (2007) Comparative genomics of elastin: sequence analysis of a highly repetitive protein. Matrix Biol 26:524–540
Heim RA, Pierce RA, Deak SB, Riley DJ, Boyd CD, Stolle CA (1991) Alternative splicing of rat tropoelastin mRNA is tissue-specific and developmentally regulated. Matrix 11:359–366
Hellsten U, Khokha MK, Grammer TC, Harland RM, Richardson P, Rokhsar DS (2007) Accelerated gene evolution and subfunctionalization in the pseudotetraploid frog *Xenopus laevis*. BMC Biol 5:31. doi:10.1186/1741-7007-5-31
Hew Y, Grzelczak Z, Lau C, Keeley FW (1999) Identification of a large region of secondary structure in the 3′-untranslated region of chicken elastin mRNA with implications for the regulation of mRNA stability. J Biol Chem 274:14415–14421
Hew Y, Lau C, Grzelczak Z, Keeley FW (2000) Identification of a GA-rich sequence as a protein-binding site in the 3′-untranslated region of chicken elastin mRNA with a potential role in the developmental regulation of elastin mRNA stability. J Biol Chem 275:24857–24864
Hsiao H, Stone PJ, Toselli P, Rosenbloom J, Franzblau C, Schreiber BM (1999) The role of the carboxy terminus of tropoelastin in its assembly into the elastic fiber. Connect Tissue Res 40:83–95
Hyllner SJ, Westerlund L, Olsson PE, Schopen A (2001) Cloning of rainbow trout egg envelope proteins: members of a unique group of structural proteins. Biol Reprod 64:805–811

Indik Z, Yeh H, Ornstein-Goldstein N, Sheppard P, Anderson N, Rosenbloom JC, Peltonen L, Rosenbloom J (1987) Alternative splicing of human elastin mRNA indicated by sequence analysis of cloned genomic and complementary DNA. Proc Natl Acad Sci USA 84:5680–5684

Indik Z, Yeh H, Ornstein-Goldstein N, Kucich U, Abrams W, Rosenbloom JC, Rosenbloom J (1989) Structure of the elastin gene and alternative splicing of elastin mRNA: implications for human disease. Am J Med Genet 34:81–90

Kawashima T, Kawashima S, Tanaka C, Murai M, Yoneda M, Putnam NH, Rokhsar DS, Kanehisa M, Satoh N, Wada H (2009) Domain shuffling and the evolution of vertebrates. Genome Res 19:1393–1403

Kielty CM (2006) Elastic fibres in health and disease. Expert Rev Mol Med 8:1–23

Knight DP, Vollrath F (2002) Biological liquid crystal elastomers. Philos Trans R Soc Lond B Biol Sci 357:155–163

Koch EA, Spitzer RH, Pithawalla RB, Castillos FA III, Parry DA (1995) Hagfish biopolymer: a type I/type II homologue of epidermal keratin intermediate filaments. Int J Biol Macromol 17:283–292

Kozel BA, Wachi H, Davis EC, Mecham RP (2003) Domains in tropoelastin that mediate elastin deposition *in vitro* and *in vivo*. J Biol Chem 278:18491–18498

Kozel BA, Rongish BJ, Czirok A, Zach J, Little CD, Davis EC, Knutsen RH, Wagenseil JE, Levy MA, Mecham RP (2006) Elastic fiber formation: a dynamic view of extracellular matrix assembly using timer reporters. J Cell Physiol 207:87–96

Leach RM Jr, Rucker RB, Van Dyke GP (1981) Egg shell membrane protein: a nonelastin desmosine/isodesmosine-containing protein. Arch Biochem Biophys 207:353–359

Levine HJ (1997) Rest heart rate and life expectancy. J Am Coll Cardiol 30:1104–1106

Lillie MA, Gosline JM (2002) The viscoelastic basis for the tensile strength of elastin. Int J Biol Macromol 30:119–127

Litscher ES, Wassarman PM (2007) Egg extracellular coat proteins: from fish to mammals. Histol Histopathol 22:337–347

Liu Y, Shao Z, Vollrath F (2008a) Elasticity of spider silks. Biomacromolecules 9:1782–1786

Liu Y, Sponner A, Porter D, Vollrath F (2008b) Proline and processing of spider silks. Biomacromolecules 9:116–121

Maruyama SR, Anatriello E, Anderson JM, Ribeiro JM, Brandão LG, Valenzuela JG, Ferreira BR, Garcia GR, Szabó MP, Patel S, Bishop R, de Miranda-Santos IK (2010) The expression of genes coding for distinct types of glycine-rich proteins varies according to the biology of three metastriate ticks, Rhipicephalus (Boophilus) microplus, Rhipicephalus sanguineus and Amblyomma cajennense. BMC Genomics 11:363

Meyer A, Van de Peer Y (2005) From 2R to 3R: evidence for a fish-specific genome duplication (FSGD). Bioessays 27:937–945

Miao M, Bellingham CM, Stahl RJ, Sitarz EE, Lane CJ, Keeley FW (2003) Sequence and structure determinants for the self-aggregation of recombinant polypeptides modeled after human elastin. J Biol Chem 278:48553–48562

Miao M, Bruce AEE, Bhanji T, Davis EC, Keeley FW (2007) Differential expression of two tropoelastin genes in zebrafish. Matrix Biol 26:115–124

Miao M, Stahl RJ, Petersen LF, Reintsch WE, Davis EC, Keeley FW (2009) Characterization of an unusual tropoelastin with truncated C-terminus in the frog. Matrix Biol 28:432–441

Mithieux SM, Weiss AS (2005) Elastin. Adv Protein Chem 70:437–461

Muiznieks LD, Keeley FW (2010) Proline periodicity modulates the self-assembly properties of elastin-like polypeptides. J Biol Chem 285:39779–39789

Muiznieks LD, Weiss AS, Keeley FW (2010) Structural disorder and dynamics of elastin. Biochem Cell Biol 88:239–250

Ott CE, Grünhagen J, Jäger M, Horbelt D, Schwill S, Kallenbach K, Guo G, Manke T, Knaus P, Mundlos S, Robinson PN (2011) MicroRNAs differentially expressed in postnatal aortic development downregulate elastin via 3′ UTR and coding-sequence binding sites. PLoS One 6:e16250. doi:10.1371/journal.pone.0016250

Parks WC, Roby JD, Wu LC, Grosso LE (1992) Cellular expression of tropoelastin mRNA splice variants. Matrix 12:156–162
Partridge SM (1962) Elastin. Adv Protein Chem 17:227–302
Patthy L (1996) Exon shuffling and other ways of module exchange. Matrix Biol 15:301–310
Pau RN (1984) Cloning of cDNA for a juvenile hormone-regulated oothecin mRNA. Biochim Biophys Acta 782:422–428
Pober BR (2010) Williams-Beuren syndrome. N Engl J Med 362:239–252
Pollock J, Baule VJ, Rich CB, Ginsburg CD, Curtiss SW, Foster JA (1990) Chick tropoelastin isoforms. From the gene to the extracellular matrix. J Biol Chem 265:3697–3702
Powell JT, Vine N, Crossman M (1992) On the accumulation of D-aspartate in elastin and other proteins of the ageing aorta. Atherosclerosis 97:201–208
Rauscher S, Pomès R (2012) Structural disorder and protein elasticity. Adv Exp Med Biol 725:159–183
Rauscher S, Baud S, Miao M, Keeley FW, Pomès R (2006) Proline and glycine control protein self-organization into elastomeric or amyloid fibrils. Structure 14:1667–1676
Ritz-Timme S, Laumeier I, Collins MJ (2003) Aspartic acid racemization: evidence for marked longevity of elastin in human skin. Br J Dermatol 149:951–959
Robson P, Wright GM, Sitarz E, Maiti A, Rawat M, Youson JH, Keeley FW (1993) Characterization of lamprin, an unusual matrix protein from lamprey cartilage. J Biol Chem 268:1440–1447
Robson P, Wright GM, Youson JH, Keeley FW (1997) A family of non-collagen-based cartilages in the skeleton of the sea lamprey, *petromyzon marinus*. Comp Biochem Physiol 118B:71–78
Robson P, Wright GM, Youson JH, Keeley FW (2000a) The structure and organization of lamprin genes: multiple-copy genes with alternative splicing and convergent evolution with insect structural proteins. Mol Biol Evol 17:1739–1752
Robson P, Wright GM, Keeley FW (2000b) Distinct non-collagen based cartilages comprising the endoskeleton of the Atlantic hagfish, *Myxine glutinosa*. Anat Embryol 202:281–290
Ronquist F, Huelsenbeck JP (2003) MrBayes 3: Bayesian phylogenetic inference under mixed models. Bioinformatics 19:1572–1574
Rudall KM (1968) Comparative biology and biochemistry of collagen. In: Gould BS (ed) Treatise on collagen, vol 2, Part A. Academic, London
Sage H (1983) The evolution of elastin: correlation of functional properties with protein structure and phylogenetic distribution. Comp Biochem Physiol 74B:373–380
Sage H, Gray WR (1979) Studies on the evolution of elastin—I. Phylogenetic distribution. Comp Biochem Physiol 64B:313–327
Sage H, Gray WR (1980) Studies on the evolution of elastin—II. Histology. Comp Biochem Physiol 66B:13–22
Sage H, Gray WR (1981) Studies on the evolution of elastin—III. The ancestral protein. Comp Biochem Physiol 68B:473–480
Savage KN, Gosline JM (2008a) The effect of proline on the network structure of major ampullate silks as inferred from their mechanical and optical properties. J Exp Biol 211:1937–1947
Savage KN, Gosline JM (2008b) The role of proline in the elastic mechanism of hydrated spider silks. J Exp Biol 211:1948–1957
Scott RW, Olson MF (2007) LIM kinases: function, regulation and association with human disease. J Mol Med (Berl) 85:555–568
Shapiro SD, Endicott SK, Province MA, Pierce JA, Campbell EJ (1991) Marked longevity of human lung parenchymal elastic fibers deduced from prevalence of D-aspartate and nuclear weapons-related radiocarbon. J Clin Invest 87:1828–1834
Starcher BC, King GS (1980) The presence of desmosine and isodesmosine in eggshell membrane protein. Connect Tissue Res 8:53–55
Stevens ED, Randall DJ (1967) Changes in blood pressure, heart rate and breathing rate during moderate swimming activity in rainbow trout. J Exp Biol 46:307–315
Szabó Z, Levi-Minzi SA, Christiano AM, Struminger C, Stoneking M, Batzer MA, Boyd CD (1999) Sequential loss of two neighboring exons of the tropoelastin gene during primate evolution. J Mol Evol 49:664–671

Tamburro AM, Bochicchia B, Pepe A (2003) Dissection of human tropoelastin: exon-by-exon chemical synthesis and related conformational studies. Biochemistry 42:13347–13362

Tamburro AM, Pepe A, Bochicchio B, Quaglino D, Ronchetti IP (2005) Supramolecular amyloid-like assembly of the polypeptide sequence coded by exon 30 of human tropoelastin. J Biol Chem 280:2682–2690

Tamburro AM, Pepe A, Bochicchio B (2006) Localizing α-helices in human tropoelastin: assembly of the elastin "puzzle". Biochemistry 45:9518–9530

Taylor JS, Van de Peer Y, Braasch I, Meyer A (2001) Comparative genomics provides evidence for an ancient genome duplication event in fish. Philos Trans R Soc Lond B Biol Sci 356:1661–1679

van Rooij E, Sutherland LB, Thatcher JE, DiMaio JM, Naseem RH, Marshall WS, Hill JA, Olson EN (2008) Dysregulation of microRNAs after myocardial infarction reveals a role of miR-29 in cardiac fibrosis. Proc Natl Acad Sci USA 105:13027–13032

Venkatachalam CM, Urry DW (1981) Development of a linear helical conformation from its cyclic correlate. β-Spiral model of the elastin poly(pentapeptide) (VPGVG)n. Macromolecules 14:1225–1229

Venkatesh B, Gilligan P, Brenner S (2000) Fugu: a compact vertebrate reference genome. FEBS Lett 476:3–7

Villarreal G Jr, Oh D-J, Kang MH, Rhee DJ (2011) Coordinated regulation of extracellular matrix synthesis by the microRNA-29 family in the trabecular meshwork. Invest Ophthalmol Vis Sci 52:3391–3397

Wagenseil JE, Mecham RP (2007) New insights into elastic fiber assembly. Birth Defects Res C 81:229–240

Waite JH, Lichtenegger HC, Stucky GD, Hansma P (2004) Exploring molecular and mechanical gradients in structural bioscaffolds. Biochemistry 43:7653–7662

Wittbrodt J, Shima A, Schartl M (2002) Evolutionary relationships between fish models. Nat Rev Genet 3:53–64

Wright GM, Keeley FW, DeMont ME (1998) Hagfish cartilage. In: Jørgensen JM, Lomholt JP, Weber RE, Malte H (eds) The biology of hagfishes. Chapman & Hall, London

Wright GM, Keeley FW, Robson P (2001) The unusual cartilaginous tissues of jawless craniates, cephalochordates and invertebrates. Cell Tissue Res 304:165–174

Wu DD, Irwin DM, Zhang YP (2008) Molecular evolution of the keratin associated protein gene family in mammals, role in the evolution of mammalian hair. BMC Evol Biol 8:241

Xu M, Lewis RV (1990) Structure of a protein superfiber: spider dragline silk. Proc Natl Acad Sci USA 87:7120–7124

Yeo GC, Keeley FW, Weiss AS (2011) Coacervation of tropoelastin. Adv Colloid Interface Sci 167:94–103

Yeo GC, Baldock C, Tuukkanen A, Roessle M, Dyksterhuis LB, Wise SG, Matthews J, Mithieux SM, Weiss AS (2012) Tropoelastin bridge region positions the cell-interactive C terminus and contributes to elastic fiber assembly. Proc Natl Acad Sci USA 21:2878–2883

Yoon K, May M, Goldstein N, Indik ZK, Oliver L, Boyd C, Rosenbloom J (1984) Characterization of a sheep elastin cDNA clone containing translated sequences. Biochem Biophys Res Commun 118:261–269

Zardoya R, Meyer A (1997) Molecular phylogenetic information on the identity of the closest living relative(s) of land vertebrates. Naturwissenschaften 84:389–397

Zhang M, Pierce RA, Wachi H, Mecham RP, Parks WC (1999) An open reading frame element mediates posttranscriptional regulation of tropoelastin and responsiveness to transforming growth factor beta1. Mol Cell Biol 19:7314–7326

Zhang P, Huang A, Ferruzzi J, Mechan RP, Starcher BC, Tellides G, Humphrey JD, Giordano FJ, Niklason LE, Sessa WC (2012) Inhibition of microRNA-29 enhances elastin levels in cells haploinsufficient for elastin and in bioengineered vessels. Arterioscler Thromb Vasc Biol 32:756–759

Chapter 5
Evolutionary Insights into Fibrillin Structure and Function in the Extracellular Matrix

Sacha Jensen, David Yadin, Ian Robertson, and Penny Handford

Abstract The 10–12 nm diameter microfibrils of the extracellular matrix are vital components of dynamic tissues, playing a major role in providing the tissues of metazoan species with their biomechanical properties. They also have functional roles in the development of elastic fibres, as a scaffold for the deposition of tropoelastin, and in the regulation of growth factors. Fibrillins are large, ~350 kDa glycoproteins that assemble to form the microfibrils. Throughout evolution, the fibrillins show a remarkably conserved domain organisation, dominated by calcium-binding epidermal growth factor domains interspersed with transforming growth factor-β-binding protein-like domains. Mutations in the fibrillin genes cause a range of human diseases including Marfan syndrome, Beals syndrome, stiff skin syndrome and the acromelic dysplasias, which affect the skin, skeleton, ocular and cardiovascular systems. In recent years, studies using molecular, cellular and animal models have yielded new information on the processes of microfibril assembly and organisation and how their regulation of growth factors is involved in disease pathogenesis. Recent advances in bioinformatics and the availability of whole genome sequences are now providing new insights into the functions of different regions of the fibrillin polypeptides, helping us to understand how fibrillin structure and function has evolved as the extracellular matrix has increased in complexity.

5.1 Introduction

The 10–12 nm diameter fibrillin microfibrils are extracellular matrix components that play a vital role in maintaining the structural integrity of highly dynamic tissues. They act as a scaffold for the deposition of elastin during the formation of elastic fibres in tissues such as the lung, blood vessels and skin. They also occur as independent

S. Jensen • D. Yadin • I. Robertson • P. Handford (✉)
Department of Biochemistry, University of Oxford, South Parks Road, Oxford OX1 3QU, UK
e-mail: Penny.handford@bioch.ox.ac.uk

F.W. Keeley and R.P. Mecham (eds.), *Evolution of Extracellular Matrix*, Biology of Extracellular Matrix, DOI 10.1007/978-3-642-36002-2_5,

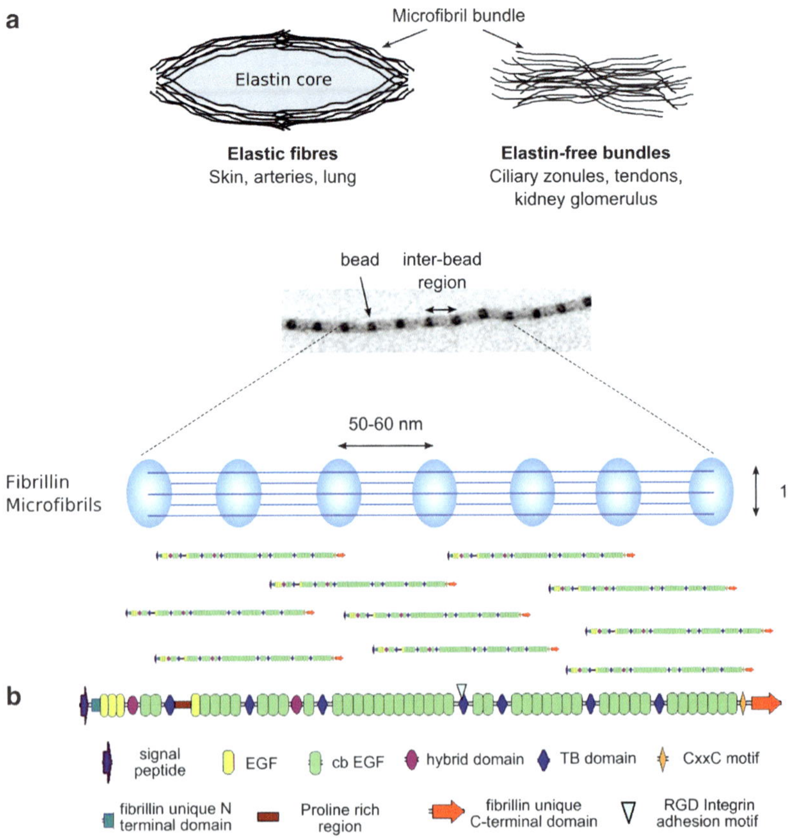

Fig. 5.1 Microfibrils and fibrillin domain structure. (**a**) The 10–12 nm fibrillin microfibrils occur in a large number of connective tissues. In elastic tissues, they surround a core of elastin, while in nonelastic tissues, they are found as bundles without elastin. When extracted from tissue and imaged by rotary shadowing electron microscopy, microfibrils have a 'beads-on-a-string' appearance, with a characteristic repeat distance between the bead structures. Microfibrils are assembled from fibrillin monomers, but their organisation within the assembly is not yet clear. Fibrillin monomers are shown as linear molecules for simplification. (**b**) The fibrillin domain structure is dominated by EGF and cbEGF domains, interspersed with 8-cysteine TB and hybrid domains. Domain boundaries are based on exon junctions and knowledge of domain structures and interdomain interfaces

networks in tissues such as tendons, the ciliary zonules of the eye and the kidney glomerulus, where they provide tensile strength (Fig. 5.1a). Interactions of fibrillins with bone morphogenetic proteins (BMPs), growth and differentiation factors (GDFs) and latent transforming growth factor-β-binding proteins (LTBPs) result in targeting of growth factors to the extracellular matrix, and interactions with integrins α5β1, αVβ3 and αvβ6 confer cell signalling and adhesion properties (Bax et al. 2003; Jovanovic et al. 2007). The fibrillin monomers that assemble to form microfibrils are large (~350 kDa) glycoproteins, with a domain architecture dominated by calcium-binding epidermal growth factor (cbEGF) domains interspersed with transforming growth factor-β-binding protein-like (TB) domains (Fig. 5.1b). Mutations in human fibrillin-encoding genes are associated with a significant burden of disease which gives rise to a spectrum of inherited connective tissue disorders. Studies of these human genetic diseases, together with analysis of mouse models, have given fundamental insights into the structural and functional roles of fibrillins. This chapter will discuss the evolutionary history and properties of these proteins, emphasising the latest findings and identifying outstanding questions.

5.2 Microfibrils

5.2.1 Microfibril Structure is Conserved Through Evolution

In vertebrate tissues, microfibrils are found at the periphery of elastic fibres, where they provide a scaffold for the deposition of elastin during elastogenesis (Jones et al. 1980; Kewley et al. 1978; Ross and Bornstein 1969). They also occur in elastin-free bundles in tissues such as the ciliary zonule, tendon, cornea, renal glomerulus and fetal membranes (Kumaratilake et al. 1989; Malak and Bell 1994; Sakai et al. 1986). The appearance of microfibrils has remained remarkably conserved throughout evolution, with structures resembling vertebrate microfibrils occurring in the tissues of several invertebrate species including jellyfish (Reber-Muller et al. 1995), sea cucumber (Thurmond and Trotter 1996), stalked crinoids (Birenheide and Motokawa 1994), lobster and whelk (Davison et al. 1995). When viewed by rotary shadowing electron microscopy, isolated microfibrils have a distinctive beaded filament appearance with an average untensioned periodicity of about 56 nm and a diameter of 10–12 nm (Fig. 5.1) (Baldock et al. 2001; Cardy and Handford 1998; Keene et al. 1991; Kielty et al. 1991). Ultrastructural and X-ray diffraction studies have shown that periodicities of up to 165 nm can occur in samples where microfibrils have become entangled (Keene et al. 1991), or when tissues are stretched (Glab and Wess 2008).

5.2.2 Microfibrils Provide Elasticity in Metazoan Tissues

The property of elasticity in vertebrate tissues is generally attributed to elastin, which is found in all vertebrate species except the agnathans (hagfish and lampreys) but has not been found in invertebrates (Sage and Gray 1977). The evolution of elastin is thought to be linked to the evolution of the closed circulatory system of vertebrates, in which there is an increased requirement for resilience in blood vessels to deal with the higher blood pressures observed compared to the open system of invertebrates (Faury 2001). Before the evolution of elastin, fibrillin microfibrils were largely responsible for tissue elasticity in many species.

The mesoglea is a simple extracellular matrix that forms between the epidermal and endodermal layers of jellyfish and is made up of several of the components found in the extracellular matrix of higher organisms, including collagens, mucopolysaccharides and fibrillin microfibrils (Reber-Muller et al. 1995; Sarras et al. 1991). As in the extracellular matrix in higher organisms, the mesoglea functions to provide structural support to the body of the animal, allows for the transport of nutrients and plays a role in cell migration and differentiation during development (Sarras et al. 1993; Shimizu et al. 2002). Extracellular fibres play an important role in maintaining the biomechanical properties of the mesoglea. Collagen is thought to provide stiffness to the tissue while fibrillin microfibrils contribute to tissue elasticity (Gambini et al. 2012; Megill et al. 2005). When hydromedusan jellyfish swim, the energy required for the refill stroke, during which water is taken up into the subumbrellar cavity, is stored by the stretched fibrillin microfibrils (Megill et al. 2005).

Non-linear elastic behaviour is functionally important in blood vessels and in vertebrate arteries. This property has been attributed to the deformation of elastin under low strains and collagen at high strains. In invertebrate species, such as lobster and horseshoe crab, abundant microfibrillar networks are found in arteries, predominantly in the middle laminae (Davison et al. 1995). The elastic properties of the blood vessels in these species are almost entirely due to microfibrils since there is no elastic component as seen in vertebrate species (Faury 2001). In a lobster (*Homarus americanus*) aorta model, it was shown that the non-linear elastic behaviour of these blood vessels could be attributed to the reorientation of fibrillin microfibrils at the lower end of the stress–strain curve and to deformation of the microfibrils at the higher end of the stress–strain curve (McConnell et al. 1997). Lobster aorta has also been used to show calcium-dependent changes in microfibril stiffness consistent with ultrastructural studies of vertebrate material (Bussiere et al. 2006; Cardy and Handford 1998; Kielty and Shuttleworth 1993). In vertebrates, the biomechanical properties of microfibrils are critical in tissues such as the ciliary zonules of the eye in which microfibril bundles are involved in anchoring the lens into position and in the aorta where they partly contribute to the elastic properties of the vessel (Faury 2001).

5.3 Fibrillin

Fibrillin was first identified as a major component of the 10–12 nm microfibrils in 1986 (Sakai et al. 1986). Cloning of the human *FBN1* and *FBN2* genes (Lee et al. 1991; Maslen et al. 1991) later showed that fibrillins are calcium-binding glycoproteins whose modular architecture is conserved across evolutionarily distant metazoan species, from jellyfish to humans (Piha-Gossack et al. 2012; Robertson et al. 2011) (Fig. 5.2). Together with the latent TGFβ (transforming growth factor-β)-binding proteins (LTBPs), they form a protein superfamily characterised by a signature '8-cysteine' or 'TB' (transforming growth factor-β-binding protein-like) domain, which is unique to this group of proteins (Robertson et al. 2011). In humans, three fibrillin isoforms (fibrillin-1, -2 and -3) have been identified, each encoded by a separate gene.

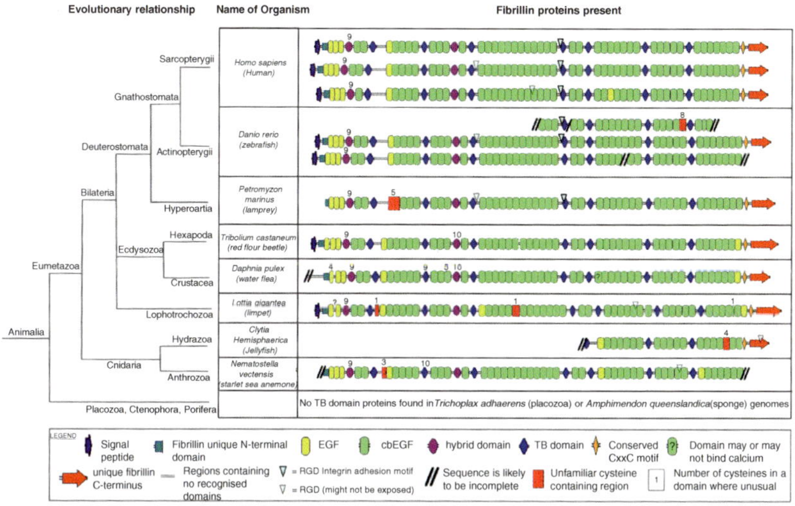

Fig. 5.2 Evolution of fibrillin domain organisation. The domain structure of fibrillins from a variety of organisms is shown. Organisms were selected to give a concise summary of metazoan evolution. The tree structure on the *left* summarises the phylogenetic relationships between the organisms, according to the Tree of Life web project (http://tolweb.org/tree/). Lines between tree nodes were drawn manually and distances do not represent any specific measure of divergence. A legend for the domains found in fibrillin is shown beneath the figure. Particular care should be taken when interpreting orange boxed regions, which are cysteine-rich regions containing an unknown domain arrangement; these may represent errors in exon prediction from genomic data

In addition to the cbEGF and TB domains that dominate the fibrillin structure, hybrid domains that have sequence and structural characteristics of both TB and cbEGF domains are also found. These appear to be unique to the fibrillin/LTBP superfamily, with two in each of the fibrillins and one in each of the LTBPs.

5.3.1 *Fibrillins Appear Early in Metazoan Evolution*

A recent analysis of fibrillin and LTBP sequences (Robertson et al. 2011) showed that the simplest eumetazoan group in which a TB domain-containing protein could be found are the cnidarians, suggesting the emergence of fibrillin at around 600 million years ago, prior to the divergence of the Cnidaria and Bilateria. At present, no TB domain-containing proteins can be found encoded by the available genome sequences of simpler metazoan species such as Porifera (sponges) or Placozoa (*Trichoplax*).

The number of fibrillin genes varies in different species, with at least two gene duplication events having been identified during evolution. Mammalian species are generally found to have three fibrillin genes, with some exceptions including the rodents (Corson et al. 2004), in which the gene encoding fibrillin-3 is thought to have been inactivated by a chromosome fission event, the armadillo and platypus (Piha-Gossack et al. 2012). In the case of the armadillo and platypus, Piha-Gossack et al. were unable to determine whether the lack of a third fibrillin gene was due to incomplete genome sequencing or species-specific gene losses. Three separate fibrillin genes were also identified in avian, lizard, amphibian and zebrafish species (Corson et al. 2004; Piha-Gossack et al. 2012; Robertson et al. 2011), while ray-finned fish species (actinopterygian), such as the Japanese pufferfish, the spotted green pufferfish and the medaka fish, were found to contain only two fibrillin genes (Piha-Gossack et al. 2012). Species with just a single fibrillin gene include the cnidarians, arthropods (insects, crustaceans and arachnids), molluscs, annelids, echinoderms, urochordates and cephalochordates. Vertebrate species with a single fibrillin gene were limited to the agnathans (lampreys and hagfish) (Piha-Gossack et al. 2012).

Although a fibrillin gene is found in several insect species, including the honey bee (*Apis mellifera*), red flour beetle (*Tribolium castaneum*), pea aphid (*Acyrthosiphon aphis*) and body louse (*Pediculus humanus corporis*), no genes encoding TB domain-containing proteins have been identified in the genome of the fruit fly (*Drosophila melanogaster*). Similarly, although a gene annotated as fibrillin has been identified in the genome of the nematode *Caenorhabditis elegans* (NCBI Reference Sequence: NM_066269.2), this sequence lacks the signature TB and hybrid domains of the fibrillin/LTBP superfamily and therefore is not a true fibrillin homologue. Using a similar argument, Piha-Gossack *et al.* (2012) suggested that the putative fibrillin sequences identified in the trematode *Schistosoma mansoni* (EnsemblMetazoa: Smp_167190, Smp_067800) are also not true fibrillin homologues. Analysis of other *Drosophila* and nematode species has shown a similar lack of fibrillin genes. Robertson et al. proposed that the loss of fibrillin genes in *Drosophila* and nematode species is due to the rapid rate of evolution observed for these genomes (Coghlan and Wolfe 2002; Ranz et al. 2001), which may be a function of their adaptation to their respective niches.

5.3.2 Humans Have 3 Fibrillin Isoforms

In humans, fibrillin-1, -2 and -3 are encoded by genes on chromosomes 15, 5 and 19, respectively (Corson et al. 2004; Lee et al. 1991). The *FBN1* and *FBN2* genes show a diphasic expression pattern, with fibrillin-2 expression generally preceding fibrillin-1 expression in most tissues except the cardiovascular system (Zhang et al. 1995). It has been suggested that fibrillin-2 may play a role in the early stages of morphogenesis, while fibrillin-1 expression correlates with the appearance of well-defined organ structures and is responsible for the structural function of microfibrils in load-bearing tissues such as the skin, aortic adventitia and ciliary zonules (Zhang et al. 1995). Although the distribution of fibrillin-1 and fibrillin-2 in embryonic and early human fetal human organs is similar, significant differences are seen in some tissues such as the kidney, liver, rib anlagen and notochord (Quondamatteo et al. 2002). Although defects in both proteins are linked to disorders affecting skeletal development, several studies have shown that fibrillin-1 and fibrillin-2 play different roles in the regulation of extracellular matrix growth factors (Arteaga-Solis et al. 2001). In bone, fibrillin-1 and -2 play separate roles in osteoblast maturation, with fibrillin-2 negatively regulating TGFβ and fibrillin-1 negatively regulating both TGFβ and BMPs (Nistala et al. 2010a,

2010b). Fibrillin-3 was first described relatively recently and its role in extracellular matrix is less well defined. Homologues of human fibrillin-3 were identified in bovine and chicken DNA sequences, although the rodent gene appears to be non-functional (Corson et al. 2004). As in the case of fibrillin-2, expression of fibrillin-3 is largely limited to fetal development, with the highest levels of expression seen in the fetal lung, brain and kidney, while only low levels were observed in adult tissues (Corson et al. 2004). Using isoform-specific antibodies, Sabatier et al. showed that the expression of fibrillin-3 does not always co-localise with fibrillin-1 or fibrillin-2 (Sabatier et al. 2011), suggesting that it may have distinct functions during development.

5.3.3 *Fibrillin-Associated Diseases (Fibrillinopathies)*

The different roles of the fibrillins in mammalian tissues are reflected in the diseases resulting from mutations in the genes encoding these proteins. In 1991, a combination of linkage studies, candidate gene analysis and mutation detection demonstrated that Marfan syndrome (MFS), a connective tissue disorder associated with cardiovascular, ocular and skeletal abnormalities was caused by mutations in the *FBN1* gene (Dietz et al. 1991; Lee et al. 1991; Maslen et al. 1991). This autosomal dominant condition occurs with a frequency of ~1:5,000 in the general population, and patients show reduced life expectancy due to aortic dissection. Subsequent studies identified a diverse range of mutations including missense, nonsense, splice-site leading to exon skipping or nonsense-mediated decay of RNA and small insertions and deletions. Databases of *FBN1* mutations associated with MFS and related disorders have shown that all protein domains may be affected, and there are no clear phenotype–genotype correlations with the exception of a subset of mutations which affect protein domains TB3-cbEGF18 (exons 24–32). These are associated with neonatal MFS, a severe form of disease that leads to death from cardiovascular complications, usually in the first 12 months. The remaining mutations may be associated with a spectrum of phenotypes ranging from MFS, where multiple organ systems are affected, to those where more limited organ systems are affected such as ectopia lentis, Shprintzen–Goldberg syndrome, progressive kyphoscoliosis, familial thoracic aneurysm and dissection, and the MASS phenotype (Table 5.1). Very recently, *FBN1* mutations have been identified in groups of patients with skin fibrosis. Patients with stiff skin syndrome, which is characterised by hard skin and joint contractures, have *FBN1* missense mutations resulting in amino acid substitutions in TB4, the integrin-binding domain of fibrillin-1 (Loeys et al. 2010). Another group of patients with mutations affecting TB5 have features belonging to the acromelic dysplasias, which includes Weill–Marchesani syndrome (WMS), geleophysic dysplasia and acromicric dysplasia (Le Goff and Cormier-Daire 2011; Le Goff et al. 2011). These phenotypes are all characterised by short stature, short hands and feet, joint limitations and thick skin caused by excessive matrix production. WMS in addition has ocular features. More recently, a WMS family has been identified with an *FBN1* deletion of the TB1, proline-rich and EGF4 domains (Sengle et al. 2012). Mutations affecting the fibrillin-2 gene (*FBN2*) are

Table 5.1 Inherited human diseases related to microfibril function

Disease	Symptoms	Genes involved	References
Marfan syndrome	Elongation of the long bones Arachnodactyly (long fingers) Pectus carinatum/pectus excavatum Abnormal joint flexibility High arched palate Ectopia lentis and early onset glaucoma Aortic aneurysm and dilatation Mitral valve prolapse	*FBN1* (dominant)	Dietz et al. (2005); Robinson et al. (2006); Robinson and Godfrey (2000)
Neonatal Marfan syndrome	Severe MFS symptoms present at birth Also exhibit joint contractures, crumpled ears and loose skin	*FBN1* (dominant)	Faivre et al. (2007); Stheneur et al. (2011); Tiecke et al. (2001)
Shprintzen–Goldberg syndrome	Craniosynostosis (early fusion of skull bones) Arachnodactyly (long fingers) and other Marfan-like skeletal features Mitral valve prolapse may occur but aortic dilatation rare Mental retardation	*FBN1* (dominant)	Hatchwell (1997)
Progressive kyphoscoliosis	Skeletal defects, particularly in the spine, no ocular or cardiac involvement	*FBN1* (dominant)	Ades et al. (2002)
Isolated ectopia lentis	Ectopia lentis	*FBN1* (dominant) *ADAMTSL4* (recessive) (Ahram et al. 2009)	Comeglio et al. (2002)
Familial thoracic aneurysm and dissection	Thoracic aneurysm and dissection without other marfanoid features	*FBN1* (dominant)	Hasham et al. (2002)
MASS phenotype	Mitral valve prolapse Aortic root dilatation without dissection Skeletal and skin abnormalities	*FBN1* (dominant)	Dietz et al. (1993)
Loeys–Dietz syndrome (LDS)	Cleft palate or bifid uvula Abnormally long limbs and fingers Pectus excavatum	*TGFBR1* *TGFBR2* (dominant)	Attias et al. (2009); Loeys et al. (2005)

(continued)

Table 5.1 (continued)

Disease	Symptoms	Genes involved	References
	Contractures of fingers and toes Abnormal joint flexibility, widely spaced eyes Aortic aneurysm and aortic dissection with tortuosity		
Stiff skin syndrome	Tight and thick skin Limited joint mobility Short stature	*FBN1* (dominant)	Loeys et al. (2010)
Weill–Marchesani syndrome	Short stature Brachydactyly (short fingers) Thick skin Myopia, glaucoma, ectopia lentis Hypermuscularity	*FBN1* (dominant) *ADAMTS10* (recessive) *ADAMTS17* (recessive) *WMS-like syndrome* *LTBP2* (recessive)	Faivre et al. (2003); Morales et al. (2009)
Geleophysic dysplasia	Severe short stature and extremities Stiff joints Skin thickening Cardiac valvular thickening Lysosomal storage like vacuoles	*FBN1* (dominant) *ADAMTSL2* (recessive GD)	Le Goff et al. (2011)
Acromicric dysplasia	Severe short stature and extremities Stiff joints Skin thickening Distinctive 'round' face Pseudomuscular build Distinctive skeletal 'notches'	*FBN1* (dominant)	Le Goff et al. (2011)
Progeria +lipodystrophy	Lipodystrophy (degeneration of adipose tissue) Ectopia lentis Marfanoid habitus (tall stature)	*FBN1* (dominant)	Goldblatt et al. (2011); Graul-Neumann et al. (2010); Horn and Robinson (2010)
Beals-Hecht syndrome or congenital contractural arachnodactyly (CCA)	Contractural arachnodactyly (long fingers bent in at the joints) and contractures of various other joints, improve with age 'Crumpled' ears Tall stature and Marfan-like appearance	*FBN2* (dominant)	Robinson et al. (2006); Robinson and Godfrey (2000)

(continued)

Table 5.1 (continued)

Disease	Symptoms	Genes involved	References
Polycystic ovary syndrome (PCOS)	Enlarged polycystic ovaries Menstrual abnormalities Infertility Other metabolic consequences	*FBN3*?	Jordan et al. (2010); Raja-Khan et al. (2010); Segars and DeCherney (2010)

associated with congenital contractural arachnodactyly (Beals syndrome) (Lee et al. 1991; Putnam et al. 1995). This shares some clinical features with MFS, but lacks ocular features and shows no overall shortened life expectancy (Godfrey 2004). Although no diseases have been firmly linked to *FBN3*, it has been suggested that an *FBN3* allele may be involved in the pathogenesis of polycystic ovary syndrome (Hatzirodos et al. 2011; Jordan et al. 2010; Urbanek et al. 2007). This is still controversial, however, with one study finding little involvement of *FBN*3 in the development of the disease (Prodoehl et al. 2009), while another found that the *FBN*3 risk allele may be associated with changes in basal glucose homeostasis in polycystic ovary syndrome (Yalamanchi et al. 2012).

A number of different *Fbn*1 mouse models have been created and given fundamental insight into the biology of fibrillin-1 and pathogenesis of MFS. Early mouse models mgR (Pereira et al. 1999) and mgΔ (Pereira et al. 1997) produced reduced levels of either normal protein or a mutant form of fibrillin with a central deletion. Neither of these models recapitulated MFS features in the heterozygous condition, but homozygous mice died prematurely from aortic dilatation and dissection. Disruption of the elastic lamellae was evident in both cases, but despite this the aorta contained many regions of contiguous elastic lamellae. Importantly, these models identified a critical threshold level of fibrillin required to maintain the vessel wall. Analysis of lung development of mgΔ mice showed they had developmental abnormalities in alveolar septation, caused by abnormal activation of TGFβ (Neptune et al. 2003). This was the first study that proved that fibrillin was not just a structurally important protein of connective tissue, but conferred regulatory properties. Subsequently, a third mouse model containing a common type of missense mutation seen in MFS patients was produced which recapitulated MFS-like features in the heterozygous state (C1039G/+), most notably mitral valve prolapse and aortic dilatation/dissection (Habashi et al. 2006; Ng et al. 2004). This model allowed the authors to test TGFβ antagonism as a strategy to treat the more serious symptoms associated with MFS. Losartan, a readily available drug that reduces TGFβ signalling, was shown to inhibit changes to the aortic wall in C1039G/+ mice, with additional beneficial effects on defective alveolar septation and muscle hypoplasia (Cohn et al. 2007). Clinical trials are in progress to test the efficacy of Losartan treatment on MFS patients. In parallel with the mouse studies, a distinct clinical condition, Loeys–Dietz syndrome, was identified (Loeys et al. 2005). This syndrome has a phenotype that overlaps with MFS (Table 5.1), but is caused by mutations in the TGFβ receptor-encoding genes, providing further evidence of the functional link between the fibrillins and TGFβ regulation.

Additional mouse models have been created to investigate mechanisms of pathogenesis underlying MFS. These include a study by Judge et al. (2004), in which YAC-based transgenesis was used to overexpress a mutant form of human fibrillin-1 containing amino acid substitution C1663R on a normal mouse background. The mice failed to show any clinical phenotype. The authors proposed this as evidence that the C1663R mutation does not exert a dominant negative effect and that haploinsufficiency must play a critical role in pathogenesis, although the ratio of mutant to wild-type protein may not have been sufficient to recreate the pathological condition. A truncated form of fibrillin-1 (GT-8) expressed as a GFP fusion protein was shown to form microfibril-like structures in the skin of GT-8/GT-8 mice at p0 and to co-localise with microfibrils in skin from GT-8/+ mice. These heterozygous mice were shown to undergo fragmentation of the aortic elastic lamellae and the authors proposed that disruption of microfibril structure by a mutant protein, i.e., a dominant negative mechanism of pathogenesis may also cause MFS (Charbonneau et al. 2010).

Recently, the mgR model has been used to demonstrate that administration of a microfibril-associated protein ADAMTSL6β can rescue the microfibril defect within the periodontal ligament, implicating this protein in microfibril formation (Saito et al. 2011). A mouse model of WMS, associated with a novel *FBN1* mutation, recapitulated the human fibrosis disorder. Biochemically this mutation was found to abrogate binding of ADAMTSL-2, -3 -6, further underscoring the importance of the ADAMTSL family for tissue-specific microfibril function (Sengle et al. 2012).

5.3.4 Fibrillins Have a Highly Conserved Domain Organisation

The domain organisation of the fibrillins has remained remarkably unchanged through evolution (Fig. 5.2) (Robertson et al. 2011). This is highlighted by the fact that the fibrillin found in the sea anemone, *Nematostella vectensis*, has a domain organisation that closely resembles human fibrillin-1, except for the replacement of two cbEGF domains with EGF domains in the sea anemone. The importance of this conservation of domain organisation is reflected in the severity of disease phenotypes associated with exon-skipping mutations that cause the deletion of whole fibrillin-1 domains (Liu et al. 1996). This would suggest that the spacing of the domains relative to each other may be functionally significant during the assembly of microfibrils. The H1Δ mouse model, in which the first hybrid domain of fibrillin-1 is deleted, was used to show that perfect register was unlikely to be required for the assembly and stability of microfibrils (Charbonneau et al. 2010). In vitro, H1Δ fibrillin-1 did not interfere with normal fibrillin-1 microfibril assembly, and homozygous H1Δ/H1Δ fibroblasts were able to assemble apparently normal microfibril networks when analysed by immunofluorescence microscopy. In contrast, a mouse model with an in-frame deletion of three domains from TB1 to EGF4, corresponding to a mutation causing Weill–Marchesani syndrome in humans (Sengle et al. 2012), showed evidence of irregularities in the periodic labelling of microfibrils that is usually observed by immunoelectron microscopy. This may suggest either that the effect of register is not always seen using low resolution methods or that deletions of different parts of the molecule have variable effects on the alignment of fibrillin within the microfibrils.

5.4 Fibrillin: Domains and Structures

5.4.1 TB Domains

TB domains are unique to the fibrillin/LTBP superfamily (Robertson et al. 2011). The domain structure (Fig. 5.3) is characterized by eight cysteines that form four disulphide bonds in a 1–3, 2–6, 4–7, 5–8 arrangement, an N-terminal globular region stabilized by a conserved aromatic residue (Lee et al. 2004; Yuan et al. 1997) and an unusual cysteine triplet comprising of cysteines 3, 4 and 5 of the domain. TB domains have been shown to mediate interactions with several cell-matrix components. The fibrillin TB4 domains of vertebrates contain an RGD sequence (Fig. 5.4) involved in the binding of integrins such as $\alpha 5\beta 1$, $\alpha V\beta 3$ and $\alpha v\beta 6$ (Bax et al. 2003, 2007; Jovanovic et al. 2007, 2008). Domain TB5 of human fibrillin-1 contains a sequence that is likely to be involved in heparin binding (Cain et al. 2005). LTBP-1, -3 and -4 have been shown to bind the propeptide of TGFβ through a disulphide exchange mechanism involving their second TB domain (sometimes referred to as the third '8-cys' domain), in which the 2–6 disulphide bond is rearranged to form two new disulphides with cysteines in the TGFβ latency-associated peptide (Gleizes et al. 1996; Saharinen and Keski-Oja 2000; Saharinen et al. 1996). The structural basis of this rearrangement has been well characterized and involves a two-amino-acid insertion between Cys6 and Cys7 of the domain (Saharinen and Keski-Oja 2000), leading to increased solvent accessibility of the Cys2–Cys6 disulphide bond (Lack et al. 2003). This two-amino-acid insertion is not seen in any of the other TB domains of the LTBPs and is not seen in the fibrillins. It is therefore unlikely that any intermolecular disulphide bond formation involving fibrillin TB domains would occur through this mechanism.

5.4.2 Hybrid Domains

DNA sequencing of *FBN1* revealed the presence of hybrid (hyb) domains, with N-terminal sequences similar to TB domains and C-terminal sequences similar to cbEGF domains (Corson et al. 1993; Pereira et al. 1993). Two hyb domains are found in each of the fibrillins and one in each of the LTBPs. Analysis of the structure of the second hybrid domain of fibrillin-1 (Fig. 5.3) revealed a 1–3, 2–5, 4–6, 7–8 disulphide bond arrangement and showed that the TB–cbEGF hybrid nature of the sequence is reflected in the 3D structure of the domain (Jensen et al. 2009). The first hyb domain of the fibrillins is unique in that it contains an extra, ninth cysteine (Charbonneau et al. 2010) (Cys204 in human fibrillin-1; Fig. 5.4b) that is absolutely conserved from humans to cnidarians (Robertson et al. 2011). This unpaired cysteine

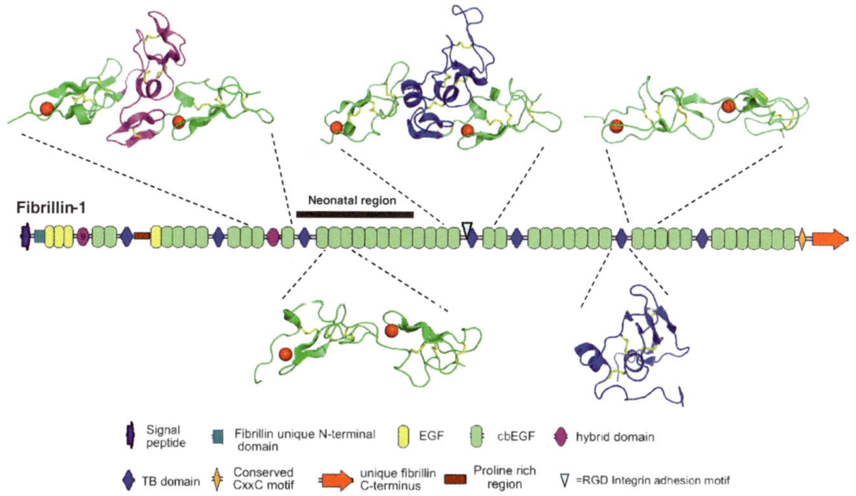

Fig. 5.3 Fibrillin domain structures. Known structures of fibrillin domains include cbEGF–cbEGF pairs, an isolated TB domain, and TB and hyb domains with flanking cbEGF domains. The region encompassing domains TB3–cbEGF18 (*black line*) is where amino acid substitutions often lead to a severe, neonatal form of Marfan syndrome and is therefore often referred to as the 'neonatal' region. Structures have been determined for fragments cbEGF9–hyb2–cbEGF10 (PDB 2W86), cbEGF12–cbEGF13 (PDB 1LMJ), cbEGF22–TB4–cbEGF23 (PDB 1UZJ), isolated domain TB6 (PDB 1APJ) and cbEGF32–cbEGF33 (PDB 1EMN)

has been shown to be solvent-exposed and may be involved in intermolecular interactions during microfibril assembly (Reinhardt et al. 2000), although the H1Δ mouse model showed normal microfibril assembly (Charbonneau et al. 2010).

5.4.3 cbEGF Domains

EGF and cbEGF domains are widely distributed extracellular modules found in proteins involved in a diverse range of functions including cell signalling, blood clotting and adhesion. They have a characteristic structure (Fig. 5.3) stabilized by three disulphide bonds in a 1–3, 2–4, 5–6 arrangement. In the fibrillins, the majority (43 out of 47 in human fibrillin-1 and -2) of the EGF domains are of the calcium-binding type, and these can be differentiated from the non-calcium-binding EGF domains by the presence of the consensus calcium-binding sequence D/N-X-D/N-E/Q-X_m-D/N*-X_n-Y/F, where m and n indicate a variable number of residues and * indicates possible post-translational β-hydroxylation. In these domains, calcium is chelated with a pentagonal bipyramidal geometry, with ligands provided by both side chain oxygen atoms and backbone carbonyl groups (Rao et al. 1995). Two classes of cbEGF domains have been described, based on the number of residues found in the linker regions between tandem cbEGF repeats (Downing et al. 1996). Class I cbEGF domains have one residue between the final cysteine of the N-terminal repeat and the first calcium-binding residue of the following domain. Class II cbEGF domains have

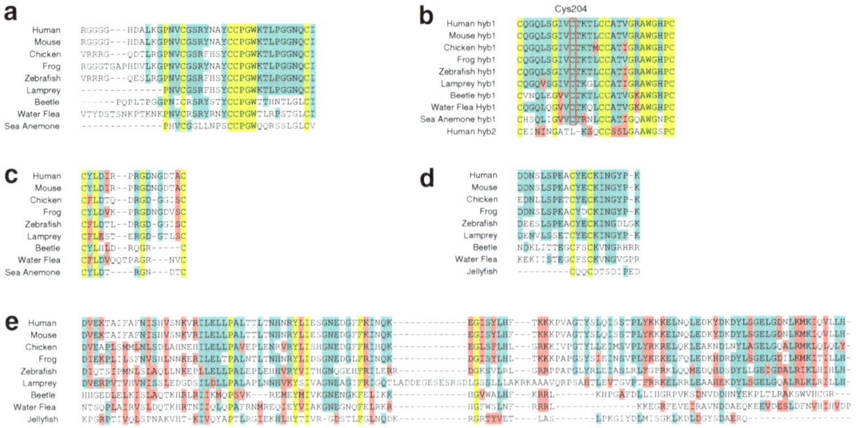

Fig. 5.4 Alignments of fibrillin domains associated with specific functional properties. (**a**) Fibrillin Unique N-terminal (FUN) domain, showing the highly conserved 4-cysteine motif and Pro-Gly-Trp sequence. (**b**) Alignment of the first hybrid domain from a range of species, in comparison with the second hybrid domain from human fibrillin-1, highlighting the absolute conservation of the 'extra' cysteine (Cys204 in humans) that may be involved in the assembly of microfibrils (Reinhardt et al. 2000). (**c**) Part of the fibrillin domain TB4, focussing on the region between Cys1 and Cys2 of the domain. The integrin-binding RGD motif is found only in the chordate species. (**d**) Alignment of part of the C-terminal unique domain, with a highly conserved Cys–Z–X–Cys motif. The function of this highly conserved sequence is still to be determined. (**e**) Alignment of part of the fibrillin C-terminal propeptide, spanning from residues 2,757–2,871 of the human fibrillin-1 sequence. This region is cleaved from the mature fibrillin polypeptide but the high degree of sequence conservation, especially in the chordate species, suggests an important functional role for this domain. Species listed: human (*Homo sapiens*) fibrillin-1, mouse (*Mus musculus*) fibrillin-1, chicken (*Gallus gallus*) fibrillin-2, frog (*Xenopus tropicalis*), zebrafish (*Danio rerio*) fibrillin-2, lamprey (*Petromyzon marinus*), beetle (*Tribolium castaneum*), water flea (*Daphnia pulex*), sea anemone (*Nematostella vectensis*) and jellyfish (*Clytia hemisphaerica*)

two residues in this position instead of one, and it has been suggested that this introduces a difference in the twist between domains in tandem cbEGF repeats (Hambleton et al. 2004). All cbEGF domains in the fibrillins are of the class I type.

Although the geometry of calcium binding by cbEGF domains is conserved, the ligands that contribute to the interaction can vary significantly between proteins. Initially it was thought that only six of seven ligands required for the pentagonal bipyramidal geometry of calcium binding were contributed by intradomain oxygen atoms, with the seventh ligand being either water or an interdomain or intermolecular protein ligand (Rao et al. 1995). It has since been shown that in some cases, such as in fibrillin-1 domain cbEGF23, all seven of the ligands required for calcium ligation are contributed by the protein (Lee et al. 2004). In other cases, such as in fibrillin-1 domain cbEGF10 and in domains EGF11–13 of human Notch-1, one or two of the ligands are donated by water molecules (Cordle et al. 2008; Jensen et al. 2009).

The stabilization of cbEGF domain structure by calcium plays an important role in microfibril function by restricting the mobility of interdomain regions. Several studies have shown that calcium binding is responsible for the rod-like arrangement of tandem cbEGF repeats. Calcium binding reduces the proteolytic susceptibility of the molecule (McGettrick et al. 2000; Reinhardt et al. 1997b) and is a prerequisite for the interaction of fibrillin with several molecules including heparin (Cain et al. 2005; Tiedemann et al. 2001); MAGP-1 (Jensen et al. 2001; Rock et al. 2004); fibulin-2, -4 and -5 (El-Hallous et al. 2007; Reinhardt et al. 1996b); and aggrecan and versican (Isogai et al. 2002). Calcium is also required for homotypic and heterotypic interactions between fibrillin-1 and fibrillin-2 (Lin et al. 2002; Marson et al. 2005).

5.4.4 Unique N- and C-Terminal Domains

The first exon of human fibrillin-1 encodes the signal peptide and N-terminal propeptide (Pereira et al. 1993). While the propeptide sequence is not conserved, a recent study found recognition sites for furin/PACE (paired basic amino acid cleaving enzyme) proteases in the propeptide region in every organism examined (Piha-Gossack et al. 2012). The vast majority of these sequences agree with the consensus motif Arg-X-(Lys/Arg)-Arg (where X is any amino acid), although there are several exceptions. Exons 2–5 of human fibrillin-1 encode the fibrillin unique N-terminal (FUN) and first three non-calcium-binding EGF-like domains, preceding the first hyb domain.

Insights into the structure of the N-terminal domains can be gained from sequence alignments. The FUN domain contains four cysteines with a conserved arrangement (Fig. 5.4a), which form intramolecular disulphide bonds (Reinhardt et al. 2000), although the pairings are not yet known. The N-terminal domain of human LTBP-1 and -2 is homologous to the FUN domain (Robertson et al. 2011) and has the same arrangement of cysteine residues. EGF-like domains 1–3 also have a conserved arrangement of cysteines and are expected to form the 1–3, 2–4, 5–6 pattern of disulphides that is found in almost every known EGF-like domain (Bork et al. 1996). These EGF-like domains lack the calcium-binding consensus residues found in cbEGFs, have shorter sequences between cysteine residues and are likely to have a more compact structure.

The FUN domain contains a highly conserved Pro-Gly-Trp sequence after the third cysteine, and there is an analogous, conserved Gly-Tyr sequence after the fourth cysteine in EGF2. A Gly-Trp-Tyr-Phe sequence is often found in cbEGF domains and in other EGF-like domains, forming part of interdomain packing sites. In non-calcium-binding EGF-like domains, the aromatic residue may pack against a histidine or asparagine on a loop in the same domain and proline in the loop between the third and fourth cysteine residue of the subsequent domain. Examples are seen in the structure of human Jagged-1 (Cordle et al. 2008), at the EGF1–EGF2 interdomain interface and in human hedgehog-interacting protein (Bishop et al. 2009), also at the EGF1–EGF2 interface. There are equivalent, highly conserved residues in fibrillin sequences at the likely FUN–EGF1 and EGF2–EGF3 interfaces (Fig. 5.5); the asparagine preceding the first cysteine in the FUN domain is replaced by a histidine in some organisms, such as the sea anemone. By contrast, these residues are not found at equivalent positions between the end of EGF1 and start of EGF2. Together with the

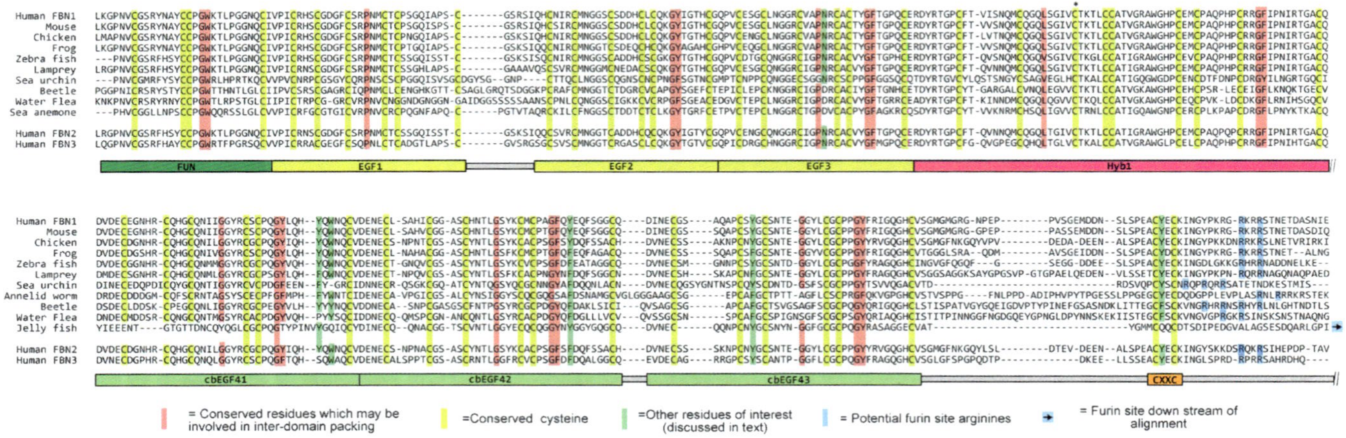

Fig. 5.5 Alignments of fibrillin N-terminus (**a**) and C-terminus (**b**). Organisms were selected to give a broad summary of metazoan evolution, although due to incomplete protein sequences, the same organisms could not be used in all alignments. Boxes beneath the alignments summarise the larger scale fibrillin domain architecture. Species used: mouse (*Mus musculus*), chicken (*Gallus gallus*), frog (*Xenopus tropicalis*), zebrafish (*Danio rerio*), lamprey (*Petromyzon marinus*), sea urchin (*Strongylocentrotus purpuratus*), beetle (*Tribolium castaneum*), water flea (*Daphnia pulex*), sea anemone (*Nematostella vectensis*), jellyfish (*Clytia hemisphaerica*) and annelid worm (*Capitella teleta*)

lack of conservation and variable length of the EGF1–EGF2 linker sequence, this suggests that there is no significant interface between these domains.

The sequence following the last cbEGF/EGF domain (cbEGF43 in human fibrillin-1) is unlike any other in the fibrillins. Moreover, there are no known homologous regions in other proteins. Contrary to previous reports (Giltay et al. 1999), there is no homology with the C-terminal domain of fibulins and therefore no reliable predictions can be made about its structure. Like the N-terminus, the unique C-terminal region has a conserved furin/PACE recognition site (Piha-Gossack et al. 2012). The sequence preceding the furin cleavage site contains the Cys–Z–X–Cys motif, where X is any amino acid and Z is tyrosine or phenylalanine (Fig. 5.4d). The cysteines and their spacing are absolutely conserved, but there are exceptions at position Z—the aromatic residue is replaced by a glutamine in the jellyfish *Clytia hemisphaerica* (Robertson et al. 2011). It is not clear whether these cysteine residues form disulphide bonds, either within or between fibrillin monomers. The C-terminal propeptide sequence, comprising 140 residues in human fibrillin-1, follows the furin recognition site. It does not contain any cysteine residues and so its structure must be independent of disulphide bonds. There are conserved elements, such as stretches of hydrophobic or charged residues (Fig. 5.4e). However, in the absence of any detailed structural or functional information, the significance of the conserved sequences remains unclear.

5.4.5 Proline-Rich and Glycine-Rich Regions

A major feature that distinguishes the three mammalian fibrillin isoforms from each other is the region between domains TB1 and EGF4, which is rich in proline residues in fibrillin-1, glycine residues in fibrillin-2 and a mixture of proline and glycine in fibrillin-3. Excluding the N- and C-terminal propeptides, which are cleaved off before incorporation of fibrillin into microfibrils, this is the only significant stretch of sequence in the fibrillins that does not contain cysteine residues. The proline-rich region of fibrillin-1 is unlikely to be structured (Ashworth et al. 1999b), but may function as a flexible region in the assembly of microfibrils (Baldock et al. 2001). The proline-rich region of fibrillin-1 and glycine-rich region of fibrillin-2 are also likely to be sites of interaction with tropoelastin during elastic fibre assembly (Rock et al. 2004; Trask et al. 2000). Analysis of these regions of the fibrillins has shown much lower homology between species than the domains flanking this region. Current data also suggest that this region could be missing entirely in species with only one fibrillin isoform, leading to speculation that the evolution of this region, in addition to the duplication of the fibrillin gene, may have coincided with the appearance of tropoelastin and been a benefit in the development of the closed circulatory system (Piha-Gossack et al. 2012).

5.4.6 Interdomain Interactions are Functionally Important

When observed by rotary shadowing electron microscopy, isolated fibrillin monomers appear thread-like and extended, with an average length of 140–150 nm (Lin et al. 2002; Reinhardt et al. 1997a; Sakai et al. 1991). Intramolecular interactions between domains play an important role in determining the overall shape and mechanical properties of fibrillin, and calcium binding plays a major role in these interactions. By rigidifying the interfaces between cbEGF–cbEGF pairs (Werner et al. 2000), calcium binding causes the structure of isolated fibrillin fragments to become more elongated compared to calcium-free or EDTA-treated fragments (Reinhardt et al. 1997a) and increases the stiffness of microfibrils (Eriksen et al. 2001). A similar effect has been observed in extracted microfibrils (Wang et al. 2009; Wess et al. 1998), and the effects of removing calcium from samples with the chelating agent ethylene glycol tetraacetic acid (EGTA) were found to be reversible on the re-addition of calcium, but not magnesium, suggesting a calcium-specific effect (Cardy and Handford 1998). Apart from influencing the overall shape of fibrillin, interdomain interactions also have an impact on how fibrillins interact with other molecules. Domain TB4 of fibrillin-1 contains an RGD motif that has been shown to be required for binding to integrins $\alpha_v\beta_3$, $\alpha_v\beta_6$ and $\alpha_5\beta_1$ (Bax et al. 2003; Jovanovic et al. 2007; Lee et al. 2004; Pfaff et al. 1996; Sakamoto et al. 1996), and the presence of domains N-terminal to domain TB4 significantly increases cell attachment to fibrillin fragments through $\alpha_v\beta_3$. This effect may be partly due to the extensive hydrophobic interface at the cbEGF22–TB4 interdomain boundary (Lee et al. 2004). The interface between domains EGF3 and hyb1 of human fibrillin-1 is a binding site for the LTBPs and fibulins (El-Hallous et al. 2007; Isogai et al. 2003; Ono et al. 2009). Interactions of the hyb1 domain with fibulin-4 and -5 depend on the context of this domain (El-Hallous et al. 2007), suggesting that interdomain interactions lead to the formation of a binding surface that spans more than one domain. Taking into account that cbEGF domains form extensive hydrophobic contacts with adjacent hyb and TB domains (Jensen et al. 2005, 2009; Lee et al. 2004) and that calcium alters the dynamics of cbEGF domain pairs (Downing et al. 1996; Jensen et al. 2005, 2009; Lee et al. 2004; Smallridge et al. 2003), these data highlight the importance of considering the native context of fibrillin domains in understanding their intra- and intermolecular interactions.

Calcium affinities of the cbEGF domains vary across the molecule and are highly dependent on domain context (Jensen et al. 2005; McGettrick et al. 2000). The wide range of calcium affinities measured across the molecule, with dissociation constant (K_d) values from the millimolar (1.6 mM for the TB6–cbEGF32 domain pair) (Kettle et al. 1999) to the nanomolar range (9 nM for the hyb1–cbEGF1 domain pair) (Jensen et al. 2009), may reflect the biomechanical properties of different regions of fibrillin. N-terminal linkage to another cbEGF domain significantly increases the calcium affinity of a cbEGF domain relative to the isolated form through the stabilisation of the calcium-binding pocket by interdomain hydrophobic interactions (Smallridge et al. 1999). Measured K_d values for calcium binding are as low as 0.3 μM in fibrillin-1 cbEGF domains in the context of a cbEGF–cbEGF pair (Suk et al. 2004).

The major sites of interdomain interaction in cbEGF–cbEGF domain pairs include a conserved aromatic residue positioned four residues C-terminal from Cys5 of the N-terminal cbEGF and an 'XG' motif involving a conserved glycine between Cys3 and Cys4 of the C-terminal cbEGF (Fig. 5.6a, b). Calcium binding to cbEGF domains is enhanced by the presence of a preceding TB or hyb domains in TB–cbEGF and hyb–cbEGF domain pairs, due to the formation of extensive hydrophobic interdomain interactions (Jensen et al. 2005; 2009; Lee et al. 2004). In these cases, the calcium affinity of the cbEGF domain in the pair is influenced by two separate hydrophobic packing sites on each of the domains. The first of these (L1) involves hydrophobic residues between Cys2 and Cys3 of the TB domain packing with residues between Cys1 and Cys2 ('loop 1') of the cbEGF domain. The second site resembles a cbEGF–cbEGF interaction, with the turn between Cys3 and Cys4 of the cbEGF domains packing against a Gly-X motif (where X is often an aromatic) found after Cys8 of the TB domain (Fig. 5.6c, d). The linker length between TB and cbEGF domain was originally thought to determine interdomain flexibility, based on the structure of the isolated fibrillin-1 TB6 domain (Yuan et al. 1997). However, it was shown later that the hydrophobic interactions at the domain interfaces have a far greater influence on interdomain interactions (Jensen et al. 2005).

TB and hyb domains involved in forming high calcium affinity domain pairs with adjacent cbEGF domains in human fibrillin-1, such as domains hyb1, hyb2 and TB4, show a high degree of sequence conservation between species at sites involved in interdomain packing (Jensen et al. 2012; Robertson et al. 2011). These domains generally have a conserved leucine or isoleucine residue positioned four residues C-terminally from the second cysteine of the domain, which is involved in packing with the 'loop 1' region of the adjacent cbEGF domain (Fig. 5.6e). The high degree of conservation seen at high calcium affinity sites may reflect a requirement for increased rigidity at these sites during microfibril assembly, for protein interactions, or for the mechanical properties of the microfibrils. The significance of the very high calcium affinities at some of these sites, however, with K_d values several orders of magnitude lower than the free calcium concentration (~1.0 to 1.2 mM) in the extracellular matrix, is unknown. A loss of hydrophobic residues at interdomain packing sites provides an explanation for the low calcium affinity and lack of interdomain interface seen in the TB6–cbEGF32 domain pair (Jensen et al. 2005, 2012; McGettrick et al. 2000), which may be functionally significant. The idea that calcium affinities drive the conservation of sequences in these regions is supported by the lack of sequence conservation in the corresponding parts of the TB1 domain (Fig. 5.6e), which is followed at its C-terminus by the proline- or glycine-rich domains, depending on the fibrillin isoform.

5.5 Models of Microfibril Organisation

At present, two broad categories of microfibril organisation have been proposed, with fibrillin monomers either folded ('pleated' models) so that each monomer fits within one interbead repeat (Baldock et al. 2006; Glab and Wess 2008; Lu et al. 2005; Wang et al. 2009) or extended, with two or more monomers per interbead distance

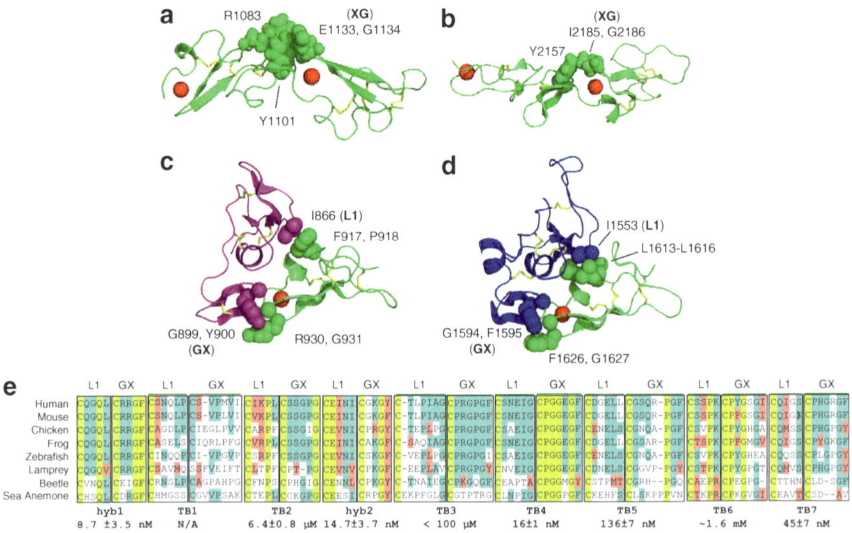

Fig. 5.6 Conservation of interdomain interactions. cbEGF–cbEGF domain pair interactions, as seen in cbEGF12–13 (**a**) and cbEGF32–33 (**b**), are stabilised by packing of conserved aromatic residues in the N-terminal domain (Y1101 in cbEGF12 and Y2157 in cbEGF32) with an 'XG' motif between the Cys3 and Cys4 of the C-terminal domain (E1133–G1134 in cbEGF13 and I2185–G2186 in cbEGF33). The cbEGF12–cbEGF13 interaction is further stabilised by the packing of methylene groups from residue R1083 against the aromatic ring of Y1101. Interfaces of heterologous domain pairs, such as TB4–cbEGF23 (**c**) and hyb2–cbEGF10 (**d**), are stabilised by extensive interdomain hydrophobic contacts involving the 'L1' and 'GX' sites. These form contacts with residues between Cys1–Cys2 (F917 and P918 in cbEGF10 and L1613–L1616 in cbEGF23) and Cys3–Cys4 (R930 and G931 in cbEGF10 and F1626 and G1627 in cbEGF23) of the adjacent cbEGF, respectively. Pairwise interactions of the domains in these structures suggest a near-linear organisation, stabilised by calcium ions (shown in red). (**e**) Interdomain interaction sites in hyb and TB domains are conserved between species, especially in domains that interact with adjacent cbEGFs to create high calcium affinity pairs. This suggests a conservation of local structure at these sites. Numbers under sequences refer to K_d values measured for calcium binding to hyb–cbEGF and TB–cbEGF pairs from human fibrillin-1 (Jensen et al. 2005, 2009). Species listed: human (*Homo sapiens*) fibrillin-1, mouse (*Mus musculus*) fibrillin-1, chicken (*Gallus gallus*) fibrillin-2, frog (*Xenopus tropicalis*), zebrafish (*Danio rerio*) fibrillin-2, lamprey (*Petromyzon marinus*), beetle (*Tribolium castaneum*), water flea (*Daphnia pulex*) and sea anemone (*Nematostella vectensis*)

(Downing et al. 1996; Kuo et al. 2007; Lee et al. 2004) (Fig. 5.7). A common feature of all models is the head-to-tail configuration of fibrillin monomers, with up to eight monomers per repeat, and interactions between the N- and C-termini being thought to be one of the first steps in assembly (Hubmacher et al. 2008; Lin et al. 2002; Marson et al. 2005; Trask et al. 1999; Sect. 5.6.2). These sites of interaction have been localised to the bead region of the beaded filament (Reinhardt et al. 1996a). The underlying mechanism of microfibril elasticity is still controversial with little agreement on the relative contributions of large-scale molecular rearrangements (pleated models) vs. more localised unfolding and refolding events at domain boundaries (extended models) to the mechanical properties of microfibrils.

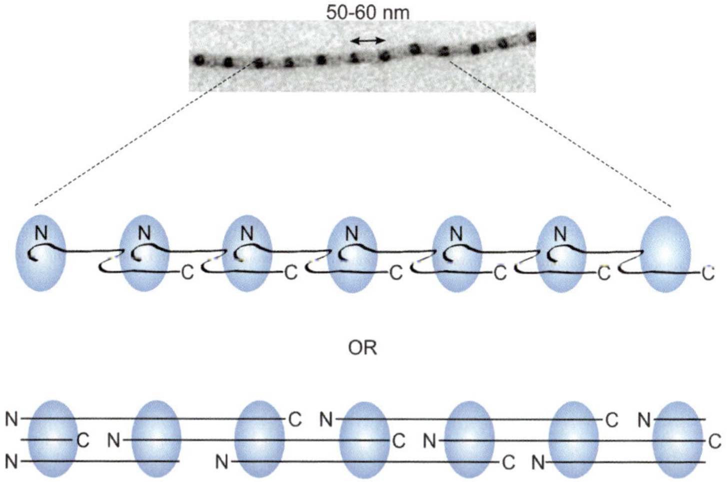

Fig. 5.7 Models of microfibril organisation. Two classes of model have been proposed to explain how fibrillin monomers, with an extended length of 150–160 nm can fit into a structure with a repeat distance of 50–60 nm. *Top*: the pleated model, which predicts kinks and turns in the fibrillin structure. *Bottom*: the staggered model, which suggests that the fibrillin molecules are near-linear and overlapped, giving rise to the observed periodicity (see text for details)

Unstaggered models of microfibril organisation have been proposed since antibody mapping was first used to map the positions of specific fibrillin epitopes within microfibrils (Reinhardt et al. 1996a). Scanning transmission electron microscopy mass mapping, antibody mapping and automated electron tomography were later used to develop a 'pleated' model of microfibril organisation in which each fibrillin monomer fit within one interbead distance (Baldock et al. 2001). In the most recent version of this model, the more globular TB and hyb domains correspond to the globular features of the microfibril observed by electron microscopy. The regions of fibrillin that interact with cells via integrins or heparan sulphate (Bax et al. 2003; Cain et al. 2005; Pfaff et al. 1996; Ritty et al. 2003; Sakamoto et al. 1996; Tiedemann et al. 2001) are predicted to cluster in the most compact part of the model. Microfibril extensibility in this model is based on conformational rearrangements at interdomain interfaces leading to the molecule becoming more linear under an imposed force. The pleated model is supported by several studies, including experimentally derived stain exclusion patterns observed by electron microscopy of microfibrils extracted from tissues (Lu et al. 2005), an X-ray diffraction study of microfibrils from zonular filaments in which computer-generated patterns were used to fit X-ray data (Glab and Wess 2008) and a small angle X-ray scattering study of purified fibrillin fragments (Baldock et al. 2006). Although the pleated model explains many of the large-scale features of microfibrils, the mechanism of structural restoration after the release of a stretching force is unclear. In addition, the model allows for bead periodicities (up to 160 nm) that have been shown to occur only under extreme extensions that may result in a loss of tissue integrity and microfibril damage (Haston et al. 2003).

Staggered, linear models of microfibril organisation present fibrillin monomers in an elongated arrangement that spans two (Downing et al. 1996; Kuo et al. 2007; Liu et al. 1996) or three (Lee et al. 2004) interbead distances and are based on nuclear magnetic resonance and X-ray structures of cbEGF-containing fibrillin-1 fragments (Downing et al. 1996; Lee et al. 2004; Smallridge et al. 2003; Yuan et al. 2002) and on the mapping of antibody binding and protease cleavage sites within fibrillin and microfibrils (Kuo et al. 2007). The most recent version of this arrangement of fibrillin in the microfibrils positions the N- and C-termini within the beaded structure of the microfibril, and the monomers are staggered so that the 'neonatal' region of fibrillin (Fig. 5.3) is also positioned near the bead. The especially severe MFS phenotypes resulting from some mutations affecting these domains suggest that this part of fibrillin-1 is functionally important, although its precise role is yet to be determined. The N- and C-termini are also positioned asymmetrically so that the N-terminal half of fibrillin overlays a core formed by the C-terminal half (Kuo et al. 2007). This version of the linear model is supported by antibody mapping data (Reinhardt et al. 1996a), by atomic resolution structures of fibrillin fragments (Downing et al. 1996; Jensen et al. 2009; Lee et al. 2004; Smallridge et al. 2003) and observations of full-length fibrillin-1 monomers by electron microscopy (Lin et al. 2002; Sakai et al. 1991). Based on the observation that fibrillin monomers have a length of ~150 nm (Lin et al. 2002; Sakai et al. 1991), an elongated, staggered organisation within microfibrils would suggest that reversible extensibility is limited to interbead periodicities of <100 nm without significant structural rearrangements. Kuo et al. suggested that periodicities of up to 160 nm seen in earlier studies of microfibrils isolated from matrix by collagenase treatment are due to cross-linked monomers held together after cleavage of sites near the fibrillin N-terminus (Kuo et al. 2007). Extensibility of microfibrils in the elongated models has been proposed to involve interdomain interactions, with the interdomain linker regions between TB and cbEGF domains acting like a molecular spring that unfolds on the application of a stretching force (Jensen et al. 2012; Lee et al. 2004). Recoil may result from the high calcium affinities seen for most fibrillin TB–cbEGF (Jensen et al. 2005) and hyb–cbEGF (Jensen et al. 2009) domain pairs, with the return to the relaxed state being driven by the calcium-dependent stabilisation and refolding of domain interfaces. The proline-rich (glycine-rich in fibrillin-2 and glycine/proline-rich in fibrillin-3) region may also contribute to extensibility, although at present there is little structural data on this region. cbEGF repeats are unlikely to contribute a large degree of extension, since untensioned microfibril periodicity is already at a maximum under conditions of saturating calcium ion concentrations (Cardy and Handford 1998).

5.6 Microfibril Assembly

The pathway of fibrillin assembly into microfibrils is not yet fully understood, but a number of important events and interactions have been identified. Unprocessed pro-fibrillin is folded in the endoplasmic reticulum and subsequently undergoes post-translational modifications during its passage through the secretory pathway.

The N- and C-terminal propeptides are removed by proteolytic cleavage, and the mature protein is secreted and then assembled near the cell surface (Reinhardt et al. 2000; Wallis et al. 2003). In particular, the N- and C-terminal regions have been implicated in assembly (Fig. 5.8). Considering the remarkably conserved fibrillin domain architecture (Fig. 5.2), the spacing of domains and length of the fibrillin molecule may also play a role in correctly aligning fibrillin monomers. Cells appear to play an important role in microfibril assembly, and there is currently no method to assay in vitro assembly outside of cell culture. Several extracellular matrix components are thought to regulate fibrillin deposition into microfibrils. Heparan sulphate is likely to be essential for assembly and can block microfibril assembly by fibroblast cells in culture (Tiedemann et al. 2001). Integrins may also play a role in controlling assembly as stiff skin syndrome, which is caused by mutations affecting the integrin-binding domain TB4 (Loeys et al. 2010), results in excessive microfibril deposition. Fibronectin is essential for microfibril assembly in mammalian cells (Sabatier et al. 2009), although it is not thought to be present in lower animals such as insects or jellyfish (Robertson et al. 2011). Finally, ADAMTS10 and ADAMTSL6β promote microfibril deposition (Kutz et al. 2011; Tsutsui et al. 2010), but the mechanism for this effect is not understood.

5.6.1 Fibrillin N- and C-Termini are Processed by Furin

Cleavage of the N- and C-terminal propeptides is likely to be an essential and conserved event in fibrillin assembly, given the high conservation of the furin/PACE recognition sites. Indeed, furin-mediated processing has been demonstrated in mammalian cells. N-terminal sequencing of peptides in the medium of fibrillin-producing cells identified Arg45, the residue after the predicted N-terminal furin/PACE recognition site, as the first residue in processed, secreted human fibrillin-1 (Kettle et al. 2000; Reinhardt et al. 2000). Similarly, the cleavage at the C-terminus was found to occur to C-terminally to Arg2731, the last basic residue of the furin consensus recognition site (Lönnqvist et al. 1998). The importance of C-terminal processing was highlighted by a patient with a *FBN1* mutation immediately upstream of the region encoding the recognition site, leading to an Arg2726-Trp substitution. Strong evidence for processing at the C-terminus was obtained from a patient with an *FBN1* mutation in the region encoding the recognition site, leading to an Arg2726-Trp substitution (Milewicz et al. 1995). The patient exhibited skeletal features of MFS, and isolated dermal fibroblast cells processed only 50% of the pro-fibrillin. Unprocessed material was present in the media, but not the insoluble matrix, suggesting that it failed to incorporate into microfibrils. It was later demonstrated that a number of furin-like proteases could cleave the fibrillin C-terminus (Ashworth et al. 1999a; Raghunath et al. 1999; Ritty et al. 1999) and that proteolysis occurred near the cell surface, as cleavage was prevented by blocking secretion (Wallis et al. 2003). Moreover, use of a specific inhibitor identified furin as the protease involved in processing, rather than another member of the furin/PACE family.

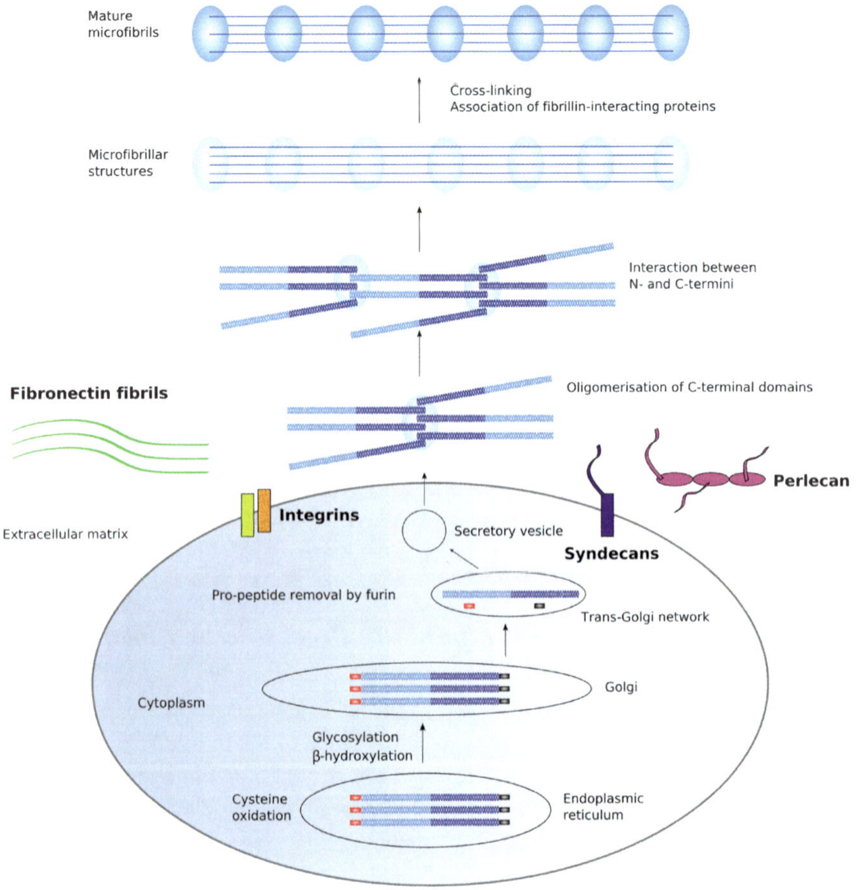

Fig. 5.8 Fibrillin assembly into microfibrils and candidate accessory proteins. Cartoon of the fibrillin microfibril assembly pathway. Pro-fibrillin is folded and modified in the secretory pathway, followed by furin-mediated proteolytic cleavage of N- (*red*) and C-terminal (*black*) regions near the cell surface. The precise locations of the fibrillin assembly events are unknown, but assembly is thought to occur near the cell surface. Assembly is likely to be initiated by oligomerisation and interactions between the N- (*light blue*) and C-terminal (*dark blue*) regions of neighbouring fibrillin molecules. The 'beads-on-a-string' appearance of extracted microfibrils only emerges after several weeks (Kielty et al. 2002). A number of factors may modulate assembly: fibronectin, integrins and heparan sulphate-conjugated proteoglycans, such as syndecans and perlecans

5.6.2 N- and C-Terminal Interactions Are an Initial Step in Assembly

A number of studies have suggested roles for the N- and C-terminal regions in assembly. Hints of an interaction between the N- and C-termini came from a study showing that the presence of an N-terminal fragment was required for processing and aggregation of a C-terminal fragment (Ashworth et al. 1999a). Later, it was demonstrated that the recombinantly produced N-terminal half of human fibrillin-1 could interact with the C-terminal half (Lin et al. 2002). Binding was also demonstrated between N- and C-terminal halves of fibrillin-1 and fibrillin-2, although the N- and C-terminal halves of fibrillin-2 did not interact. This suggested a basis for the formation of heterotypic microfibrils that contain more than one fibrillin isoform. The regions involved in binding were then narrowed down: Marson et al. identified an interaction between the FUN–cbEGF2 region at the N-terminus and the TB7-unique-C-terminal region at the C-terminus (Marson et al. 2005). Subsequently, it was shown that the minimal regions for the interaction were the FUN–EGF3 region at the N-terminus (El-Hallous et al. 2007) and cbEGF41–43 at the C-terminus (Hubmacher et al. 2008).

The N–C interaction alone is not sufficient for microfibril assembly. Hubmacher et al. demonstrated the importance of the multimerisation of the C-terminus (Hubmacher et al. 2008). The last three cbEGF domains (cbEGF41–43) were shown to promote oligomerisation, which increased the binding affinity of the C-terminal half for the N-terminal half. Purified multimers blocked microfibril assembly in cell culture and had beadlike structures, as judged by electron microscopy. These 'beads' could be analogous to those seen in the 'beads-on-a-string' structures of microfibrils extracted from tissues. The N-terminus may also oligomerise—interactions between N-terminal fragments have been observed (Marson et al. 2005). Moreover, the conserved, unpaired cysteine in the hyb1 domain is likely to be involved in intermolecular disulphide bond formation. This residue is reactive in fibrillin produced by cultures of chick embryo aortae (Reinhardt et al. 2000) and appears to be surface-exposed in a homology model of the hybrid 1 structure (Jensen et al. 2009).

Alignments of the N- and C-terminal sequences across species may give clues about the mechanism of fibrillin assembly. In the FUN–EGF3 region, there are conserved residues that are not likely to be structurally important, for example, the loop between the first and second cysteines in the FUN domain and the loop between the third and fourth cysteines in the EGF1 domain (Fig. 5.5). However, interpretations are complicated by the fact that microfibril-associated proteins, specifically BMP prodomains, also interact with these domains (Sengle et al. 2008a). Since BMPs were present before the emergence of fibrillins, it is possible that conserved sequences may reflect heterotypic rather than homotypic interaction sites.

Sequences of the cbEGF41–43 region show some common features, but also differences between fibrillin-containing species. In addition to the residues common to all cbEGF domains ('packing' and 'consensus'), there is a notably high abundance of aromatic residues, many of which are conserved (Fig. 5.5). Comparisons with known cbEGF domain structures suggests that they may be clustered (Downing et al. 1996; Smallridge et al. 2003). If so, they could create hydrophobic surfaces that might be involved in multimerisation, the interaction between the fibrillin N- and C-termini, or heterotypic interactions with other matrix molecules. One striking variable in cbEGF41–43 sequences is the cbEGF42–cbEGF43 interface. All vertebrates have calcium-binding consensus residues in the domain cbEGF43, but many invertebrates do not (Robertson et al. 2011). This could reflect a difference in the assembly mechanism, but until more details about the molecular nature of the interaction between the fibrillin N- and C-termini are known, firm conclusions cannot be drawn.

5.6.3 Fibronectin Is Essential for Microfibril Assembly

Fibronectin is a major component of the extracellular matrix involved in several processes including cell adhesion, wound healing and development. The importance of fibronectin is demonstrated by the embryonic lethality of fibronectin null mice (George et al. 1993). In recent years, it has been shown that the fibronectin network is essential for the assembly of fibrillin into microfibrils (Kinsey et al. 2008; Sabatier et al. 2009). Using siRNAs specific for fibronectin and peptides that block the function of fibronectin, Sabatier et al. were able to block fibrillin network formation by human dermal fibroblasts (Sabatier et al. 2009). Similar results were obtained in fibronectin RNAi experiments by Kinsey et al. (2008), who also suggested that integrin $\alpha5\beta1$ was involved in microfibril assembly through its role in the development of the fibronectin network. The interaction of fibrillin with fibronectin was localised to the C-termini of fibrillin-1, -2 and -3, with multimeric forms of the recombinant constructs used in these assays interacting more strongly with fibronectin than the monomeric forms (Sabatier et al. 2009). This suggested a model of increased fibronectin binding after the multimerisation of the fibrillin C-terminus observed in an earlier study (Hubmacher et al. 2008), which is likely to be an important early step in microfibril assembly. A second interaction was observed between fibronectin and the N-terminus of fibrillin-1, but not of fibrillin-2, although the significance of this difference is not clear (Sabatier et al. 2009). The binding site for the fibrillins on fibronectin was localised to FNI_{6-9} (Sabatier et al. 2009).

Although fibronectin was found to be essential for the assembly of a fibrillin network in human fibroblast cultures, a recent review of the evolutionary history of the fibrillins found no evidence of genes encoding fibronectins in the available genomes of lower metazoan species, including the cnidarians and arthropods (Robertson et al. 2011). The absence of a fibronectin homologue may reflect a requirement for a greater degree of organisation in the assembly of microfibrils in the more complex extracellular matrices of higher metazoans. Others, however, have reported that a fibronectin-like molecule has existed since the emergence of the sponges (Labat-Robert 2012), and immunological studies have shown that fibronectin-like proteins occur in early metazoan species. Extracts of mesoglea from the cnidarian *Hydra vulgaris* analysed by Western blotting show anti-fibronectin antibody-reactive bands at 200 and 335 kDa, which could be specifically blocked by human plasma fibronectin (Sarras et al. 1991). The presence of a fibronectin-like protein has also been demonstrated in the sponge *Ephydatia muderi* (Labat-Robert et al. 1981). No sequence information is available to indicate that any of these early metazoan proteins have a domain structure homologous to fibronectin, however, so at present the role of these fibronectin-like molecules in microfibril assembly remains speculative.

5.6.4 Integrins

Fibrillin-1 has been shown by several studies to be one of the main extracellular matrix components involved in cell adhesion via integrins (Bax et al. 2003; Jovanovic et al. 2007; 2008; Pfaff et al. 1996; Sakamoto et al. 1996). Interactions between the fibrillins and integrins have been implicated in the development of disease. The tight skin phenotype of the *Tsk* mouse model is caused by a naturally occurring in-frame duplication resulting in the expression of a fibrillin-1 polypeptide containing an additional TB4 domain (Bona et al. 1997). In humans, all known mutations that cause stiff skin syndrome have been found to affect the integrin-binding domain TB4 (Loeys et al. 2010). Fibrillin-1 fragments containing the RGD site can also upregulate the expression and production of the matrix metalloproteases MMP-1 and MMP-3, indicating a potential role in the pathogenesis of MFS (Booms et al. 2005). There is also some evidence to suggest that, in addition to its role in cell-matrix interactions, the interaction between fibrillins and integrins may be important in the assembly of the microfibril network. Blocking integrin $\alpha v\beta 3$ function with a cyclic peptide antagonist inhibits microfibril assembly by periodontal ligament cells (Tsuruga et al. 2009), although this effect was not observed in similar experiments by Kinsey et al. (2008), who showed that assembly was influenced more by integrin $\alpha 5\beta 1$, possibly through its interactions with fibronectin.

Apart from the integrin-binding site found in domain TB4 of human fibrillin-1, additional RGD motifs have been found in domains TB4 of fibrillin-2 and -3, domain TB3 of fibrillin-2 and cbEGF18 of fibrillin-3 (Corson et al. 2004; Lee et al. 1991; Maslen et al. 1991; Pereira et al. 1993). At present, it is not clear whether the sites in domains TB3 and cbEGF18 are functional in tissues. Although RGD motifs are mostly found in vertebrates (Robertson et al. 2011), recent bioinformatics data has shown that the domain TB4 RGD motif is found in earlier species including the cephalochordate *Branchiostoma lanceolatum* (lancelet or amphioxus), extending the presence of this site to other chordates (Piha-Gossack et al. 2012).

An RGD motif is thought to have first appeared in domains TB3 and TB4 in an ancestral fibrillin around the time of the divergence of vertebrates and other chordates. After the duplication events that lead to the development of the current fibrillin isoforms, the RGD motif in domain TB3 was eventually lost from fibrillin-1 around the divergence of mammals from other tetrapods and from fibrillin-3 around the appearance of marsupials (Piha-Gossack et al. 2012). An RGD motif in fibrillin-3 domain cbEGF18 has only been observed in sequences from humans and chimpanzees, suggesting that this has appeared very recently in evolution (Piha-Gossack et al. 2012). Based on previous fibrillin-1 cbEGF domain structures (Downing et al. 1996; Smallridge et al. 2003), the position of this sequence within the cbEGF domain, between the second and third cysteines, suggests that it may be exposed and available for interactions with integrins.

5.6.5 *Heparin/Heparan Sulphate*

Interactions of fibrillin with cell-surface heparan sulphate proteoglycans are likely to be an important part of the microfibril assembly process. Microfibril assembly by cultured skin fibroblasts can be inhibited by the addition of heparin or a highly sulphated and iduronated form of heparan sulphate (Tiedemann et al. 2001). Inhibition of microfibril assembly by the compounds 4-methylumbelliferyl-β-D-xylopyranoside and *p*-nitrophenyl-β-D-xylopyranoside, which reduce the amount of glycosaminoglycans attached to proteoglycans, further suggested that glycosaminoglycans attached to proteoglycans are involved in the process of microfibril assembly (Tiedemann et al. 2001). Similar inhibition of microfibril assembly by heparin was seen using osteosarcoma (MG63) and pigmented epithelial (PE) cell lines (Ritty et al. 2003). The binding of heparin/heparan sulphate has been localised to four regions on human fibrillin-1: the N-terminus up to domain cbEGF1, the C-terminal region including domains cbEGF38–41 and the domain immediately preceding the furin cleavage site, domains cbEGF12–14 in the 'neonatal' region, and the TB5–cbEGF25 region (Cain et al. 2005; 2008; Ritty et al. 2003; Tiedemann et al. 2001).

Fibrillin polypeptides can encounter heparan sulphate chains attached either to secreted proteoglycans, such as perlecan, or cell membrane-bound proteoglycans such as syndecans and glypicans. Fibrillin-1 interacts with perlecan, and this interaction is likely to be important in anchoring microfibrils to basement membranes and possibly microfibril assembly (Tiedemann et al. 2005). Perlecans are evolutionarily ancient molecules, found throughout metazoan species, including the placozoan *Trichoplax* (Robertson et al. 2011). At present there is no information on whether perlecan interacts with fibrillin in the simplest microfibril-containing species, such as cnidarians, so the significance of any interaction between perlecan and microfibrils in these species is unknown. Similarly, syndecans are an evolutionarily conserved family of cell-surface proteoglycans, found in early metazoans including the cnidarians (Choi et al. 2011). They have been suggested to have a role in microfibril function (Bax et al. 2007; Cain et al. 2008); however, at present, a direct interaction between fibrillin and a syndecan is yet to be demonstrated.

5.7 Role of Microfibrils in Matrix Regulation

As previously discussed, mouse models of MFS and the identification of Loeys–Dietz syndrome have demonstrated a clear physiological link between disruption of the fibrillin network and dysregulation of TGFβ signalling. However, the mechanisms behind this connection are still unclear. Fibrillin interactions with TGFβ are mediated by the LTBPs (Fig. 5.9a), which form a covalently linked complex with the latency-associated peptide of TGFβ. The interaction of the TGFβ latency-associated peptide involves the rearrangement of the disulphide bond formed between Cys2 and Cys6 of the second TB domain of the LTBPs, through a mechanism that does not occur in any of the TB domains found in any of the fibrillins (Chen et al. 2005; Saharinen and

Keski-Oja 2000). This complex of TGFβ, its propeptide and LTBP is referred to as the large latent complex (LLC), and it has been shown that the presence of LTBP-1, and formation of this complex, significantly enhances TGFβ secretion in cell culture and ensures correct disulphide bond formation in the small latent complex (Miyazono et al. 1991). However, TGFβ remains in a latent state while bound to its propeptide and is unable to bind to its receptors. As well as LTBP expression enhancing secretion of TGFβ, it has been shown that LTBP-3 is dependent on binding to TGFβ for secretion from the cell (Penttinen et al. 2002). Despite this, significant amounts of LTBP-1 and -4 are secreted from cells without bound TGFβ.

LTBPs may be more than just chaperones for TGFβ secretion. LTBPs are large extracellular matrix proteins, with a multi domain organisation similar to but distinct from fibrillin (Robertson et al. 2011). Only the second TB domain of LTBPs has so far been demonstrated to interact with TGFβ, while other regions of the LTBPs have instead been shown to interact with insoluble components of the extracellular matrix (Fig. 5.9b) (Unsold et al. 2001). The C-termini of LTBP-1, -2 and -4 have been shown to bind non-covalently to the N-terminus of fibrillin-1 (Isogai et al. 2003; Massam-Wu et al. 2010; Ono et al. 2009) and the N-termini of LTBP-1 and -4 associate with fibronectin (Fontana et al. 2005; Kantola et al. 2008). Figure 5.9b summarises the known interactions of the LTBPs, and proposed organisation of the LLC in the extracellular matrix.

The interaction between LTBP and fibrillin is particularly interesting, as it may provide clues for a mechanistic basis for the link between the fibrillin network and TGFβ signalling seen in MFS. Initial studies of the LTBP/fibrillin interaction localised the binding regions to the N-terminal domains of fibrillin-1 and C-terminal domains of LTBP-1 and -4 (Isogai et al. 2003). It was also shown that the equivalent region of LTBP-3 did not interact with the same N-terminal fragments of fibrillin-1. A separate study demonstrated that LTBP-2 interacts with a similar region of fibrillin-1 to LTBP-1 and -4, but not with fibrillin-2 (Hirani et al. 2007).

5.7.1 LTBP–Fibrillin Interaction

Although no specific LTBP residues have yet been implicated in the LTBP–fibrillin interaction, in LTBP-1 and -4 the binding sites have been localised to a C-terminal region (Isogai et al. 2003; Ono et al. 2009). The C-terminus of LTBP-1 demonstrates striking conservation of the third EGF domain across species (Robertson et al. 2011), suggesting it may be important for fibrillin binding. Furthermore, most residues within this domain are also conserved in LTBP-2 and -4, but not LTBP-3, which does not bind fibrillin.

The fibrillin N-terminus (Fig. 5.5) has been shown to interact with LTBPs in the region around its first hybrid domain. Substitution of Asn164 in EGF3 with serine has a significant effect on the binding of LTBP-4, but not LTBP-1 (Ono et al. 2009). Figure 5.5 shows that this residue is conserved between vertebrates and sea urchin, which is consistent with the distribution of LTBPs. While this does not preclude

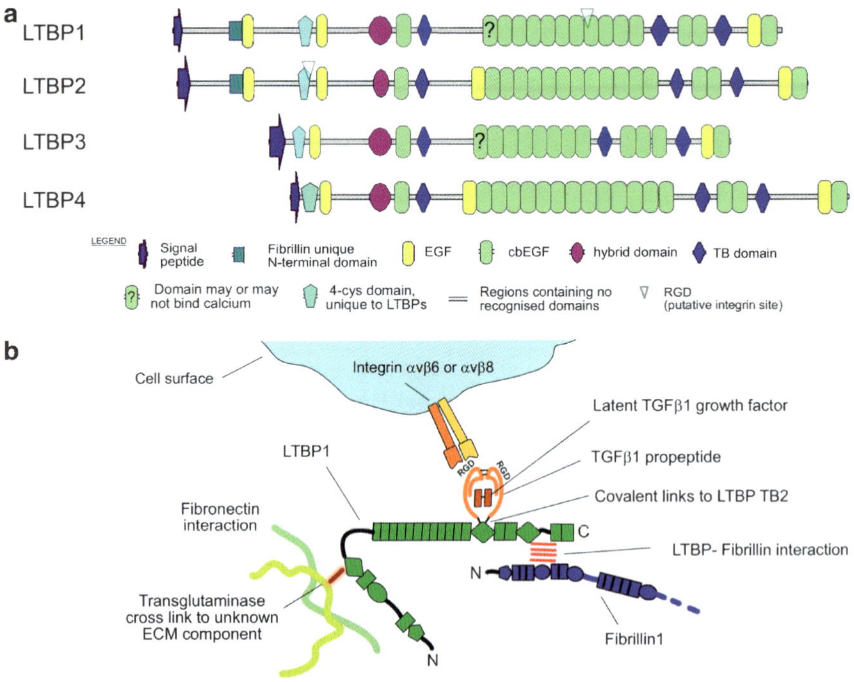

Fig. 5.9 Latent transforming growth factor-beta binding proteins (LTBPs). (**a**) Domain structures of the four human LTBP isoforms. (**b**) The proposed organisation of LTBP-1 within the extracellular matrix, bound to the TGFβ propeptide via a disulphide linkage, which is itself non-covalently bound to the dimeric growth factor. LTBP-1 also interacts with the extracellular matrix via its N- and C-termini, and specific interactions have been demonstrated with fibrillin at the C-terminus and fibronectin at the N-terminus. In the case of TGFβ-1 and -3, the propeptide contains integrin-binding sites that are important for traction-mediated activation by integrins at the cell surface

LTBP-4 type interactions occurring in these organisms, further experiments are needed to determine whether or not LTBP–fibrillin interactions occur in all vertebrates.

5.7.2 Evolution of the LTBP–TGFβ Complex

The first LTBP and the first true TGFβ-like proteins appear to have emerged in the deuterostomes as they can be found in the *S. purpuratus* sea urchin genome, but not elsewhere. The sea urchin LTBP also possesses the key amino acid insertion for disulphide bond exposure and TGFβ binding (Robertson et al. 2011). The TGFβ superfamily, which includes the BMPs and GDFs, precedes the evolution of the metazoa (Herpin et al. 2004; Kahlem and Newfeld 2009), but close homologues of TGFβ-1, -2 and -3 that possess the necessary cysteine for covalent binding LTBP are restricted to the deuterostomes. This might suggest that the functions of LTBP and TGFβ co-evolved, although the sea squirt *C. intestinalis* possesses a TGFβ-like growth factor but no LTBP. This may be because TGFβ is able to function independently of an LTBP in this organism, or that LTBP sequences may have been missed.

5.7.3 *Interactions with Other Growth Factors*

While TGFβ does not interact directly with fibrillin, a number of related growth factors can, including BMP-2, BMP-4, BMP-5, BMP-7, BMP-10 and GDF-5 (Gregory et al. 2005; Sengle et al. 2008a; 2011). Initial SPR studies showed that all five of the prodomains of these growth factors could interact with the N-terminal regions (from the unique N-terminus to EGF4) of both fibrillin-1 and -2, with similar affinities (Gregory et al. 2005; Sengle et al. 2008a). BMP-2, -4 and -10 and GDF-5 also interacted with the EGF4–TB1 region of fibrillin-1 with lower affinity, and BMP-10 also interacted with a fragment of fibrillin-1 from domain TB7 to the C-terminus. The binding site of BMP-7 was localised more precisely to the N-terminus, and shown to be dependent on the presence of the FUN domain (Sengle et al. 2008a). Rotary shadowing electron microscopy of the complex was consistent with BMP-7 binding to the N-terminus of fibrillin. However, not all TGFβ superfamily prodomains interact with this region of fibrillin. The GDF-8 prodomain, for example, binds to the glycosaminoglycans side chains of perlecan domain V (Sengle et al. 2011).

BMP-7 and BMP-4 have been shown to co-localise with fibrillin-1 in tissue sections (Gregory et al. 2005; Sengle et al. 2008a), demonstrating that these interactions do occur in vivo. One hypothesis is that binding to fibrillin may somehow localise and direct growth factor signals to specific regions of the extracellular matrix, and in so doing direct specific features of tissue development (Ramirez and Rifkin 2009).

While the TGFβ and GDF-8 growth factor/prodomain complexes are latent and unable to activate signalling (Lee and McPherron 2001), BMP-4, -5, -7 and -9 growth factor/prodomain complexes are not latent and are able to bind to their receptors and activate intracellular signalling events with their propeptides still attached (Brown et al. 2005; Sengle et al. 2008b, 2011). The BMP-10 prodomain on the other hand was shown to confer latency when bound to BMP-10, and BMP-10 can be released when the prodomain is cleaved by BMP-1 (Sengle et al. 2011). One possibility is that the interaction with fibrillin confers latency to these growth factors, by localising them to the extracellular matrix, away from the cell surface. Signalling would then be initiated by displacement from fibrillin, potentially by proteases, or contact of microfibrils with the cell surface.

5.7.4 *Evolutionary Aspects of Growth Factor Interactions*

As no specific residues involved in the BMP-fibrillin or GDF-fibrillin interactions have yet been defined, it is not possible to examine the evolution of their interaction using sequence alignments. The FUN domain is found consistently in fibrillins, from sea anemone through to humans (Figs. 5.4 and 5.5). Since BMP-5/7 growth factors are also present in these organisms (Robertson et al. 2011), BMP-5/7

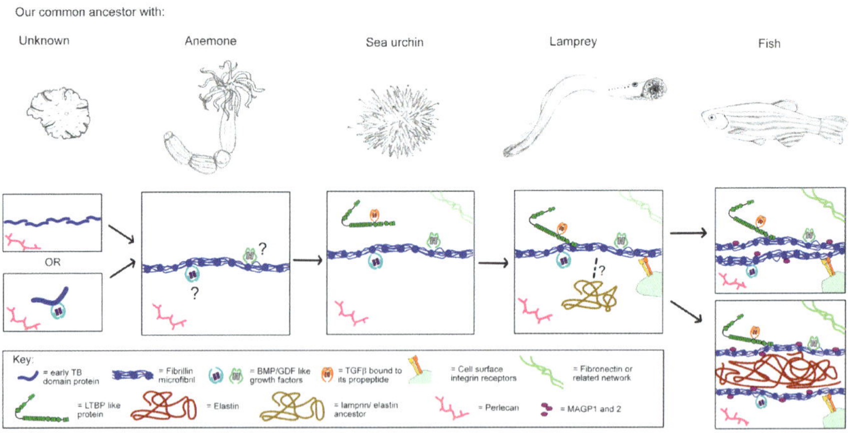

Fig. 5.10 Summary of microfibril evolution. Boxes contain possible functions acquired by fibrillin microfibrils in the common ancestors of humans shown above. The unknown organism represents the first organism to utilise TB domain proteins, where it is unclear if they might have first acquired a structural role, or initially played a role in growth factor binding

interactions are also possible. However BMP-like growth factors appear to precede the fibrillins, and so it is likely that their initial signalling function evolved first, and the modulatory function provided by fibrillin/prodomain interactions evolved later. What is not clear, however, is whether the fibrillins first evolved as a structural protein, or as a growth factor regulator. A greater understanding of the interactions involved, and more genome sequences from early metazoans, will be needed to answer this question.

5.8 Conclusion and Outstanding Questions

Since the discovery of fibrillin as the main component of the 10–12 nm diameter microfibrils in 1986 (Sakai et al. 1986), studies of fibrillin-1 have employed the molecular, cellular and whole organism toolkits to gain fundamental insights into its structure, interactions and molecular pathology. However, several gaps in our knowledge of fibrillin microfibrils still remain including:

- Details of the higher order arrangement of fibrillin within microfibrils
- An ability to assay microfibril assembly to allow the dissection of the process in order to identify specific interaction sites and the effects of mutations
- Direct evidence of the dominant negative effects in disease pathogenesis
- How protein–protein and protein–glycosaminoglycan interactions involving microfibrils vary in different tissues, leading to diseases, such as stiff skin syndrome and MFS, with different phenotypes
- How LTBP heterogeneity contributes to growth factor regulation by microfibrils

The recent advances in bioinformatics and release of whole genome sequences allow us to now assess the molecular interactions involved in microfibril assembly, and matrix regulation by fibrillins, from a new point of view. We can now start to understand how these molecules evolved to acquire their structure and regulatory functions (Fig. 5.10).

Acknowledgements We acknowledge funding from the Wellcome Trust (grant no. 086613), Arthritis Research UK (grant no. 19810) and the BBSRC.

References

Ades LC, Sreetharan D, Onikul E, Stockton V, Watson KC, Holman KJ (2002) Segregation of a novel FBN1 gene mutation, G1796E, with kyphoscoliosis and radiographic evidence of vertebral dysplasia in three generations. Am J Med Genet 109:261–270

Ahram D, Sato TS, Kohilan A, Tayeh M, Chen S, Leal S, Al-Salem M, El-Shanti H (2009) A homozygous mutation in ADAMTSL4 causes autosomal-recessive isolated ectopia lentis. Am J Hum Genet 84:274–278

Arteaga-Solis E, Gayraud B, Lee SY, Shum L, Sakai L, Ramirez F (2001) Regulation of limb patterning by extracellular microfibrils. J Cell Biol 154:275–281

Ashworth JL, Kelly V, Rock MJ, Shuttleworth CA, Kielty CM (1999a) Regulation of fibrillin carboxy-terminal furin processing by N-glycosylation, and association of amino- and carboxy-terminal sequences. J Cell Sci 112(Pt 22):4163–4171

Ashworth JL, Kelly V, Wilson R, Shuttleworth CA, Kielty CM (1999b) Fibrillin assembly: dimer formation mediated by amino-terminal sequences. J Cell Sci 112(Pt 20):3549–3558

Attias D, Stheneur C, Roy C, Collod-Beroud G, Detaint D, Faivre L, Delrue MA, Cohen L, Francannet C, Beroud C et al (2009) Comparison of clinical presentations and outcomes between patients with TGFBR2 and FBN1 mutations in Marfan syndrome and related disorders. Circulation 120:2541–2549

Baldock C, Koster AJ, Ziese U, Rock MJ, Sherratt MJ, Kadler KE, Shuttleworth CA, Kielty CM (2001) The supramolecular organization of fibrillin-rich microfibrils. J Cell Biol 152: 1045–1056

Baldock C, Siegler V, Bax DV, Cain SA, Mellody KT, Marson A, Haston JL, Berry R, Wang MC, Grossmann JG et al (2006) Nanostructure of fibrillin-1 reveals compact conformation of EGF arrays and mechanism for extensibility. Proc Natl Acad Sci U S A 103:11922–11927

Bax DV, Bernard SE, Lomas A, Morgan A, Humphries J, Shuttleworth CA, Humphries MJ, Kielty CM (2003) Cell adhesion to fibrillin-1 molecules and microfibrils is mediated by alpha 5 beta 1 and alpha v beta 3 integrins. J Biol Chem 278:34605–34616

Bax DV, Mahalingam Y, Cain S, Mellody K, Freeman L, Younger K, Shuttleworth CA, Humphries MJ, Couchman JR, Kielty CM (2007) Cell adhesion to fibrillin-1: identification of an Arg-Gly-Asp-dependent synergy region and a heparin-binding site that regulates focal adhesion formation. J Cell Sci 120:1383–1392

Birenheide R, Motokawa T (1994) Morphological Basis and Mechanics of Arm Movement in the Stalked Crinoid Metacrinus-Rotundus (Echinodermata, Crinoida). Mar Biol 121:273–283

Bishop B, Aricescu AR, Harlos K, O'Callaghan CA, Jones EY, Siebold C (2009) Structural insights into hedgehog ligand sequestration by the human hedgehog-interacting protein HHIP. Nat Struct Mol Biol 16:698–703

Bona CA, Murai C, Casares S, Kasturi K, Nishimura H, Honjo T, Matsuda F (1997) Structure of the mutant fibrillin-1 gene in the tight skin (TSK) mouse. DNA Res 4:267–271

Booms P, Pregla R, Ney A, Barthel F, Reinhardt DP, Pletschacher A, Mundlos S, Robinson PN (2005) RGD-containing fibrillin-1 fragments upregulate matrix metalloproteinase expression in cell culture: a potential factor in the pathogenesis of the Marfan syndrome. Hum Genet 116: 51–61

Bork P, Downing AK, Kieffer B, Campbell ID (1996) Structure and distribution of modules in extracellular proteins. Q Rev Biophys 29:119–167

Brown MA, Zhao Q, Baker KA, Naik C, Chen C, Pukac L, Singh M, Tsareva T, Parice Y, Mahoney A et al (2005) Crystal structure of BMP-9 and functional interactions with pro-region and receptors. J Biol Chem 280:25111–25118

Bussiere CT, Wright GM, DeMont ME (2006) The mechanical function and structure of aortic microfibrils in the lobster Homarus americanus. Comp Biochem Physiol A Mol Integr Physiol 143:417–428

Cain SA, Baldock C, Gallagher J, Morgan A, Bax DV, Weiss AS, Shuttleworth CA, Kielty CM (2005) Fibrillin-1 interactions with heparin. Implications for microfibril and elastic fiber assembly. J Biol Chem 280:30526–30537

Cain SA, Baldwin AK, Mahalingam Y, Raynal B, Jowitt TA, Shuttleworth CA, Couchman JR, Kielty CM (2008) Heparan sulfate regulates fibrillin-1 N- and C-terminal interactions. J Biol Chem 283:27017–27027

Cardy CM, Handford PA (1998) Metal ion dependency of microfibrils supports a rod-like conformation for fibrillin-1 calcium-binding epidermal growth factor-like domains. J Mol Biol 276:855–860

Charbonneau NL, Carlson EJ, Tufa S, Sengle G, Manalo EC, Carlberg VM, Ramirez F, Keene DR, Sakai LY (2010) In vivo studies of mutant fibrillin-1 microfibrils. J Biol Chem 285: 24943–24955

Chen Y, Ali T, Todorovic V, O'Leary JM, Kristina Downing A, Rifkin DB (2005) Amino acid requirements for formation of the TGF-beta-latent TGF-beta binding protein complexes. J Mol Biol 345:175–186

Choi Y, Chung H, Jung H, Couchman JR, Oh ES (2011) Syndecans as cell surface receptors: Unique structure equates with functional diversity. Matrix Biol 30:93–99

Coghlan A, Wolfe KH (2002) Fourfold faster rate of genome rearrangement in nematodes than in Drosophila. Genome Res 12:857–867

Cohn RD, van Erp C, Habashi JP, Soleimani AA, Klein EC, Lisi MT, Gamradt M, ap Rhys CM, Holm TM, Loeys BL et al (2007) Angiotensin II type 1 receptor blockade attenuates TGF-beta-induced failure of muscle regeneration in multiple myopathic states. Nat Med 13:204–210

Comeglio P, Evans AL, Brice G, Cooling RJ, Child AH (2002) Identification of FBN1 gene mutations in patients with ectopia lentis and marfanoid habitus. Br J Ophthalmol 86: 1359–1362

Cordle J, Johnson S, Tay JZ, Roversi P, Wilkin MB, de Madrid BH, Shimizu H, Jensen S, Whiteman P, Jin B et al (2008) A conserved face of the Jagged/Serrate DSL domain is involved in Notch trans-activation and cis-inhibition. Nat Struct Mol Biol 15:849–857

Corson GM, Chalberg SC, Dietz HC, Charbonneau NL, Sakai LY (1993) Fibrillin binds calcium and is coded by cDNAs that reveal a multidomain structure and alternatively spliced exons at the 5′ end. Genomics 17:476–484

Corson GM, Charbonneau NL, Keene DR, Sakai LY (2004) Differential expression of fibrillin-3 adds to microfibril variety in human and avian, but not rodent, connective tissues. Genomics 83:461–472

Davison IG, Wright GM, DeMont ME (1995) The structure and physical properties of invertebrate and primitive vertebrate arteries. J Exp Biol 198:2185–2196

Dietz HC, Cutting GR, Pyeritz RE, Maslen CL, Sakai LY, Corson GM, Puffenberger EG, Hamosh A, Nanthakumar EJ, Curristin SM et al (1991) Marfan syndrome caused by a recurrent de novo missense mutation in the fibrillin gene. Nature 352:337–339

Dietz HC, Loeys B, Carta L, Ramirez F (2005) Recent progress towards a molecular understanding of Marfan syndrome. Am J Med Genet C Semin Med Genet 139C:4–9

Dietz HC, McIntosh I, Sakai LY, Corson GM, Chalberg SC, Pyeritz RE, Francomano CA (1993) Four novel FBN1 mutations: significance for mutant transcript level and EGF-like domain calcium binding in the pathogenesis of Marfan syndrome. Genomics 17:468–475

Downing AK, Knott V, Werner JM, Cardy CM, Campbell ID, Handford PA (1996) Solution structure of a pair of calcium-binding epidermal growth factor-like domains: implications for the Marfan syndrome and other genetic disorders. Cell 85:597–605

El-Hallous E, Sasaki T, Hubmacher D, Getie M, Tiedemann K, Brinckmann J, Batge B, Davis EC, Reinhardt DP (2007) Fibrillin-1 interactions with fibulins depend on the first hybrid domain and provide an adaptor function to tropoelastin. J Biol Chem 282:8935–8946

Eriksen TA, Wright DM, Purslow PP, Duance VC (2001) Role of Ca(2+) for the mechanical properties of fibrillin. Proteins 45:90–95

Faivre L, Collod-Beroud G, Loeys BL, Child A, Binquet C, Gautier E, Callewaert B, Arbustini E, Mayer K, Arslan-Kirchner M et al (2007) Effect of mutation type and location on clinical outcome in 1,013 probands with Marfan syndrome or related phenotypes and FBN1 mutations: an international study. Am J Hum Genet 81:454–466

Faivre L, Gorlin RJ, Wirtz MK, Godfrey M, Dagoneau N, Samples JR, Le Merrer M, Collod-Beroud G, Boileau C, Munnich A, Cormier-Daire V (2003) In frame fibrillin-1 gene deletion in autosomal dominant Weill-Marchesani syndrome. J Med Genet 40:34–36

Faury G (2001) Function-structure relationship of elastic arteries in evolution: from microfibrils to elastin and elastic fibres. Pathol Biol 49:310–325

Fontana L, Chen Y, Prijatelj P, Sakai T, Fassler R, Sakai LY, Rifkin DB (2005) Fibronectin is required for integrin alphavbeta6-mediated activation of latent TGF-beta complexes containing LTBP-1. FASEB J 19:1798–1808

Gambini C, Abou B, Ponton A, Cornelissen AJ (2012) Micro- and macrorheology of jellyfish extracellular matrix. Biophys J 102:1–9

George EL, Georges-Labouesse EN, Patel-King RS, Rayburn H, Hynes RO (1993) Defects in mesoderm, neural tube and vascular development in mouse embryos lacking fibronectin. Development 119:1079–1091

Giltay R, Timpl R, Kostka G (1999) Sequence, recombinant expression and tissue localization of two novel extracellular matrix proteins, fibulin-3 and fibulin-4. Matrix Biol 18:469–480

Glab J, Wess T (2008) Changes in the molecular packing of fibrillin microfibrils during extension indicate intrafibrillar and interfibrillar reorganization in elastic response. J Mol Biol 383: 1171–1180

Gleizes PE, Beavis RC, Mazzieri R, Shen B, Rifkin DB (1996) Identification and characterization of an eight-cysteine repeat of the latent transforming growth factor-beta binding protein-1 that mediates bonding to the latent transforming growth factor-beta1. J Biol Chem 271: 29891–29896

Godfrey M (2004) Fibrillin-2 mutations in congenital contractural arachnodactyly. In: Robinson PN, Godfrey M (eds) Marfan syndrome: a primer for clinicians and scientists. Kluwer, New York, p 123

Goldblatt J, Hyatt J, Edwards C, Walpole I (2011) Further evidence for a marfanoid syndrome with neonatal progeroid features and severe generalized lipodystrophy due to frameshift mutations near the 3′ end of the FBN1 gene. Am J Med Genet 155A:717–720

Graul-Neumann LM, Kienitz T, Robinson PN, Baasanjav S, Karow B, Gillessen-Kaesbach G, Fahsold R, Schmidt H, Hoffmann K, Passarge E (2010) Marfan syndrome with neonatal progeroid syndrome-like lipodystrophy associated with a novel frameshift mutation at the 3′ terminus of the FBN1-gene. Am J Med Genet 152A:2749–2755

Gregory KE, Ono RN, Charbonneau NL, Kuo CL, Keene DR, Bachinger HP, Sakai LY (2005) The prodomain of BMP-7 targets the BMP-7 complex to the extracellular matrix. J Biol Chem 280:27970–27980

Habashi JP, Judge DP, Holm TM, Cohn RD, Loeys BL, Cooper TK, Myers L, Klein EC, Liu G, Calvi C et al (2006) Losartan, an AT1 antagonist, prevents aortic aneurysm in a mouse model of Marfan syndrome. Science 312:117–121

Hambleton S, Valeyev NV, Muranyi A, Knott V, Werner JM, McMichael AJ, Handford PA, Downing AK (2004) Structural and functional properties of the human notch-1 ligand binding region. Structure 12:2173–2183

Hasham SN, Guo DC, Milewicz DM (2002) Genetic basis of thoracic aortic aneurysms and dissections. Curr Opin Cardiol 17:677–683

Haston JL, Engelsen SB, Roessle M, Clarkson J, Blanch EW, Baldock C, Kielty CM, Wess TJ (2003) Raman microscopy and X-ray diffraction, a combined study of fibrillin-rich microfibrillar elasticity. J Biol Chem 278:41189–41197

Hatchwell E (1997) Shprintzen-Goldberg syndrome results from mutations in fibrillin-1, not monosomy 22q11. J Pediatr 131:164–165

Hatzirodos N, Bayne RA, Irving-Rodgers HF, Hummitzsch K, Sabatier L, Lee S, Bonner W, Gibson MA, Rainey WE, Carr BR et al (2011) Linkage of regulators of TGF-beta activity in the fetal ovary to polycystic ovary syndrome. FASEB J 25:2256–2265

Herpin A, Lelong C, Favrel P (2004) Transforming growth factor-beta-related proteins: an ancestral and widespread superfamily of cytokines in metazoans. Dev Comp Immunol 28: 461–485

Hirani R, Hanssen E, Gibson MA (2007) LTBP-2 specifically interacts with the amino-terminal region of fibrillin-1 and competes with LTBP-1 for binding to this microfibrillar protein. Matrix Biol 26:213–223

Horn D, Robinson PN (2010) Progeroid facial features and lipodystrophy associated with a novel splice site mutation in the final intron of the FBN1 gene. Am J Med Genet 155A:721–724

Hubmacher D, El-Hallous EI, Nelea V, Kaartinen MT, Lee ER, Reinhardt DP (2008) Biogenesis of extracellular microfibrils: Multimerization of the fibrillin-1 C terminus into bead-like structures enables self-assembly. Proc Natl Acad Sci U S A 105:6548–6553

Isogai Z, Aspberg A, Keene DR, Ono RN, Reinhardt DP, Sakai LY (2002) Versican interacts with fibrillin-1 and links extracellular microfibrils to other connective tissue networks. J Biol Chem 277:4565–4572

Isogai Z, Ono RN, Ushiro S, Keene DR, Chen Y, Mazzieri R, Charbonneau NL, Reinhardt DP, Rifkin DB, Sakai LY (2003) Latent transforming growth factor beta-binding protein 1 interacts with fibrillin and is a microfibril-associated protein. J Biol Chem 278:2750–2757

Jensen SA, Corbett AR, Knott V, Redfield C, Handford PA (2005) Ca2+–dependent interface formation in fibrillin-1. J Biol Chem 280:14076–14084

Jensen SA, Iqbal S, Lowe ED, Redfield C, Handford PA (2009) Structure and interdomain interactions of a hybrid domain: a disulphide-rich module of the fibrillin/LTBP superfamily of matrix proteins. Structure 17:759–768

Jensen SA, Reinhardt DP, Gibson MA, Weiss AS (2001) Protein interaction studies of MAGP-1 with tropoelastin and fibrillin-1. J Biol Chem 276:39661–39666

Jensen SA, Robertson IB, Handford PA (2012) Dissecting the fibrillin microfibril: structural insights into organization and function. Structure 20:215–225

Jones CJ, Sear CH, Grant ME (1980) An ultrastructural study of fibroblasts derived from bovine ligamentum nuchae and their capacity for elastogenesis in culture. J Pathol 131:35–53

Jordan CD, Bohling SD, Charbonneau NL, Sakai LY (2010) Fibrillins in human adult ovary and polycystic ovary syndrome: is fibrillin-3 affected in PCOS? J Histochem Cytochem 58:903–915

Jovanovic J, Iqbal S, Jensen S, Mardon H, Handford P (2008) Fibrillin-integrin interactions in health and disease. Biochem Soc Trans 36:257–262

Jovanovic J, Takagi J, Choulier L, Abrescia NG, Stuart DI, van der Merwe PA, Mardon HJ, Handford PA (2007) alphaVbeta6 is a novel receptor for human fibrillin-1. Comparative studies of molecular determinants underlying integrin-rgd affinity and specificity. J Biol Chem 282:6743–6751

Judge DP, Biery NJ, Keene DR, Geubtner J, Myers L, Huso DL, Sakai LY, Dietz HC (2004) Evidence for a critical contribution of haploinsufficiency in the complex pathogenesis of Marfan syndrome. J Clin Invest 114:172–181

Kahlem P, Newfeld SJ (2009) Informatics approaches to understanding TGFbeta pathway regulation. Development 136:3729–3740

Kantola AK, Keski-Oja J, Koli K (2008) Fibronectin and heparin binding domains of latent TGF-beta binding protein (LTBP)-4 mediate matrix targeting and cell adhesion. Exp Cell Res 314: 2488–2500

Keene DR, Maddox BK, Kuo HJ, Sakai LY, Glanville RW (1991) Extraction of extendable beaded structures and their identification as fibrillin-containing extracellular matrix microfibrils. J Histochem Cytochem 39:441–449

Kettle S, Card CM, Hutchinson S, Sykes B, Handford PA (2000) Characterisation of fibrillin-1 cDNA clones in a human fibroblast cell line that assembles microfibrils. Int J Biochem Cell Biol 32:201–214

Kettle S, Yuan X, Grundy G, Knott V, Downing AK, Handford PA (1999) Defective calcium binding to fibrillin-1: consequence of an N2144S change for fibrillin-1 structure and function. J Mol Biol 285:1277–1287

Kewley MA, Williams G, Steven FS (1978) Studies of elastic tissue formation in the developing bovine ligamentum nuchae. J Pathol 124:95–101

Kielty CM, Cummings C, Whittaker SP, Shuttleworth CA, Grant ME (1991) Isolation and ultrastructural analysis of microfibrillar structures from foetal bovine elastic tissues. Relative abundance and supramolecular architecture of type VI collagen assemblies and fibrillin. J Cell Sci 99(Pt 4):797–807

Kielty CM, Shuttleworth CA (1993) The role of calcium in the organization of fibrillin microfibrils. FEBS Lett 336:323–326

Kielty CM, Wess TJ, Haston L, Ashworth JL, Sherratt MJ, Shuttleworth CA (2002) Fibrillin-rich microfibrils: elastic biopolymers of the extracellular matrix. J Muscle Res Cell Motil 23: 581–596

Kinsey R, Williamson MR, Chaudhry S, Mellody KT, McGovern A, Takahashi S, Shuttleworth CA, Kielty CM (2008) Fibrillin-1 microfibril deposition is dependent on fibronectin assembly. J Cell Sci 121:2696–2704

Kumaratilake JS, Gibson MA, Fanning JC, Cleary EG (1989) The tissue distribution of microfibrils reacting with a monospecific antibody to MAGP, the major glycoprotein antigen of elastin-associated microfibrils. Eur J Cell Biol 50:117–127

Kuo CL, Isogai Z, Keene DR, Hazeki N, Ono RN, Sengle G, Peter Bachinger H, Sakai LY (2007) Effects of fibrillin-1 degradation on microfibril ultrastructure. J Biol Chem 282:4007–4020

Kutz WE, Wang LW, Bader HL, Majors AK, Iwata K, Traboulsi EI, Sakai LY, Keene DR, Apte SS (2011) ADAMTS10 protein interacts with fibrillin-1 and promotes its deposition in extracellular matrix of cultured fibroblasts. J Biol Chem 286:17156–17167

Labat-Robert J (2012) Cell-Matrix interactions, the role of fibronectin and integrins. A survey. Pathol Biol (Paris) 60:15–19

Labat-Robert J, Robert L, Auger C, Lethias C, Garrone R (1981) Fibronectin-like protein in Porifera: its role in cell aggregation. Proc Natl Acad Sci U S A 78:6261–6265

Lack J, O'Leary JM, Knott V, Yuan X, Rifkin DB, Handford PA, Downing AK (2003) Solution structure of the third TB domain from LTBP1 provides insight into assembly of the large latent complex that sequesters latent TGF-beta. J Mol Biol 334:281–291

Le Goff C, Cormier-Daire V (2011) The ADAMTS(L) family and human genetic disorders. Hum Mol Genet 20:R163–R167

Le Goff C, Mahaut C, Wang LW, Allali S, Abhyankar A, Jensen S, Zylberberg L, Collod-Beroud G, Bonnet D, Alanay Y et al (2011) Mutations in the TGFbeta binding-protein-like domain 5 of FBN1 are responsible for acromicric and geleophysic dysplasias. Am J Hum Genet 89:7–14

Lee B, Godfrey M, Vitale E, Hori H, Mattei MG, Sarfarazi M, Tsipouras P, Ramirez F, Hollister DW (1991) Linkage of Marfan syndrome and a phenotypically related disorder to two different fibrillin genes. Nature 352:330–334

Lee SJ, McPherron AC (2001) Regulation of myostatin activity and muscle growth. Proc Natl Acad Sci U S A 98:9306–9311

Lee SS, Knott V, Jovanovic J, Harlos K, Grimes JM, Choulier L, Mardon HJ, Stuart DI, Handford PA (2004) Structure of the integrin binding fragment from fibrillin-1 gives new insights into microfibril organization. Structure 12:717–729

Lin G, Tiedemann K, Vollbrandt T, Peters H, Batge B, Brinckmann J, Reinhardt DP (2002) Homo- and heterotypic fibrillin-1 and −2 interactions constitute the basis for the assembly of microfibrils. J Biol Chem 277:50795–50804

Liu W, Qian C, Comeau K, Brenn T, Furthmayr H, Francke U (1996) Mutant fibrillin-1 monomers lacking EGF-like domains disrupt microfibril assembly and cause severe marfan syndrome. Hum Mol Genet 5:1581–1587

Loeys BL, Chen J, Neptune ER, Judge DP, Podowski M, Holm T, Meyers J, Leitch CC, Katsanis N, Sharifi N et al (2005) A syndrome of altered cardiovascular, craniofacial, neurocognitive and skeletal development caused by mutations in TGFBR1 or TGFBR2. Nat Genet 37:275–281

Loeys BL, Gerber EE, Riegert-Johnson D, Iqbal S, Whiteman P, McConnell V, Chillakuri CR, Macaya D, Coucke PJ, De Paepe A et al (2010) Mutations in fibrillin-1 cause congenital scleroderma: stiff skin syndrome. Sci Transl Med 2:23ra20

Lönnqvist, L, Reinhardt, D, Sakai, L, Peltonen, L (1998) Evidence for furin-type activity-mediated C-terminal processing of profibrillin-1 and interference in the processing by certain mutations. Hum Mol Genet 7(13):2039–2044

Lu Y, Holmes DF, Baldock C (2005) Evidence for the intramolecular pleating model of fibrillin microfibril organisation from single particle image analysis. J Mol Biol 349:73–85

Malak TM, Bell SC (1994) Distribution of fibrillin-containing microfibrils and elastin in human fetal membranes: a novel molecular basis for membrane elasticity. Am J Obstet Gynecol 171:195–205

Marson A, Rock MJ, Cain SA, Freeman LJ, Morgan A, Mellody K, Shuttleworth CA, Baldock C, Kielty CM (2005) Homotypic fibrillin-1 interactions in microfibril assembly. J Biol Chem 280: 5013–5021

Maslen CL, Corson GM, Maddox BK, Glanville RW, Sakai LY (1991) Partial sequence of a candidate gene for the Marfan syndrome. Nature 352:334–337

Massam-Wu T, Chiu M, Choudhury R, Chaudhry SS, Baldwin AK, McGovern A, Baldock C, Shuttleworth CA, Kielty CM (2010) Assembly of fibrillin microfibrils governs extracellular deposition of latent TGF beta. J Cell Sci 123:3006–3018

McConnell CJ, DeMont ME, Wright GM (1997) Microfibrils provide non-linear elastic behaviour in the abdominal artery of the lobster Homarus americanus. J Physiol 499(Pt 2):513–526

McGettrick AJ, Knott V, Willis A, Handford PA (2000) Molecular effects of calcium binding mutations in Marfan syndrome depend on domain context. Hum Mol Genet 9:1987–1994

Megill WM, Gosline JM, Blake RW (2005) The modulus of elasticity of fibrillin-containing elastic fibres in the mesoglea of the hydromedusa Polyorchis penicillatus. J Exp Biol 208:3819–3834

Milewicz DM, Grossfield J, Cao SN, Kielty C, Covitz W, Jewett T (1995) A mutation in FBN1 disrupts profibrillin processing and results in isolated skeletal features of the Marfan syndrome. J Clin Invest 95:2373–2378

Miyazono K, Olofsson A, Colosetti P, Heldin CH (1991) A role of the latent TGF-beta 1-binding protein in the assembly and secretion of TGF-beta 1. EMBO J 10:1091–1101

Morales J, Al-Sharif L, Khalil DS, Shinwari JM, Bavi P, Al-Mahrouqi RA, Al-Rajhi A, Alkuraya FS, Meyer BF, Al Tassan N (2009) Homozygous mutations in ADAMTS10 and ADAMTS17 cause lenticular myopia, ectopia lentis, glaucoma, spherophakia, and short stature. Am J Hum Genet 85:558–568

Neptune ER, Frischmeyer PA, Arking DE, Myers L, Bunton TE, Gayraud B, Ramirez F, Sakai LY, Dietz HC (2003) Dysregulation of TGF-beta activation contributes to pathogenesis in Marfan syndrome. Nat Genet 33:407–411

Ng CM, Cheng A, Myers LA, Martinez-Murillo F, Jie C, Bedja D, Gabrielson KL, Hausladen JM, Mecham RP, Judge DP, Dietz HC (2004) TGF-beta-dependent pathogenesis of mitral valve prolapse in a mouse model of Marfan syndrome. J Clin Invest 114:1586–1592

Nistala H, Lee-Arteaga S, Siciliano G, Smaldone S, Ramirez F (2010a) Extracellular regulation of transforming growth factor beta and bone morphogenetic protein signaling in bone. Ann N Y Acad Sci 1192:253–256

Nistala H, Lee-Arteaga S, Smaldone S, Siciliano G, Carta L, Ono RN, Sengle G, Arteaga-Solis E, Levasseur R, Ducy P et al (2010b) Fibrillin-1 and −2 differentially modulate endogenous TGF-beta and BMP bioavailability during bone formation. J Cell Biol 190:1107–1121

Ono RN, Sengle G, Charbonneau NL, Carlberg V, Bachinger HP, Sasaki T, Lee-Arteaga S, Zilberberg L, Rifkin DB, Ramirez F et al (2009) Latent transforming growth factor beta-binding proteins and fibulins compete for fibrillin-1 and exhibit exquisite specificities in binding sites. J Biol Chem 284:16872–16881

Penttinen C, Saharinen J, Weikkolainen K, Hyytiainen M, Keski-Oja J (2002) Secretion of human latent TGF-beta-binding protein-3 (LTBP-3) is dependent on co-expression of TGF-beta. J Cell Sci 115:3457–3468

Pereira L, D'Alessio M, Ramirez F, Lynch JR, Sykes B, Pangilinan T, Bonadio J (1993) Genomic organization of the sequence coding for fibrillin, the defective gene product in Marfan syndrome. Hum Mol Genet 2:961–968

Pereira, L, Andrikopoulos, K, Tian, J, Lee, SY, Keene, DR, Ono, R, Reinhardt, DP, Sakai, LY, Biery, NJ, Bunton, T, Dietz, HC, Ramirez, F (1997) Targetting of the gene encoding fibrillin-1 recapitulates the vascular aspect of Marfan syndrome. Nat Genet 17:218–22

Pereira L, Lee SY, Gayraud B, Andrikopoulos K, Shapiro SD, Bunton T, Biery NJ, Dietz HC, Sakai LY, Ramirez F (1999) Pathogenetic sequence for aneurysm revealed in mice underexpressing fibrillin-1. Proc Natl Acad Sci U S A 96:3819–3823

Pfaff M, Reinhardt DP, Sakai LY, Timpl R (1996) Cell adhesion and integrin binding to recombinant human fibrillin-1. FEBS Lett 384:247–250

Piha-Gossack A, Sossin W, Reinhardt DP (2012) The evolution of extracellular fibrillins and their functional domains. PLoS One 7:e33560

Prodoehl MJ, Hatzirodos N, Irving-Rodgers HF, Zhao ZZ, Painter JN, Hickey TE, Gibson MA, Rainey WE, Carr BR, Mason HD et al (2009) Genetic and gene expression analyses of the polycystic ovary syndrome candidate gene fibrillin-3 and other fibrillin family members in human ovaries. Mol Hum Reprod 15:829–841

Putnam EA, Zhang H, Ramirez F, Milewicz DM (1995) Fibrillin-2 (FBN2) mutations result in the Marfan-like disorder, congenital contractural arachnodactyly. Nat Genet 11:456–458

Quondamatteo F, Reinhardt DP, Charbonneau NL, Pophal G, Sakai LY, Herken R (2002) Fibrillin-1 and fibrillin-2 in human embryonic and early fetal development. Matrix Biol 21: 637–646

Raghunath M, Putnam EA, Ritty T, Hamstra D, Park ES, Tschodrich-Rotter M, Peters R, Rehemtulla A, Milewicz DM (1999) Carboxy-terminal conversion of profibrillin to fibrillin at a basic site by PACE/furin-like activity required for incorporation in the matrix. J Cell Sci 112(Pt 7):1093–1100

Raja-Khan N, Kunselman AR, Demers LM, Ewens KG, Spielman RS, Legro RS (2010) A variant in the fibrillin-3 gene is associated with TGF-beta and inhibin B levels in women with polycystic ovary syndrome. Fertil Steril 94:2916–2919

Ramirez F, Rifkin DB (2009) Extracellular microfibrils: contextual platforms for TGFbeta and BMP signaling. Curr Opin Cell Biol 21:616–622

Ranz JM, Casals F, Ruiz A (2001) How malleable is the eukaryotic genome? Extreme rate of chromosomal rearrangement in the genus Drosophila. Genome Res 11:230–239

Rao Z, Handford P, Mayhew M, Knott V, Brownlee GG, Stuart D (1995) The structure of a Ca(2+)-binding epidermal growth factor-like domain: its role in protein-protein interactions. Cell 82:131–141

Reber-Muller S, Spissinger T, Schuchert P, Spring J, Schmid V (1995) An extracellular matrix protein of jellyfish homologous to mammalian fibrillins forms different fibrils depending on the life stage of the animal. Dev Biol 169:662–672

Reinhardt DP, Gambee JE, Ono RN, Bachinger HP, Sakai LY (2000) Initial steps in assembly of microfibrils. Formation of disulfide-cross-linked multimers containing fibrillin-1. J Biol Chem 275:2205–2210

Reinhardt DP, Keene DR, Corson GM, Poschl E, Bachinger HP, Gambee JE, Sakai LY (1996a) Fibrillin-1: organization in microfibrils and structural properties. J Mol Biol 258:104–116

Reinhardt DP, Mechling DE, Boswell BA, Keene DR, Sakai LY, Bachinger HP (1997a) Calcium determines the shape of fibrillin. J Biol Chem 272:7368–7373

Reinhardt DP, Ono RN, Sakai LY (1997b) Calcium stabilizes fibrillin-1 against proteolytic degradation. J Biol Chem 272:1231–1236

Reinhardt DP, Sasaki T, Dzamba BJ, Keene DR, Chu ML, Gohring W, Timpl R, Sakai LY (1996b) Fibrillin-1 and fibulin-2 interact and are colocalized in some tissues. J Biol Chem 271: 19489–19496

Ritty TM, Broekelmann T, Tisdale C, Milewicz DM, Mecham RP (1999) Processing of the fibrillin-1 carboxyl-terminal domain. J Biol Chem 274:8933–8940

Ritty TM, Broekelmann TJ, Werneck CC, Mecham RP (2003) Fibrillin-1 and −2 contain heparin-binding sites important for matrix deposition and that support cell attachment. Biochem J 375: 425–432

Robertson I, Jensen S, Handford P (2011) TB domain proteins: evolutionary insights into the multifaceted roles of fibrillins and LTBPs. Biochem J 433:263–276

Robinson PN, Arteaga-Solis E, Baldock C, Collod-Beroud G, Booms P, De Paepe A, Dietz HC, Guo G, Handford PA, Judge DP et al (2006) The molecular genetics of Marfan syndrome and related disorders. J Med Genet 43:769–787

Robinson PN, Godfrey M (2000) The molecular genetics of Marfan syndrome and related microfibrillopathies. J Med Genet 37:9–25

Rock MJ, Cain SA, Freeman LJ, Morgan A, Mellody K, Marson A, Shuttleworth CA, Weiss AS, Kielty CM (2004) Molecular basis of elastic fiber formation. Critical interactions and a tropoelastin-fibrillin-1 cross-link. J Biol Chem 279:23748–23758

Ross R, Bornstein P (1969) The elastic fiber. I. The separation and partial characterization of its macromolecular components. J Cell Biol 40:366–381

Sabatier L, Chen D, Fagotto-Kaufmann C, Hubmacher D, McKee MD, Annis DS, Mosher DF, Reinhardt DP (2009) Fibrillin assembly requires fibronectin. Mol Biol Cell 20:846–858

Sabatier L, Miosge N, Hubmacher D, Lin G, Davis EC, Reinhardt DP (2011) Fibrillin-3 expression in human development. Matrix Biol 30:43–52

Sage EH, Gray WR (1977) Evolution of elastin structure. Adv Exp Med Biol 79:291–312

Saharinen J, Keski-Oja J (2000) Specific sequence motif of 8-Cys repeats of TGF-beta binding proteins, LTBPs, creates a hydrophobic interaction surface for binding of small latent TGF-beta. Mol Biol Cell 11:2691–2704

Saharinen J, Taipale J, Keski-Oja J (1996) Association of the small latent transforming growth factor-beta with an eight cysteine repeat of its binding protein LTBP-1. EMBO J 15:245–253

Saito M, Kurokawa M, Oda M, Oshima M, Tsutsui K, Kosaka K, Nakao K, Ogawa M, Manabe R, Suda N et al (2011) ADAMTSL6beta protein rescues fibrillin-1 microfibril disorder in a Marfan syndrome mouse model through the promotion of fibrillin-1 assembly. J Biol Chem 286:38602–38613

Sakai LY, Keene DR, Engvall E (1986) Fibrillin, a new 350-kD glycoprotein, is a component of extracellular microfibrils. J Cell Biol 103:2499–2509

Sakai LY, Keene DR, Glanville RW, Bachinger HP (1991) Purification and partial characterization of fibrillin, a cysteine-rich structural component of connective tissue microfibrils. J Biol Chem 266:14763–14770

Sakamoto H, Broekelmann T, Cheresh DA, Ramirez F, Rosenbloom J, Mecham RP (1996) Cell-type specific recognition of RGD- and non-RGD-containing cell binding domains in fibrillin-1. J Biol Chem 271:4916–4922

Sarras MP Jr, Madden ME, Zhang XM, Gunwar S, Huff JK, Hudson BG (1991) Extracellular matrix (mesoglea) of Hydra vulgaris. I. Isolation and characterization. Dev Biol 148:481–494

Sarras MP Jr, Zhang X, Huff JK, Accavitti MA, St John PL, Abrahamson DR (1993) Extracellular matrix (mesoglea) of Hydra vulgaris III. Formation and function during morphogenesis of hydra cell aggregates. Dev Biol 157:383–398

Segars JH, DeCherney AH (2010) Is there a genetic basis for polycystic ovary syndrome? J Clin Endocrinol Metabol 95:2058–2060

Sengle G, Charbonneau NL, Ono RN, Sasaki T, Alvarez J, Keene DR, Bachinger HP, Sakai LY (2008a) Targeting of bone morphogenetic protein growth factor complexes to fibrillin. J Biol Chem 283:13874–13888

Sengle G, Ono RN, Lyons KM, Bachinger HP, Sakai LY (2008b) A new model for growth factor activation: type II receptors compete with the prodomain for BMP-7. J Mol Biol 381: 1025–1039

Sengle G, Ono RN, Sasaki T, Sakai LY (2011) Prodomains of transforming growth factor beta (TGFbeta) superfamily members specify different functions: extracellular matrix interactions and growth factor bioavailability. J Biol Chem 286:5087–5099

Sengle G, Tsutsui K, Keene DR, Tufa SF, Carlson EJ, Charbonneau NL, Ono RN, Sasaki T, Wirtz MK, Samples JR et al (2012) Microenvironmental regulation by fibrillin-1. PLoS Genet 8:e1002425

Shimizu H, Zhang X, Zhang J, Leontovich A, Fei K, Yan L, Sarras MP Jr (2002) Epithelial morphogenesis in hydra requires de novo expression of extracellular matrix components and matrix metalloproteinases. Development 129:1521–1532

Smallridge RS, Whiteman P, Doering K, Handford PA, Downing AK (1999) EGF-like domain calcium affinity modulated by N-terminal domain linkage in human fibrillin-1. J Mol Biol 286:661–668

Smallridge RS, Whiteman P, Werner JM, Campbell ID, Handford PA, Downing AK (2003) Solution structure and dynamics of a calcium binding epidermal growth factor-like domain pair from the neonatal region of human fibrillin-1. J Biol Chem 278:12199–12206

Stheneur C, Faivre L, Collod-Beroud G, Gautier E, Binquet C, Bonithon-Kopp C, Claustres M, Child AH, Arbustini E, Ades LC et al (2011) Prognosis factors in probands with an FBN1 mutation diagnosed before the age of 1 year. Pediatr Res 69:265–270

Suk JY, Jensen S, McGettrick A, Willis AC, Whiteman P, Redfield C, Handford PA (2004) Structural consequences of cysteine substitutions C1977Y and C1977R in calcium-binding epidermal growth factor-like domain 30 of human fibrillin-1. J Biol Chem 279:51258–51265

Thurmond F, Trotter J (1996) Morphology and biomechanics of the microfibrillar network of sea cucumber dermis. J Exp Biol 199:1817–1828

Tiecke F, Katzke S, Booms P, Robinson PN, Neumann L, Godfrey M, Mathews KR, Scheuner M, Hinkel GK, Brenner RE et al (2001) Classic, atypically severe and neonatal Marfan syndrome: twelve mutations and genotype-phenotype correlations in FBN1 exons 24–40. Eur J Hum Genet 9:13–21

Tiedemann K, Batge B, Muller PK, Reinhardt DP (2001) Interactions of fibrillin-1 with heparin/heparan sulfate, implications for microfibrillar assembly. J Biol Chem 276:36035–36042

Tiedemann K, Sasaki T, Gustafsson E, Gohring W, Batge B, Notbohm H, Timpl R, Wedel T, Schlotzer-Schrehardt U, Reinhardt DP (2005) Microfibrils at basement membrane zones interact with perlecan via fibrillin-1. J Biol Chem 280:11404–11412

Trask TM, Ritty TM, Broekelmann T, Tisdale C, Mecham RP (1999) N-terminal domains of fibrillin 1 and fibrillin 2 direct the formation of homodimers: a possible first step in microfibril assembly. Biochem J 340(Pt 3):693–701

Trask TM, Trask BC, Ritty TM, Abrams WR, Rosenbloom J, Mecham RP (2000) Interaction of tropoelastin with the amino-terminal domains of fibrillin-1 and fibrillin-2 suggests a role for the fibrillins in elastic fiber assembly. J Biol Chem 275:24400–24406

Tsuruga E, Sato A, Ueki T, Nakashima K, Nakatomi Y, Ishikawa H, Yajima T, Sawa Y (2009) Integrin alphavbeta3 regulates microfibril assembly in human periodontal ligament cells. Tissue Cell 41:85–89

Tsutsui K, Manabe R, Yamada T, Nakano I, Oguri Y, Keene DR, Sengle G, Sakai LY, Sekiguchi K (2010) ADAMTSL-6 is a novel extracellular matrix protein that binds to fibrillin-1 and promotes fibrillin-1 fibril formation. J Biol Chem 285:4870–4882

Unsold C, Hyytiainen M, Bruckner-Tuderman L, Keski-Oja J (2001) Latent TGF-beta binding protein LTBP-1 contains three potential extracellular matrix interacting domains. J Cell Sci 114:187–197

Urbanek M, Sam S, Legro RS, Dunaif A (2007) Identification of a polycystic ovary syndrome susceptibility variant in fibrillin-3 and association with a metabolic phenotype. J Clin Endocrinol Metabol 92:4191–4198

Wallis DD, Putnam EA, Cretoiu JS, Carmical SG, Cao SN, Thomas G, Milewicz DM (2003) Profibrillin-1 maturation by human dermal fibroblasts: proteolytic processing and molecular chaperones. J Cell Biochem 90:641–652

Wang MC, Lu Y, Baldock C (2009) Fibrillin microfibrils: a key role for the interbead region in elasticity. J Mol Biol 388:168–179

Werner JM, Knott V, Handford PA, Campbell ID, Downing AK (2000) Backbone dynamics of a cbEGF domain pair in the presence of calcium. J Mol Biol 296:1065–1078

Wess TJ, Purslow PP, Sherratt MJ, Ashworth J, Shuttleworth CA, Kielty CM (1998) Calcium determines the supramolecular organization of fibrillin-rich microfibrils. J Cell Biol 141: 829–837

Yalamanchi SK, Sam S, Cardenas MO, Holaday LW, Urbanek M, Dunaif A (2012) Association of fibrillin-3 and transcription factor-7-like 2 gene variants with metabolic phenotypes in PCOS. Obesity 20(6):1273–1278

Yuan X, Downing AK, Knott V, Handford PA (1997) Solution structure of the transforming growth factor beta-binding protein-like module, a domain associated with matrix fibrils. EMBO J 16:6659–6666

Yuan X, Werner JM, Lack J, Knott V, Handford PA, Campbell ID, Downing AK (2002) Effects of the N2144S mutation on backbone dynamics of a TB-cbEGF domain pair from human fibrillin-1. J Mol Biol 316:113–125

Zhang H, Hu W, Ramirez F (1995) Developmental expression of fibrillin genes suggests heterogeneity of extracellular microfibrils. J Cell Biol 129:1165–1176

Chapter 6
Molecular Evolution of the Microfibril-Associated Proteins: The Fibulins and the MAGPs

Fernando Segade

Abstract This chapter reviews the evolutionary histories of two representative families of microfibril-associated proteins as distinctive examples of adding evolutionary novelty to the ancient microfibril genetic module. The microfibril-associated glycoproteins (MAGPs), a recent addition to the microfibril, evolved from a *MAGP1-like* gene in basal vertebrates. The extant *MAGP1* and *MAGP2* genes arose in a multigene duplication event, with functional divergence mediated by relaxation of natural selection and correlated evolution between *MAGP2* and *Notch1*. Unique MAGP1 functions in bony vertebrates likely evolved by gain of function at the N-terminus. The fibulins (FBLNs) appeared at the base of the metazoans, contemporaneously with the *Fibrillin* gene. Diversification accelerated in chordates and vertebrates, to a total of nine *FBLN* genes. The vertebrate *FBLN3/4/5* clade of elastogenic FBLNs evolved from a chordate ancestor by two gene duplication events in quick succession. Accelerated evolution in sequence and expression patterns occurred preferentially on the *FBLN5* branch. Coevolution of functional motifs in *FBLN3/4/5* with their partners in elastic fiber deposition contributed to the nonredundant specialization of the elastogenic FBLNs.

6.1 Introduction

In multicellular animals, the microfibrils (MFs) are ubiquitous features in the extracellular matrix (ECM) of most tissues (Hubmacher and Reinhardt 2011; Wagenseil and Mecham 2007). MFs make an essential contribution to the structural integrity of tissues, anchor cells to matrix components through cell adhesion motifs, sequester growth factors in the ECM, and signal to cells by interacting with

---

F. Segade (✉)
Department of Anatomy and Cell Biology, University of Pennsylvania School of Dental Medicine, 240 South 40th Street, Philadelphia, PA 19104, USA
e-mail: fernando.segade@gmail.com

F.W. Keeley and R.P. Mecham (eds.), *Evolution of Extracellular Matrix*, Biology of Extracellular Matrix, DOI 10.1007/978-3-642-36002-2_6,

cell-surface receptors (Hubmacher and Reinhardt 2011; Wagenseil and Mecham 2007). The major structural proteins of the MF are the fibrillins, large modular glycoproteins consisting of tandem arrays of calcium-binding epidermal growth factor (cbEGF)-like modules interrupted by distinctive 8-cysteine domains (Hubmacher and Reinhardt 2011; Kielty et al. 2005; Wagenseil and Mecham 2007). Ultrastructurally, the MFs present an extraordinarily conserved morphology of 10–12 nm filaments with periodic "knots" that give the structure its characteristic "beads-on-a-string" appearance (Hubmacher and Reinhardt 2011). In vertebrates, MFs are especially abundant in elastic tissues, bone, and cartilage (Mecham and Davis 1994), where they serve as the scaffold for elastin deposition to build elastic fibers (Kielty et al. 2002; Wagenseil and Mecham 2007). Importantly, the fibrillin MFs possess intrinsic elastic properties even in the absence of elastin or elastin-like proteins (McConnell et al. 1996, 1997). This ability to endow tissues with mechanical recoil is almost certainly the MF ancestral function. The evolutionary origins of the *Fibrillin* gene and of the MF structure predate the divergence of Cnidaria (coelenterates; jellyfish, sea anemones, and allies) and Bilateria (Putnam et al. 2007; Reber-Muller et al. 1995).

In addition to their elastic function, the MFs serve as the functional locus for a number of proteins, known collectively as microfibril-associated proteins, of which more than 35 have been described (Kielty et al. 2002). Some MF-associated proteins contribute to the biomechanics of the MFs and elastic fibers, with specialized roles in deposition, post-secretion modification, and higher-order assembly of the complex. Among the better-characterized examples of this class of microfibrillar proteins, we find the fibulins (FBLNs), emilins, and lysyl oxidases (LOXs) (Kielty et al. 2002). A second class of MF-associated proteins associates to the MF without a direct involvement in its structural function. Growth factor-binding proteins such as the microfibril-associated glycoproteins (MAGPs) and latent transforming growth factor β-binding proteins (LTBPs), which modulate the availability of growth factors in tissues, belong to this category (Kielty et al. 2002; Ramirez and Dietz 2007; Wagenseil and Mecham 2007).

From both the functional and evolutionary standpoints, the MF is a genetic module, a suite of genes whose products form a semiautonomous unit with a high density of functional linkages (Fraser 2006). Genetic modules are characterized by a relatively small set of "core" genes coding for essential components and a variable number of ancillary genes that in varying degrees may be dispensable for the genetic module's core function. Genetic modules commonly follow a dual mode of evolution, with the core genes evolving at a slow rate to ensure continuity of structure and function, while the associated components may evolve at accelerated rates when there is a requirement for adaptive flexibility (Hartwell et al. 1999). This is clearly seen in the case of the MFs, where the fibrillin gene shows low rates of sequence evolution and gene diversification (Robertson et al. 2011), resulting in highly conserved morphology and biochemical properties in lineages separated for more than half a billion years of divergent evolution. In contrast, typical MF-associated protein gene families like the *FBLN*s show elevated rates of sequence evolution, domain shuffling, and gene gains and losses (Segade 2010).

Since genes within a functional module tend to coevolve (Chen and Dokholyan 2006; Fraser et al. 2004), the study of the molecular evolutionary history of the auxiliary components of the MF should provide a much welcomed view of the evolution of the MF as a whole, including the elucidation of the critical evolutionary steps that allowed this ancient module to adapt to the organizational complexity of the vertebrates. To this end, we review next the evolutionary history of two representative MF-associated protein gene families, the *MAGP*s and the *FBLN*s.

The *MAGP*s are a comparatively recent evolutionary novelty, found only in vertebrates, whereas the *FBLN* family is evolutionarily ancient, with an origin contemporaneous with the *Fibrillin* gene and the MF itself. Both MAGPs and FBLNs have acquired novel roles in vertebrates while maintaining a close functional relationship with the MF. Together they represent two very different instances of how the process of apposition of new gene products to the ancestral fibrillin core characterized the evolution of the MF.

6.2 The Microfibril-Associated Glycoprotein Gene Family

6.2.1 Introduction to the MAGPs

The MAGPs, MAGP1 (Gibson et al. 1991) and MAGP2 (Gibson et al. 1996), are low-molecular weight components of the MF (Gibson et al. 1991; Henderson et al. 1996) forming a unique family of microfibrillar proteins with no significant similarity to other components of the ECM. I must note here that the nomenclature of the MAGPs is confusing and in need of review. Two other proteins with the designation "MAGP," namely, MAGP-3 and MAGP-36, bear no structural or phylogenetic relation to MAGP1 or MAGP2. Equally confusing, the genes for MAGP1 and MAGP2 have been assigned the designations *MFAP2* (from *Microfibril-associated protein 2*) and *MFAP5*, respectively, although they are not phylogenetically related to three other *MFAP* genes (*MFAP1*, *MFAP3*, and *MFAP4*). Here I refer to the genes coding for MAGP1 and MAGP2 as the "*MAGP* gene family," and, to avoid unnecessary confusion, I will use the designation *MAGP1* and *MAGP2* for their genes (including in rodents and zebrafish where it is customary not to use all-capital gene names).

MAGP1 is a 183-residue protein with two distinctive domains (Fig. 6.1): an N-terminal region spanning almost half of the mature protein sequence, highly acidic and enriched in proline and glutamine (Fig. 6.1), and a cysteine-rich C-terminal half (Fig. 6.1) that contains the matrix-binding domain (MBD) which in mammals anchors MAGP1 to the ECM (Segade et al. 2002) and mediates its interaction with fibrillin (Werneck et al. 2004). In MAGP1, the MBD is 54 residues in length and includes seven essential cysteines (Segade et al. 2002). Many ECM proteins contain cysteine-stabilized folds, like the tandem arrays of cbEGF modules present in FBLNs, fibrillins, and LTBPs, domains which provide their characteristic rigidity and fibrillar conformation. However, the MAGP MBD represents a unique invention of the MAGPs (Segade 2009). MAGP1 is widely expressed in mammalian tissues, mainly in mesenchymal and connective tissue (Chen et al. 1993), and associates with the MFs except in those located adjacent to the plasma membrane of aortic endothelial cells and at the junction of the zonule and the lens capsule of the eye (Davis 1994; Henderson et al. 1996; Kumaratilake et al. 1989). MAGP1 localizes to the bead structure of the MF (Henderson et al. 1996) and binds tropoelastin and fibrillin, among other proteins. This led to the idea that MAGP1 might function as a bridge between fibrillin and elastin during elastic fiber assembly. However, recent reports have widened our view of MAGP1. Manipulation of *MAGP1* gene expression in zebrafish produced malformations in cranial and caudal blood vessels (Chen et al. 2006) and reduced growth of the hyaline-retinal vasculature (Alvarez et al. 2007), indicative of a wider role of MAGP1 in development. The vascular phenotype was not replicated in the null *MAGP1* mouse, where developmental defects included bone abnormalities, increased fat deposition, bleeding diathesis, and impaired response to injury (Craft et al. 2010, 2011; Weinbaum et al. 2008; Werneck et al. 2008). No elastic phenotype was present. Together, these observations suggested that MAGP1 is not required for elastin deposition, but it is mainly involved in processes of tissue homeostasis or differentiation.

MAGP2 is a 170–173-residue protein with high similarity to MAGP1 in the region corresponding to the MBD (Fig. 6.1) (Gibson et al. 1996; Hatzinikolas and Gibson 1998; Segade 2009; Segade et al. 2000, 2002). Outside of the MBD, MAGP2 is rich in serine and threonine and contains an RGD motif that binds the $\alpha v\beta 3$ integrin (Fig. 6.1) (Gibson et al. 1999). MAGP2 lacks the proline-, glutamine-, and tyrosine-rich sequences characteristic of MAGP1. Although MAGP2 was found in both elastin-associated and -free MFs, its tissue and developmental pattern of expression are more restricted than those of MAGP1 (Frankfater et al. 2000; Gibson et al. 1998). The presence of an RGD motif (Gibson et al. 1999) and the ability to stimulate elastic fiber assembly (Lemaire et al. 2007) suggested that MAGP2 anchors the MF to the cell surface (Wagenseil and Mecham 2007). That MAGP2 is also involved in tissue growth and development was demonstrated by elevated proliferation rate and invasiveness shown by MAGP2-expressing endothelial cells (Albig et al. 2007), an effect mediated by the blocking of Notch signaling (Albig et al. 2008). MAGP2 binds to and induces the shedding of Jagged1 and Notch1 (Miyamoto et al. 2006; Nehring et al. 2005), acting as a negative (Albig et al. 2008) or positive (Nehring et al. 2005) regulator of the Notch signaling pathway in an apparent cell-type-dependent manner.

6.2.2 Origin and Evolution of the MAGP Genes

In relative terms, the *MAGP* genes represent a recent addition to the MF complex. Based on the identification of a *MAGP1* homolog with a recognizable MAGP1 MBD in the genome of the sea lamprey, an agnathan (jawless) vertebrate (Segade

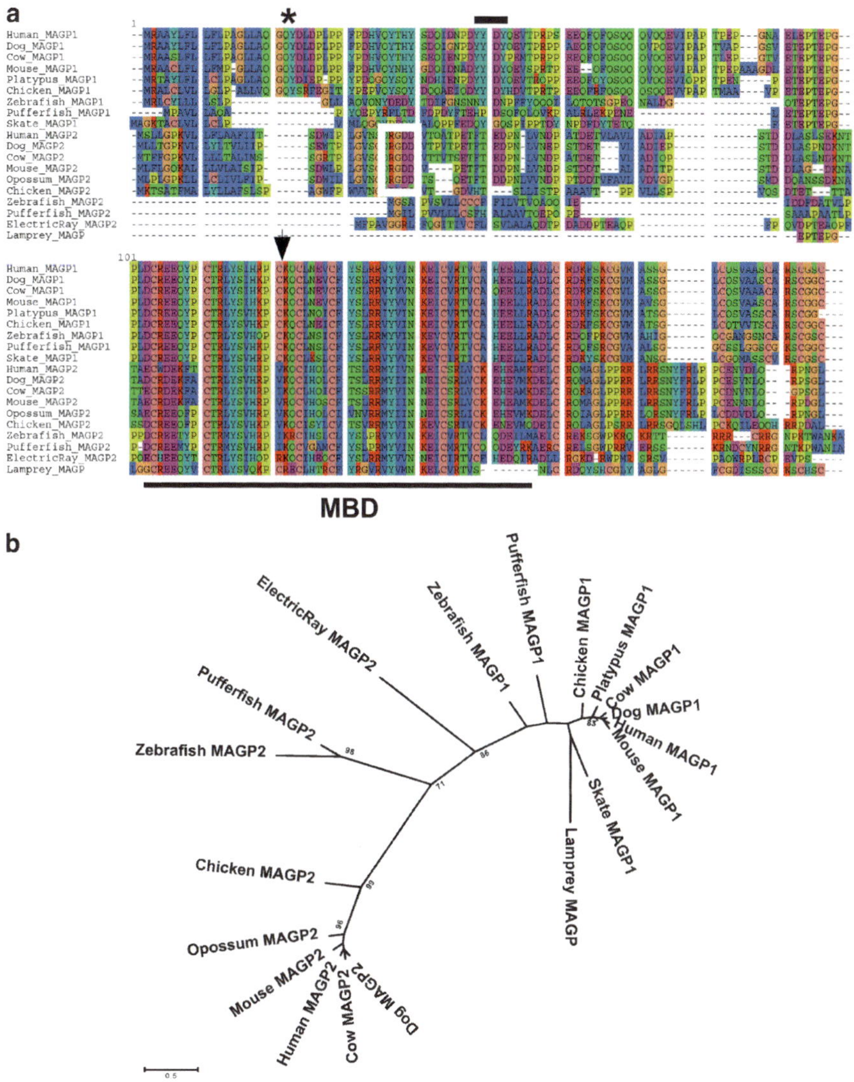

Fig. 6.1 (**a**) Clustal W alignment of amino acid sequences of vertebrate MAGP1 and MAGP2 proteins. The *asterisk* indicates the position of the Gln-20 transglutamination site in MAGP1; the *horizontal bar* above the *upper panel* indicates the tyrosination sites in MAGP1; the *arrowhead* shows the conserved Cys-104 in MAGP1 and lost in MAGP2; the *white-bordered rectangle* indicates the RGD motif in mammalian MAGP2; and a horizontal line under the *lower panel* highlights the matrix-binding domain (MBD). (**b**) Phylogenetic reconstruction of the evolutionary history of the *MAGP* genes inferred by maximum likelihood on the Tamura 3-parameter model (Tamura 1992). The tree with the highest log likelihood (−7321.3903) is shown. Reliability of branching was assessed by bootstrapping with 250 pseudo-replicates. A discrete gamma distribution was used to model evolutionary rate differences among sites ($\alpha = 2.47$). The scale bar represents 0.5 expected substitutions per site. The tree is rooted on the lamprey *MAGP1* branch. The figure was adapted from Segade (2009) and used with permission

2009) (Fig. 6.1), the origin of the *MAGP1* (*MFAP2*) gene can be dated to the stem of the vertebrate lineage in the interval between the appearance of the Craniates (the chordate clade that includes agnathans and gnathostomes) and before the Agnatha-Gnathostomata split, or 650–800 My ago (Blair and Hedges 2005). From a comparison of lamprey and tetrapod MBD sequences, including the conservation of functional residues, like the cysteine homologous to human Cys-101 that defines the ECM function of the MBD (Fig. 6.1) (Segade et al. 2002), it was proposed that ancestral MAGP1 was able to interact with the ECM and, likely, with a fibrillin-like EGF-rich protein (Segade 2009). Based on the pattern of MAGP1 gene expression, mostly in tissues derived from the embryonic neural crest, like smooth muscle, dermal connective tissue, cartilage, and bone (Chen et al. 1993), it was postulated that the appearance of MAGP1 correlated with the evolution of the neural crest (Segade 2009). Evolution of the neural crest is generally considered one of the essential vertebrate innovations (Shimeld and Holland 2000).

The reconstructed phylogeny of the *MAGP* genes and a synteny analysis of their loci in mammalian genomes showed that *MAGP2* (*MFAP5*) originated in a multigene duplication of the *MAGP1* genomic neighborhood (Segade 2009). The duplication was dated to the period between the divergence of the cartilaginous fishes (Chondrichthyes; sharks, rays, and allies) and the bony vertebrates (Osteichthyes), based on the apparent lack of *MAGP2* in both lampreys and elasmobranchs (Segade 2009). However, a reevaluation of certain features in the electric ray *Torpedo californica MAGP* sequence clearly shows that the gene is a *MAGP2* ortholog, and not an aberrant *MAGP1* gene, as originally reported (Segade 2009). A revised maximum likelihood phylogenetic reconstruction in Fig. 6.1 places the electric ray *MAGP2* sequence within the *MAGP2* clade with a strong statistical support. The updated analysis places the *MAGP1*/*MAGP2* gene duplication earlier in vertebrate evolution, between the divergence of the agnathans (which possess only *MAGP1*) and the Chondrichthyes-Osteichthyes split, approximately 525–600 My ago (Hedges 2009).

After duplication, the *MAGP* genes followed distinct evolutionary paths, shown by their patterns of expression (Albig et al. 2007; Chen et al. 1993; Faraco et al. 1995; Frankfater et al. 2000; Gibson et al. 1991, 1996; Segade et al. 2000), lineage-specific motifs (Brown-Augsburger et al. 1996; Gibson et al. 1996; Segade et al. 2002; Trask et al. 2001), function (Albig et al. 2007, 2008; Chen et al. 2006; Weinbaum et al. 2008; Werneck et al. 2008), and promoter elements (Frankfater et al. 2000; Segade and Mecham 2005). While the MBD is always present, sites otherwise functional in human and other mammals are not necessarily conserved (Fig. 6.1). Thus, in MAGP1, glutamine 20 (full-length human MAGP1 numbering convention) that serves as substrate for transglutamination and tyrosines 47, 48, and 50, the sites for posttranslational sulfation (Trask et al. 2001), are found only in amniotes (although glutamine 20 is present in skate, a chondrichthyan). Likewise, the N-terminus in nonmammalian MAGP2 is poor in negatively charged amino acids compared (Segade 2009), while the previously mentioned RGD potential cell attachment site (positions 30–32 in human MAGP2) (Gibson et al. 1999) represents a uniquely mammalian feature (Segade 2009).

6.2.3 Evolutionary Mechanisms in the Functional Diversification of the MAGP Genes

Early evolutionary studies of the *MAGP* genes found that, in placental mammals, *MAGP2* shows a higher rate of non-synonymous substitution per non-synonymous site (Ka) than *MAGP1*, interpreted as indicative of the faster evolution of *MAGP2* (Segade et al. 2000). A more extensive relative rate test, using *MAGP* coding sequences across the major vertebrate taxa (Segade 2009), demonstrated that the Ka rate on the branch leading to *MAGP2* (Ka = 0.37) was significantly higher than the Ka rate on the *MAGP1* branch (Ka = 0.20) and seemed to confirm that, in fact, *MAGP2* evolves faster than *MAGP1*. Note in Fig. 6.1 the appreciable longer branches leading to *MAGP2* sequences, representing higher numbers of substitutions per site, when compared with the shorter *MAGP1* branches, especially within the mammalian clade, consistent with a more conservative evolution of *MAGP1*. However, maximum likelihood tests for modes of selection that best fit the evolution of the *MAGP* genes uncovered no evidence of positive (Darwinian) natural selection, showing instead that both *MAGP1* and *MAGP2* have evolved under a regime of strong purifying (negative) selection (Segade 2009). The relative rate test can be interpreted as a transitory relaxation of selective constraints on *MAGP2* that elevated its Ka rate. It is well known that, after gene duplication events, relaxation of purifying selection is a common phenomenon in the early stages of the divergent evolution of the two duplicates, in what is known as Type I divergence (Gu 1999, 2001). During type I divergence, in a given position, site-specific shifts in substitution rates between paralogs will be detected by a characteristic variability of amino acids in one duplicate and a fixed residue in the other duplicate. Variable residues are an indication that the stranglehold of negative selection has weakened, which allows the gene to explore an expanded sequence universe and, therefore, to eventually evolve a new function (neofunctionalization). Another kind of functional divergence, type II divergence, results from site-specific shifts in amino acid properties (Gu 2006). Type II divergence occurs in later stages in the process when residues adapted to new functionalities become fixed within a gene lineage. A significant type II coefficient ($\theta_{\mathrm{II}} > 0$) is consistent with the presence of Darwinian selection, but detection of a higher type I than type II divergence ($\theta_{\mathrm{I}} = \theta_{\mathrm{II}}$; $\theta_{\mathrm{I}} > 0$) is an indication of gain of function through relaxation of selective constraints rather than from positive selection of fixed variants (Gu 2006). Results from the divergence test on a collection of vertebrate gene sequences showed significant type I divergence but nonexistent type II divergence between *MAGP1* and *MAGP2*, and it confirmed that temporarily lower selective constraints were involved in the functional separation of the *MAGP* genes. Moreover, Bayesian analysis of the sequences found that in 10 out of 11 candidate sites for functional divergence, the amino acid is fixed on MAGP1 and variable in MAGP2 (Segade 2009), demonstrating that the relaxed selective regime took place preferentially on the *MAGP2* lineage and, therefore, that MAGP2 functionally diverged from the ancestral role shared with MAGP1, in agreement with previous tests and the phylogenetic tree reconstruction. The candidate sites mapped to the C-terminal end of the protein, including to the MBD (Segade 2009). Significantly, the biological effect of the substitution at position 104, which is always a cysteine in MAGP1 and valine/isoleucine in MAGP2 (Fig. 6.1), was previously examined experimentally and found to be partly responsible for the lower affinity of the MAGP2 MBD for the ECM (Segade et al. 2002), thus validating the results of the divergence test.

At one point along its evolutionary history, MAGP2 acquired the ability to interact with Jagged1 and Notch1 (Albig et al. 2008; Miyamoto et al. 2006; Nehring et al. 2005). The timing of this gain of function is still unclear since we lack experimental data on the binding of MAGP2 to Notch-like proteins in nonmammalian systems. Nevertheless, a correlated evolution test, serving as a proxy for the detection of coevolution between two interacting partners (Pazos et al. 2008), showed that the evolution of the *Notch1* gene and the MAGP2 MBD, through which MAGP2 interacts with Notch1 (Miyamoto et al. 2006), were significantly correlated ($r = 0.984$, $P = 0.004$) (Segade 2009). In view of these results, a model of coevolution-induced elevation in Ka rate was put forward to explain the accelerated Ka rate on the *MAGP2* branch. In basic terms, when interacting motifs coevolve, sequence substitutions in one partner will require the positive (Darwinian) selection for compensating substitutions in the second partner to maintain continuity of function (Adkins 2008), leading to a detectable elevation in Ka rate. Since the Notch pathway is evolutionarily older than *MAGP2* (Kortschak et al. 2001) and Notch and Jagged are central in vertebrate development, it is conceivable that after the Notch-binding activity of MAGP2 evolved, a coevolutionary process led to the fixation of non-synonymous substitution in *MAGP2* by positive selection for a more efficient interaction. In this scenario, the *Notch* genes drive the effect, and therefore, we should expect to see a one-sided acceleration in Ka rate in MAGP2. However, rapid evolution of supposedly constrained development genes is not uncommon (Dorus et al. 2006; Fares et al. 2003; Kortschak et al. 2001) and may, in fact, account for the specific anatomical, physiological, or behavioral differences between lineages. An especially relevant case is the fast evolution of *Notch4* in the rodent lineage (Kortschak et al. 2001). If MAGP2 binds not just Notch1 but other faster-evolving Notch proteins such as Notch4, the coevolutionary hypothesis is even more likely to explain the accelerated evolution of the *MAGP2* gene.

Tests of evolution showed that *MAGP1* evolves under negative selection (Segade 2009). However, the statistical analyses that were used all have a strict requirement for ungapped multiple sequence alignments in order to obtain meaningful conclusions. This precluded the analysis of the evolutionary dynamics of the highly divergent N-terminal domain, leading to their exclusion from the original study (Segade 2009). However, the low conservation of the N-terminal sequences may actually reflect their evolutionary flexibility and offer clues as to the origin of functions mediated by regions outside of the MBD. I have mentioned that in mouse, experimental ablation of the *MAGP1* gene induced a complex phenotype which was ascribed to perturbations in TGFβ signaling (Weinbaum et al. 2008) and that in zebrafish, genetic manipulation of *MAGP1* provoked the appearance of vascular defects (Alvarez et al. 2007; Chen et al. 2006) phenotypically similar to those in a number of Mendelian diseases of the ECM that alter TGFβ homeostasis (Ramirez et al. 2007). These observations are consistent with a shared and conceivable ancestral role of MAGP1 in a TGFβ-modulating activity. The bone function of MAGP1 in mammals would then be a newly evolved mammalian character (or amniote, since data from birds are lacking) character. Since the MBDs in mammalian and teleost MAGP1 are ~90 % identical (Fig. 6.1), it is unlikely the MBD has undergone such a dramatic gain of function, thus leaving the N-terminus as the most logical candidate. In this sense, the mammalian N-terminus is significantly longer than in teleosts (average of 77 ± 2 residues in 13 mammal species; average of 56 ± 6 residues in ten teleost species; $P < 0.0001$) and contains more negatively charged and proline and glutamine residues (Fig. 6.1). Importantly, mammalian MAGP1 binds TGFβ/BMP growth factors through regions of its N-terminus that share no significant homology to teleost MAGP1 (T. J. Broekelmann and R. P. Mecham, personal communication). It is important to remember that two of the most characteristic features of the N-terminal half of mammalian MAGP1, namely, its negative charge and the presence of P/Q-rich sequences, are also found in proteins active in tissue mineralization. The *SCPP* (secretory calcium-binding phosphoprotein) gene family is a striking example (Kawasaki 2011). Two *SCPP* clades are present in vertebrates, the acidic SCPPs (also known as SIBLINGs or small integrin-binding ligand N-linked glycoproteins) (Bellahcene et al. 2008) and the P/Q-rich SCPPs (Kawasaki 2011). Whereas the acidic SCPPs help in the mineralization of the bone and dental collagen matrix by binding calcium ions, the P/Q-rich SCPPs create spaces between the mineralizing tissue and the adjacent mesenchymal cells (Kawasaki 2011) through protein-protein interactions and self-assembly induced by the P/Q regions (Williamson 1994). It is noteworthy that the *SCPP* genes evolved by duplication and functional diversification of a *SPARC-like 1* (*SPARCL1*) gene (Kawasaki 2011). *SPARCL1* is paralogous to *SPARC* (Secreted Protein Acidic and Rich in Cysteine, or osteonectin) (Yan and Sage 1999), a matricellular protein with which MAGP1 has been compared (Segade et al. 2007). The unusually acidic MAGP1 has a strong propensity to form supramolecular aggregates (Gibson et al. 1986), seemingly mediated by the P/Q-containing N-terminus and, thus, combining in a single sequence physicochemical properties that are separate in the two *SCPP* subfamilies. Although not directly involved in tissue mineralization (Craft et al. 2010), I speculate that, by similarity to the SCPPs, the acidic and P/Q-rich N-terminus of MAGP1 in mammals and other tetrapods helps to localize the protein to the proximity of the mineral bone matrix, thus ensuing its correct localization for the sequestration of active forms of the BMP-like growth factors. The P/Q-poor N-termini of teleost MAGP1 would be expected not to localize to mineral tissues and not to bind BMP-like growth factors. In short, evolution of the MAGP1 bone function may have occurred through P/Q enrichment and elongation of an ancestral TGFβ-binding sequence. The short teleost and elasmobranch MAGP1 proteins may represent reasonable facsimiles of the common ancestor. A detailed investigation of TGFβ/BMP-binding activities in teleost, elasmobranch, and lamprey MAGP1 should clarify the evolutionary steps that carried MAGP1 from the vasculature to the vertebrate bone.

6.3 The Fibulins

6.3.1 Introduction to the Fibulins

The *FBLN* genes code for a family of secreted glycoproteins in the 50–200 kDa range, defined by the presence of a unique globular domain (Fibulin C-terminal [FC] domain or domain III) at the C-terminal end. The FC domain mediates protein-protein interaction, including with a number of ECM components (Argraves et al. 2003; Cooley and Argraves 2011; de Vega et al. 2009; Timpl et al. 2003). A tandem array with a variable number of cbEGF modules (domain II) always precedes the FC domain. Additional modules are present in the N-terminal sequences (domain I), except in the short or "elastic" FBLNs (FBLN3, 4, and 5) which contain only the tandem cbEGF modules and an atypical EGF motif (Cooley and Argraves 2011; Yanagisawa and Davis 2010). Eight FBLNs have been described in placental mammals (Cooley and Argraves 2011; de Vega et al. 2009), and up to 14 orthologous genes were identified in vertebrates (Segade 2010). The *FBLN* family includes the two vertebrate hemicentin genes, *HMCN1* and *HMCN2*, which code for FC-domain proteins and are, therefore, bona fide *FBLN* genes. HMCN1 and HMCN2 are sometimes referred to as FBLN6 and FBLN8, respectively (Cooley and Argraves 2011). However, since (1) phylogenetic studies have shown that the *HMCN* genes cluster in a monophyletic branch which is a sister clade to the other *FBLN*s (Segade 2010), (2) the official gene names for the vertebrate hemicentins are *HMCN1* and *HMCN2* (Flicek et al. 2012), and (3) the designation *FBLN8* has been previously used for a paralogous gene to *FBLN7* present in most vertebrates except placental mammals (Segade 2010), I will exclusively use the appellations "hemicentin" and "*HMCN* genes" to refer to members of this clade.

FBLNs are cell adhesion proteins. Among other roles, in mammals, FBLN1 regulates cell migration and guidance by suppressing motility, FBLN2 promotes smooth muscle cell migration, FBLN3 enhances cell adhesion, and FBLN5 inhibits cell migration (Cooley and Argraves 2011). Mediation of cell-cell and cell-ECM interactions is the most probable ancestral FBLN function. This is supported by the report that in nematodes, which possess two *FBLN* genes (*fbl1* and *him1*) orthologous to vertebrate *FBLN1* and *HMCN1*, respectively, FBLN1 determines cell and organ shape (Hesselson et al. 2004; Kubota et al. 2004; Muriel et al. 2005), and HMCN collaborates in the establishment of temporary cell-cell contacts during cell migration (Vogel and Hedgecock 2001; Vogel et al. 2006). Interestingly, a close association between FBLNs and the formation of an elastic ECM was found in nematodes, which may represent a primordial FBLN function. In these animals, the FBLN1D splice form, which is conserved in vertebrates (Barth et al. 1998), assembles into flexible structures that connect the pharyngeal and body-wall muscle basement membranes (Muriel et al. 2005). These flexible tracks are unique nematode features that may be considered functional (though not phylogenetical) homologs to the elastic fibers of vertebrates (Muriel et al. 2005). In this sense, vertebrate basement membrane FBLNs also show an affinity for elastic structures: FBLN1 to the amorphous elastin core (Roark et al. 1995) and FBLN2 to the MF/elastin interface (Kobayashi et al. 2007; Reinhardt et al. 1996). Only in vertebrates a specialized *FBLN* subfamily, represented by the *FBLN3/4/5* genes, evolved specialized functions in elastogenesis (Argraves et al. 2003; Cooley and Argraves 2011; de Vega et al. 2009; Timpl et al. 2003; Wagenseil and Mecham 2007). The essential role that the *FBLN3/4/5* genes play in the deposition and assembly of elastic fibers is demonstrated by the defects seen in the architecture of elastic fibers in knockout mice for *FBLN3* (McLaughlin et al. 2007; Rahn et al. 2009), *FBLN4* (Hanada et al. 2007; McLaughlin et al. 2006; Moltzer et al. 2011) and *FBLN5* (Nakamura et al. 2002; Spencer et al. 2005; Yanagisawa et al. 2002). While the function of FBLN3 in elastogenesis is still unclear, FBLN4 and FBLN5 seem to limit the size of the tropoelastin coacervates during secretion, assist in the transfer of elastin to the MF, and facilitate cross-linking by LOXs (Choudhury et al. 2009; Wagenseil and Mecham 2007), acting as dynamic "bridges" between the growing elastin polymer and the MF scaffold (Cooley and Argraves 2011; Wagenseil and Mecham 2007; Yanagisawa and Davis 2010).

6.3.2 Evolutionary Relationships Between FBLN Genes

Historically, the FBLNs were classified according to their domain architectures, which only approximated their phylogenetic relationships. Reconstruction of the *FBLN* gene phylogeny using a maximum likelihood method with FBLN coding sequences from 180 metazoan species clarified the intrafamily relationships (Segade 2010) and showed that the *FBLN* genes cluster into two major clades, one branch containing the *FBLN* genes sensu stricto and a second branch with the two *HMCN* genes (Vogel and Hedgecock 2001) (Fig. 6.2). The placement of the *HMCN* genes in a sister group to the remaining *FBLN* genes reflects the unique sequence features of the HMCN proteins, with a domain I formed by one N-terminal von Willebrand factor module followed by a tandem array of up to 50 immunoglobulin modules, six thrombospondin domains, and one G2 nidogen fold preceding eight cbEGF modules in domain II (Argraves et al. 2003; de Vega et al. 2009; Vogel and Hedgecock 2001; Vogel et al. 2006).

Within the *FBLN* clade, each of the previously recognized structural types of FBLN was placed in statistically robust monophyletic groups. *FBLN1* and *FBLN2* were reconstructed as sister groups (Fig. 6.2). In mammals, the FBLN1 and FBLN2 proteins share a domain II of nine cbEGF modules, with three anaphylatoxin modules in domain I, whilst FBLN2 contains an extra ~400-residue N-terminal domain, composed of cysteine-rich (150 amino acids) and cysteine-free segments

a

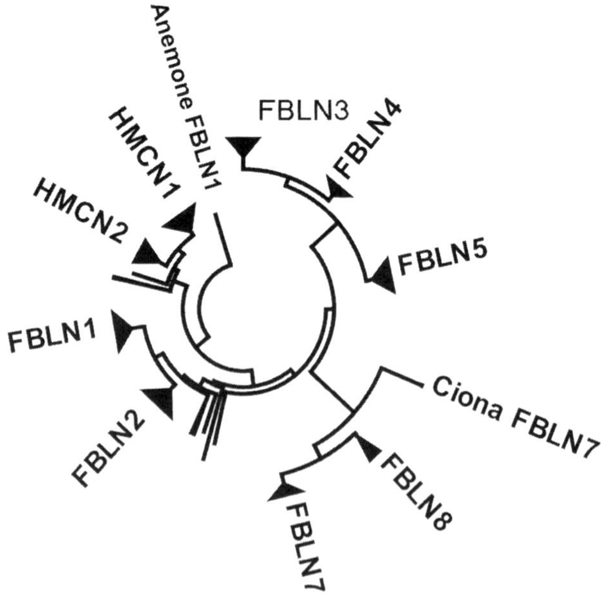

b

Cnidarian	Ecdysozoan	Urochordate	Teleost	Amphibian	Avian	Monotreme	Marsupial	Eutherian
FBLN1	FBLN1	FBLN1	FBLN1A	FBLN1	FBLN1	FBLN1	FBLN1	FBLN1
			FBLN1B					
			FBLN2A	FBLN2	FBLN2	FBLN2	FBLN2	FBLN2
			FBLN2B					
		FBLN3/4/5	FBLN3A	FBLN3	FBLN3	FBLN3	FBLN3	FBLN3
			FBLN3B					
			FBLN4A	FBLN4	†	FBLN4	FBLN4	FBLN4
			FBLN4B					
			FBLN5	FBLN5	FBLN5	FBLN5	FBLN5	FBLN5
			‡					
		FBLN7/8	FBLN7	FBLN7	FBLN7	FBLN7	FBLN7	FBLN7
			‡					
			FBLN8A	†	FBLN8	FBLN8	†	†
			FBLN8B					
HMCN	HMCN	HMCN	HMCN1	HMCN1	HMCN1	HMCN1	HMCN1	HMCN1
			‡					
			HMCN2	†	†	HMCN2	HMCN2	HMCN2
			‡					

Fig. 6.2 Evolution of the *FBLN* genes. (**a**) Radiant view of the phylogenetic relationships of the *FBLN* gene family inferred by maximum likelihood of protein sequences with a JTT substitution matrix and a gamma distribution ($\alpha = 0.98$). The tree with the highest log likelihood (−21724.5)

(Pan et al. 1993) (Fig. 6.3). It is likely that, after duplication, the extra sequences upstream of the anaphylatoxin domains in FBLN2 were gained in a process of functional divergence after duplication. The *FBLN1/FBLN2* chordate genes and the invertebrate *FBLN1* gene are sister taxa to the chordate *FBLN3/4/5* and *FBLN7/8*.

The vertebrate *FBLN3/4/5*, the short or "elastic *FBLN*" genes, form a natural clade with 100 % bootstrap support (Fig. 6.2). The FBLN3/4/5 proteins (FBLN3/EFEMP1, FBLN4/EFEMP2, and FBLN5/DANCE/EVEC) are the smallest and more highly conserved of the vertebrate fibulins (Argraves et al. 2003; Cooley and Argraves 2011; de Vega et al. 2009; Yanagisawa and Davis 2010). They lack domain I, and domain II contains five cbEGF modules in tandem preceded by an atypical EGF-like module with a 28–88 amino acid link sequence between the fourth and fifth cysteines (Giltay et al. 1999) (Fig. 6.3). Within the vertebrate *FBLN3/4/5* cluster, *FBLN5* nests in a high-support sister clade to the *FBLN3* and *FBLN4* sequences, consistent with an early divergence from the ancestral *FBLN3/4/5* common ancestor. The short branches at the base of each gene clade are indicative of rapid diversification from two gene duplications in quick succession (Fig. 6.2).

Finally, the *FBLN7/8* genes cluster in a monophyletic clade positioned as a sister clade to the elastic *FBLN* genes (Fig. 6.2), indicative of their common ancestry. FBLN7 is the most recently characterized human FBLN, expressed mostly in cartilage and dentition-related tissues (de Vega et al. 2007). FBLN7 has a domain I formed by one sushi/CCP protein-protein interaction segment and a domain II with three cbEGF modules (de Vega et al. 2007). The recently identified *FBLN8* gene, found in most vertebrates except placental and marsupial mammals (Fig. 6.2), where it may play an eye-related function, codes for a protein with a similar structure to FBLN7 (Segade 2010).

6.3.3 *Origin and Evolutionary History of the FBLN Gene Family*

Fibulins are ancient proteins. *FBLN* gene sequences are present in the genome of species from every major metazoan phylum (Barth et al. 1998; Hesselson et al. 2004; Kubota et al. 2004; Segade 2010; Vogel et al. 2006), with their origin dated between the divergence of the Porifera (sponges), which lack *FBLN* sequences

Fig. 6.2 (continued) is shown. Reliability of branching was assessed by bootstrapping with 150 pseudo-replicates. Where a large number of sequences were collapsed to unclutter the figure, area of the triangles is proportional to the number of species in the subtrees. The scale bar represents 1.0 expected substitution per site. (**b**) Simplified view of the evolutionary diversification of the *FBLN* gene family. Major taxa are represented in the top row of the table. Inside the table cells, orthologous genes are listed horizontally. The dagger represents lineage-specific gene losses; the double dagger signifies losses of probable duplicates from the fish-specific gene duplication. The figure was adapted from Segade (2010) and used with permission

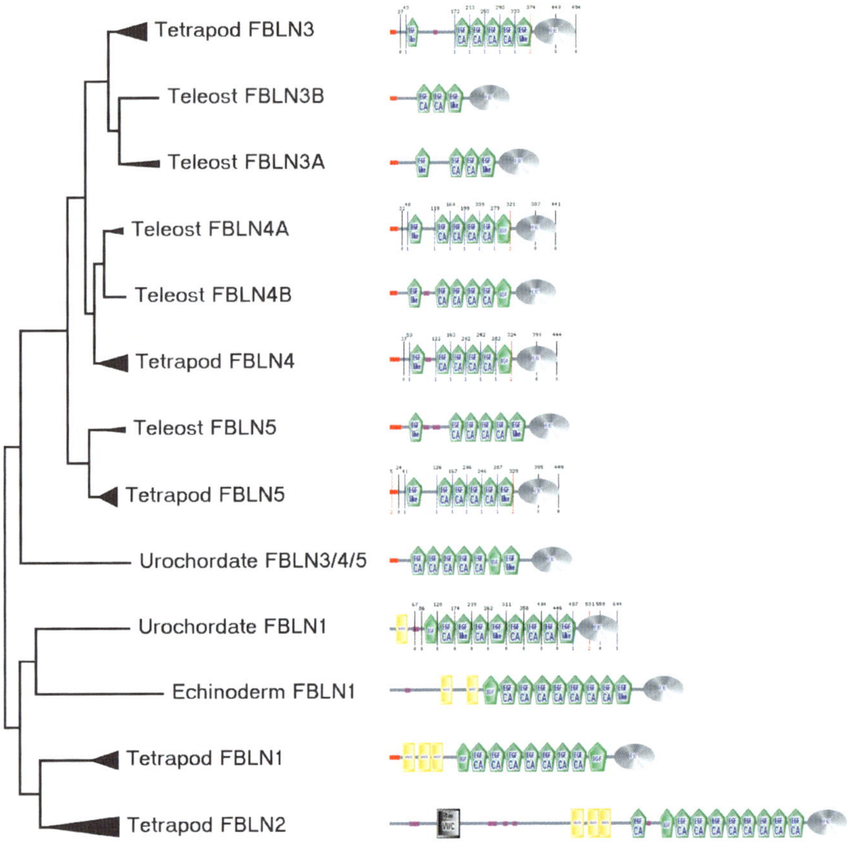

Fig. 6.3 Domain architecture of the chordate FBLN1/2 and FBLN3/4/5 proteins. SMART (Simple Modular Architecture Research Tool) visualization of domains in representative FBLN proteins. The FC domain is not recognized by SMART, so its presence and boundaries were manually located by sequence comparison using Clustal W. The tree was computed using the Minimum Evolution algorithm and the JTT substitution matrix with a gamma distribution. *Numbers* correspond to the beginning and end of exon sequences in those sequences where exon organization is known. *ANATO* anaphylatoxin domain, *EGF* epidermal growth factor-like module, *EGF-CA* calcium-binding EGF, *FC* fibulin-like C-terminal domain, *vWC* von Willebrand factor module. The figure was adapted from Segade (2010) and used with permission

(Segade 2010), and the Cnidaria-Bilateria split, or around 1 Gy ago (Blair 2009). This period in Earth's history, deep in the Proterozoic, saw the evolution of multicellularity and with it, a leap in the genetic, histological, and developmental complexity of organisms. Many of the extant multidomain protein families became established at this time through extensive gene duplications and domain shuffling events (Miyata and Suga 2001). For ECM molecules, new combinations of preexisting functional modules evolved from cell adhesion proteins present in the unicellular ancestors (King et al. 2003). A complex ECM as the medium for cell migration, cell-cell communication, and reservoir of growth factors was decisive in the evolution of the cell-type specialization and embryology with differentiated tissue layers characteristic of Eumetazoans (Morris 1993). A window on the probable role of *FBLNs* in early animals is provided by protostomes like the nematode *Caenorhabditis elegans*, where FBLN and HMCN are functional partners that are codependent for their deposition in flexible tracks, mechanosensory neurons, and uterine attachments (Hesselson et al. 2004; Kubota et al. 2004; Muriel et al. 2006; Vogel et al. 2006). Even early-branching Eumetazoans (Cnidaria, Nematodes) possess a two-gene *FBLN* complement (Fig. 6.2). This suggests that an ancestral *FBLN* gene with the basic FBLN architecture was present in basal Eumetazoans. This *FBLN* concestor then underwent a gene duplication that established the primordial *FBLN1* and *HMCN* branches before the Cnidaria-Bilateria split. Subsequent paralogous divergence through recombination and exon shuffling added the immunoglobulin and von Willebrand factor modules characteristic of extant HMCNs (Vogel and Hedgecock 2001).

For a long period of their evolutionary history, the *FBLN* family entered a phase of stasis. Cnidarians, nematodes, arthropods, and echinoderms encode one ortholog each of *FBLN1* and *HMCN*, and no significant expansion of the *FBLN* gene repertoire took place before the appearance of the chordates (Fig. 6.2). The chordate ancestor is widely believed to have possessed a notochord, a hollow dorsal nerve cord, pharyngeal slits, an endostyle, and a post anal tail (Rychel et al. 2006). The rise of this complex organism was concomitant with the evolution of new genes (Dehal et al. 2002; McLysaght et al. 2002). The tunicate *Ciona*, a urochordate, sister taxon to the vertebrates (Putnam et al. 2008), can be used to ascertain the complement of *FBLN* genes in the prochordate ancestor. The genome of *Ciona* includes four *FBLN* genes, respectively, orthologous to vertebrate *FBLN1/2*, *FBLN3/4/5*, *FBLN7/8*, and *HMCN1/2* (Segade 2010) (Fig. 6.2). Comparative synteny analysis showed that *Ciona FBLN3/4/5* and *FBLN7/8* genes evolved by duplications of the *FBLN1* gene before the divergence of urochordates and vertebrates (Segade 2010). The origin of the *FBLN3/4/5* and *FBLN7/8* families can then be dated to the base of the chordate lineage, or around 775–850 My ago (Blair 2009).

Shown in Fig. 6.2 is the increase in *FBLN* genes in the vertebrate lineage with the duplication of the *FBLN1/FBLN2*, *FBLN3/FBLN4/FBLN5* (two rounds), *FBLN7/FBLN8*, and *HMCN1/HMCN2* genes (Segade 2010). The origin of the vertebrates from the prochordate ancestor has been linked to an expansion in genomic complexity that provided the raw genetic material for the anatomical and developmental innovations of the vertebrate body plan, including neural crest cells, bone and cartilage, dermal connective tissue, and a high-pressure circulatory system (Shimeld and Holland 2000). The presence of tetralogons (four gene copies per genome) is consistent with two whole-genome duplications early in the history of vertebrates before the divergence of tetrapods and teleosts (bony fishes) (Kasahara 2007). Whether the duplication of the prochordate *FBLN* genes happened before or after the split of the Agnatha (jawless vertebrates; lampreys and hagfish) is unclear. Only *FBLN1* and a single gene belonging to the *FBLN3/4/5* clade were found in the lamprey genome (Segade 2010), indicating that the duplications took place after the divergence of gnathostomes and agnathans. Alternatively, the low quality of the lamprey genome assembly may still conceal the presence of other *FBLN* genes since no *FBLN7-like* gene was identified either. As shown in Fig. 6.2, no further gain of *FBLN* genes occurred in tetrapods. However, extensive lineage-specific *FBLN* gene losses did happen. Thus, *FBLN4* was lost in birds; *FBLN8* in therian mammals (placentals and marsupials) and the amphibian *Xenopus tropicalis*; and *HMCN2* from sauropsids (birds plus squamate [lizard] reptiles) and *X. tropicalis* (Segade 2010). These losses are particularly intriguing in the context of the essentiality of some of the *FBLN* genes and, more generally, on the evolution of gene redundancy.

In teleost fishes, a third whole-genome duplication (the fish-specific whole-genome duplication, FSGD) occurred early in their history (Brunet et al. 2006; Christoffels et al. 2004; Jaillon et al. 2004; Meyer and Van de Peer 2005; Santini et al. 2009; Vandepoele et al. 2004). The presence of the recently diverged paralog pairs *FBLN1A/B*, *FBLN 2A/B*, *FBLN3A/B*, *FBLN4A/B*, and *FBLN8A/B* in the genomes of several teleost species (Segade 2010) argues in favor of their origin in the FSGD. Teleost-specific paralogs of *FBLN5*, *FBLN7*, *HMCN1*, and *HMCN2* were likely lost after duplication (Segade 2010), the usual outcome of most duplicated genes, including up to 85 % of the paralogs from the FSGD (Brunet et al. 2006).

6.3.4 Evolution of the Elastogenic Function in the FBLN3/4/5 Gene Family

In bony vertebrates, FBLN4 and FBLN5 are essential for the deposition and assembly of elastin onto the MF scaffold. Identification of a *Ciona* ortholog of the *FBLN3/4/5* genes implies that the origin of FBLNs with elastogenic properties predates the evolution of their functional partner, tropoelastin. Polymeric elastin is present in every vertebrate taxon except agnathans (Kozel et al. 2011), which dates the evolution of the tropoelastin gene (*ELN*) to between 500 and 600 My ago, or more than 250 My after the evolution of the *FBLN3/4/5* ancestor (Hedges 2009). Two questions immediately come to mind: (1) what was the function of the *FBLN3/4/5* ortholog in a basal chordate, and (2) how did the elastogenic activity of *FBLN3/4/5* in vertebrates evolve in a protein family that predates the origin of tropoelastin?

We can only speculate on the *FBLN3/4/5* functions in *Ciona*, since we lack experimental data on its biochemistry and gene expression data from tunicates are still sparse (Satou et al. 2005). Tunicates, like *Ciona*, are relatively simple chordates characterized by the presence of a mesenchymal (cartilage-like) notochord; a tunic, which adds support to the animal once the free-swimming larva fixes itself to a solid substrate; cartilaginous gills; and a beating heart (Gans 1989). Many of these organs are elastic, and in vertebrates, their homologs are rich in MFs and elastic fibers (Mecham and Davis 1994). I speculate that FBLN3/4/5 also participates in the assembly of MFs and similar elastic ECM features in organs with intrinsic elasticity perhaps through a hypothetical interaction with peptide sequences with similar physicochemical characteristics to tropoelastin (e.g., hydrophobicity or self-aggregation), a biochemical property that was conceivably inherited from FBLN1, from which FBLN3/4/5 evolved since, as we have seen in Sect. 6.3.1, an elastic ECM function may be ancestral to the FBLNs. This idea is supported by the essential role of nematode FBLN1 in the assembly of flexible tracks and the proto-elastogenic activity still present in vertebrate FBLN1 as shown by its localization to MFs and elastic fibers (Kobayashi et al. 2007; Nakamura et al. 2002; Wagenseil and Mecham 2007, 2009; Yanagisawa et al. 2002).

After the two rapid gene duplications that diversified the *FBLN3/4/5* genes (Fig. 6.2), the architecture of the FBLN3/4/5 proteins quickly congealed into their modern domain structure after the FBLN1 anaphylatoxin domains were lost and the number of cbEGF modules was reduced concomitantly with the insertion of a link sequence into the first EGF module (Segade 2010) (Fig. 6.3).

An analysis of gene-specific variation in selection pressure within the FBLN3/4/5 clade found, as expected, that the three genes evolved within a strong negative selection regime (Segade 2010). However, three stem branches leading to the *FBLN3/4* clade, the *FBLN5* gene, and the *FBLN3/FBLN4* split node showed accelerated evolution, which was interpreted as the result of a temporary relaxation of the prevailing purifying selection after each duplication event (Segade 2010), similar to what we have seen caused the divergence of the *MAGP* genes. We have reviewed in Sect. 6.2.2 that relaxed natural selection after gene duplication is usually associated with type I functional divergence (Gu 1999). A divergence test found significant type I divergence after the split of *FBLN5* and the *FBLN3/4* common ancestor and, later, after the duplication of *FBLN3* and *FBLN4* (Fig. 6.3) (Segade 2010). Divergence occurred preferentially on the *FBLN5* branch, as shown by the presence of variable amino acids in FBLN5 and unique in FBLN3 and FBLN4 at positions with the highest likelihood of driving the functional divergence between members of the family (Segade 2010). Divergence of *FBLN5* was not restricted to sequence changes as demonstrated by the computation of an expression divergence parameter that measured how quantitatively assessed patterns of gene expression (using numerical data from gene chip experiments) between FBLN3/4/5 genes correlated with the time elapsed since their split. The highest expression divergence was found between *FBLN3* and *FBLN5* and the lowest between *FBLN3* and *FBLN4* (Segade 2010). Together, the data predict that FBLN4 and FBLN5 are the most divergent and not functionally redundant, suggesting that gain of functions, conceivably for elastic fiber assembly, evolved de novo in *FBLN4* and *FBLN5*, while the *FBLN3* gene may still represent the ancestral function of the family, namely, cell adhesion. This is consistent with the phenotypes observed in *FBLN3/4/5* knockout animals where elastic fiber defects were found in null *FBLN4* or *FBLN5* mice, but not in the null *FBLN3* mouse (Cooley and Argraves 2011).

Mapping of the regions departing from the prevailing rate of sequence evolution was performed to identify regions with the highest potential for the evolution of new functions. A sliding window analysis of variation in substitution rates for every position along the sequences of the FBLN3/4/5 proteins identified peaks in Ka rate (non-synonymous substitutions) that spanned the atypical first EGF module in all three genes, and over the FC domain of *FBLN5* (Segade 2010) (Fig. 6.4). The atypical EGF domain is a strong candidate to account for the differences in function. Interestingly, the high-Ka regions mediate the elastogenic functions of the short FBLNs (Choudhury et al. 2009; Horiguchi et al. 2009; Zheng et al. 2007). Thus, elastin binds FBLN5 at the first EGF module and FC domain (Choudhury et al. 2009; Zheng et al. 2007), and the LOX prodomain interacts with a region in FBLN4 that includes its first EGF module (Horiguchi et al. 2009). In most genes, the measured Ka rate is low due to the negative selection pressure on functional motifs where most non-synonymous mutations would be deleterious and rapidly eliminated by natural selection. To account for the elevated evolutionary rate of the FBLN3/4/5 functional motifs, a coevolutionary model was proposed in which the Ka peaks reflect coevolutionary adaptation of the *FBLN* genes to sequence substitutions in their functional partners, tropoelastin, and the LOX prodomain throughout vertebrate evolution (Segade 2010).

6.4 Perspectives

For decades now, researchers have shed light on numerous aspects of the biology of the microfibril-associated proteins. Hundreds of articles have been published describing the identification of new functions, critical motifs, or disease-causing mutations. We now know that no gene is an island, that every gene evolves in concert, and that interactions between gene products within a functional module create selective constraints. The MF complex that we try to understand today is the result of a grand and ultimately successful experiment in mutagenesis. Uncounted numbers of sequence combinations and gene expression patterns likely arose throughout evolution, only to be discarded by natural selection except for a vanishingly small set preserved in modern genes. Learning to identify those successful mutations, from simple residue changes to genome-scale duplications, should provide inspiration for further research. Throughout this chapter, I examined our current knowledge on the evolutionary history of two of the better-characterized families of microfibril-associated proteins, the FBLNs and the MAGPs. While the particular details differ, their evolutionary pathways are, in many respects, similar. We have seen that major increases in their gene number, mediated by the appearance of gene duplicates, most probably took place in the genome duplications early in the history of vertebrates. I also highlighted the importance that stochastic events such as sequence substitutions in a small number of positions during short periods of relaxed natural selection allowed the exploration of new functionalities and working partners, leading to the specialization of once identical paralogous genes. Coevolution between functional partners has apparently played a major role in the evolution of many (all?) microfibrillar genes. The coevolution models were based on statistical tests of correlated evolution between *MAGP2* and *Notch*

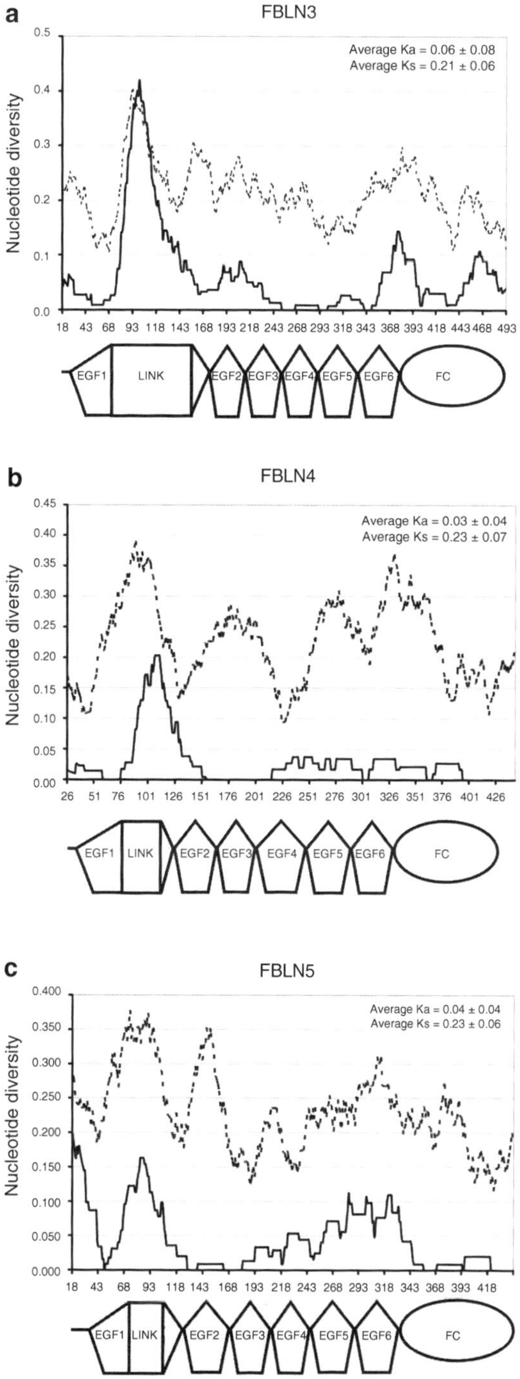

Fig. 6.4 Sliding window analysis of the synonymous (Ks, stippled line) and non-synonymous (Ka, continuous line) substitutions per codon along the *FBLN3* (**a**), *FBLN4* (**b**), and *FBLN5* (**c**) coding sequences in placental mammals with a 25-codon window and one-codon step. Below each panel, a simplified overview of the domain architectures of the FBLN3/4/5 proteins is shown. The figure was adapted from Segade (2010) and used with permission

genes (Segade 2009) and experimental assays of MAGP/Notch, FBLN/ELN, and FBLN/LOXL interaction (Albig et al. 2008; Choi et al. 2009; Choudhury et al. 2009; Horiguchi et al. 2009; Kobayashi et al. 2007; Miyamoto et al. 2006; Yanagisawa and Davis 2010; Zheng et al. 2007). To find evidence of actual coevolution, as opposed to statistically correlated evolution, experiments will need to verify the occurrence of constrained evolution due to biochemical interactions, for example, by measuring their functional affinities in a larger set of nonmammalian species.

Several questions arise from a close examination of the molecular evolution of the microfibrillar proteins, starting with their specificity of function across animal lineages and through eons of time. In the case of the *FBLN*s, we are highly confident that the primordial *FBLN1* and *HMCN*, the founder genes of the family, mediated cell adhesion. However, if the repertoire of *FBLN* genes in the tunicate *Ciona* is representative of that of a basal chordate, our prochordate ancestor already possessed duplicated copies of the invertebrate *FBLN1*-like gene represented by the orthologs of vertebrate *FBLN1/2*, *FBLN3/4/5*, and *FBLN7/8*. We know nothing about the roles of the three FBLNs in tunicates, in particular whether the FBLN3/4/5 co-ortholog participates in the construction of elastic ECM or if, alternatively, it is exclusively involved in cell adhesion. If the former, what chordate lineage saw the specialization of the FBLN3/4/5 protein(s) towards an elastic function? Perhaps it had to wait until after the duplication and divergence of the three vertebrate genes, or, perhaps, though not likely, it required the appearance of tropoelastin first. We simply do not have data on whether an "elastogenic" FBLN is needed for the deposition of other elastic proteins that are present outside of gnathostomes and even whether these elastin-like proteins require a MF scaffold to assemble. Using the lone FBLN3/4/5 protein from lamprey (Segade 2010), its histological localization, search of binding partners by two-hybrid approaches, and biochemical evidence of its involvement in the assembly of the elastin-like (but evolutionarily unrelated) lamprins (Robson et al. 2000) should clarify the origin of the elastic function in the FBLN3/4/5 subfamily, experimentation which could then be conceivably extended to the basal chordates, starting with a search of the tunicate genome for MF-associated elastin-like genes.

The MFs, as a functional genetic module, must be able to absorb changes in their environment, including the constantly evolving cellular and extracellular milieu of multicellular organisms. They also need to possess a certain degree of adaptability in order to maintain the basic function of the MF even after novel developmental, biochemical, and even behavioral traits evolve in different species (Hartwell et al. 1999). It has been often proposed that an increase in gene diversity (e.g., through gene duplication and division of functions [subfunctionalization]) promotes resilience. The concept of "gene redundancy," often mentioned to explain the frequently baffling mild phenotypes seen in null animal models for "essential" genes, including a number of MF-related genes, seems to represent a clear case of evolutionary robustness by gene diversification. Although fully redundant genes are not likely to exist, the *MAGP* and *FBLN* families provide natural experiments of apparent gene redundancy by paralogous genes, like the natural knockouts represented by the losses of *FBLN4* in birds or *FBLN8* in therians. In experimental systems and in human genetic diseases, total or partial loss of function of microfibrillar genes is quite often debilitating if not lethal (Ramirez et al. 2007). It is, therefore, puzzling that the absence of FBLN4 does not impair elastogenesis in birds. Likewise, nothing is known about the function of FBLN8 in non-therian mammals and other vertebrates and how the loss of this protein was compensated (conceivably by its close relative FBLN7) in our direct ancestors. Studying the effects of gene death, and not just brought by the evolution of new genes, should provide valuable data on the resilience and adaptability of the MF complex and, in a more general sense, on the compensatory mechanisms at the genetic and developmental level that organisms follow to overcome the inactivation of such critically important genes, ultimately providing insights on how to manage pathological loss of function in human patients.

We should not forget that the molecular evolution of the protein-coding sequence is just a small portion of the total evolutionary history of a gene. Perhaps even more important than changes in amino acid sequence is the evolution of regulatory sequences due to their lower sequence constraints, short length of the functional motifs, and their combinatorial nature. To fully understand the evolution of the microfibrillar proteins, molecular evolution studies must now be extended to those regulatory sequences, including promoters, enhancers, and splice regulators. Between closely related species, changes in gene expression are likely to be the most common mechanism of evolutionary diversification, especially in genes that are part of a selectively constrained module where non-synonymous mutations are, with high probability, deleterious. Even between distant lineages, functional variation is as likely to originate from regulatory changes as from protein sequence substitutions (Hoekstra and Coyne 2007). Unfortunately, the statistical tools and evolutionary models for the analysis of noncoding sequences are not as advanced as those for protein evolution. However, by focusing on closely related species within the same taxon, like primates or rodents, where noncoding sequence divergence is low, we should be able to identify regions or even motifs that depart from the prevailing purifying selection regime and are, therefore, candidates for divergence in gene expression and, probably, in functionality of the microfibrillar complex.

Finally, most of the research on the MFs and the microfibrillar proteins is guided by a biomedical perspective given the clinical implications that dysfunction in the elastic organs brings about. Many genes for microfibril-associated proteins are loci for genetic diseases, and it is now thought that the same genes responsible for monogenic defects are at the origin of the more common polygenic diseases. By exploring the molecular evolution of the microfibrillar genes in humans and other primates, including the constraints posed by their functional partners, we should expand our view of the mechanisms underlying those genetic diseases and of the ultimate bases for the complex phenotypes that defective microfibrillar genes cause in apparently unrelated target organs.

References

Adkins RM (2008) Coevolution: molecular. In: Cooper DN, Kehrer-Sawatzki H (eds) Handbook of human molecular evolution. Wiley, Chichester, pp 322–328

Albig AR, Roy TG, Becenti DJ, Schiemann WP (2007) Transcriptome analysis of endothelial cell gene expression induced by growth on matrigel matrices: identification and characterization of MAGP-2 and lumican as novel regulators of angiogenesis. Angiogenesis 10:197–216

Albig AR, Becenti DJ, Roy TG, Schiemann WP (2008) Microfibril-associate glycoprotein-2 (MAGP-2) promotes angiogenic cell sprouting by blocking notch signaling in endothelial cells. Microvasc Res 76:7–14

Alvarez Y, Cederlund ML, Cottell DC, Bill BR, Ekker SC, Torres-Vazquez J, Weinstein BM, Hyde DR, Vihtelic TS, Kennedy BN (2007) Genetic determinants of hyaloid and retinal vasculature in zebrafish. BMC Dev Biol 7:114

Argraves WS, Greene LM, Cooley MA, Gallagher WM (2003) Fibulins: physiological and disease perspectives. EMBO Rep 4:1127–1131

Barth JL, Argraves KM, Roark EF, Little CD, Argraves WS (1998) Identification of chicken and C. elegans fibulin-1 homologs and characterization of the C. elegans fibulin-1 gene. Matrix Biol 17:635–646

Bellahcene A, Castronovo V, Ogbureke KU, Fisher LW, Fedarko NS (2008) Small integrin-binding ligand N-linked glycoproteins (SIBLINGs): multifunctional proteins in cancer. Nat Rev Cancer 8:212–226

Blair JE (2009) Animals (Metazoa). In: Hedges SB, Kumar S (eds) The timetree of life. Oxford University Press, Oxford, pp 223–230

Blair JE, Hedges SB (2005) Molecular phylogeny and divergence times of deuterostome animals. Mol Biol Evol 22:2275–2284

Brown-Augsburger P, Broekelmann T, Rosenbloom J, Mecham RP (1996) Functional domains on elastin and microfibril-associated glycoprotein involved in elastic fibre assembly. Biochem J 318(pt 1):149–155

Brunet FG, Crollius HR, Paris M, Aury JM, Gibert P, Jaillon O, Laudet V, Robinson-Rechavi M (2006) Gene loss and evolutionary rates following whole-genome duplication in teleost fishes. Mol Biol Evol 23:1808–1816

Chen Y, Dokholyan NV (2006) The coordinated evolution of yeast proteins is constrained by functional modularity. Trends Genet 22:416–419

Chen Y, Faraco J, Yin W, Germiller J, Francke U, Bonadio J (1993) Structure, chromosomal localization, and expression pattern of the murine Magp gene. J Biol Chem 268:27381–27389

Chen E, Larson JD, Ekker SC (2006) Functional analysis of zebrafish microfibril-associated glycoprotein-1 (Magp1) in vivo reveals roles for microfibrils in vascular development and function. Blood 107:4364–4374

Choi J, Bergdahl A, Zheng Q, Starcher B, Yanagisawa H, Davis EC (2009) Analysis of dermal elastic fibers in the absence of fibulin-5 reveals potential roles for fibulin-5 in elastic fiber assembly. Matrix Biol 28:211–220

Choudhury R, McGovern A, Ridley C, Cain SA, Baldwin A, Wang MC, Guo C, Mironov A Jr, Drymoussi Z, Trump D, Shuttleworth A, Baldock C, Kielty CM (2009) Differential regulation of elastic fiber formation by fibulin-4 and -5. J Biol Chem 284:24553–24567

Christoffels A, Koh EG, Chia JM, Brenner S, Aparicio S, Venkatesh B (2004) Fugu genome analysis provides evidence for a whole-genome duplication early during the evolution of ray-finned fishes. Mol Biol Evol 21:1146–1151

Cooley MA, Argraves WS (2011) The fibulins. In: Mecham RP (ed) The extracellular matrix: an overview. Springer, Berlin, pp 337–367

Craft CS, Zou W, Watkins M, Grimston S, Brodt MD, Broekelmann TJ, Weinbaum JS, Teitelbaum SL, Pierce RA, Civitelli R, Silva MJ, Mecham RP (2010) Microfibril-associated glycoprotein-1, an extracellular matrix regulator of bone remodeling. J Biol Chem 285:23858–23867

Craft CS, Broekelmann TJ, Zou W, Chappel JC, Teitelbaum SL, Mecham RP (2011) Oophorectomy-induced bone loss is attenuated in MAGP1-deficient mice. J Cell Biochem 113:93–99

Davis EC (1994) Immunolocalization of microfibril and microfibril-associated proteins in the subendothelial matrix of the developing mouse aorta. J Cell Sci 107(pt 3):727–736

de Vega S, Iwamoto T, Nakamura T, Hozumi K, McKnight DA, Fisher LW, Fukumoto S, Yamada Y (2007) TM14 is a new member of the fibulin family (fibulin-7) that interacts with extracellular matrix molecules and is active for cell binding. J Biol Chem 282:30878–30888

de Vega S, Iwamoto T, Yamada Y (2009) Fibulins: multiple roles in matrix structures and tissue functions. Cell Mol Life Sci 66:1890–1902

Dehal P, Satou Y, Campbell RK, Chapman J, Degnan B, De Tomaso A, Davidson B, Di Gregorio A, Gelpke M, Goodstein DM, Harafuji N, Hastings KE, Ho I, Hotta K, Huang W, Kawashima T, Lemaire P, Martinez D, Meinertzhagen IA, Necula S, Nonaka M, Putnam N, Rash S, Saiga H, Satake M, Terry A, Yamada L, Wang HG, Awazu S, Azumi K, Boore J, Branno M, Chin-Bow S, DeSantis R, Doyle S, Francino P, Keys DN, Haga S, Hayashi H, Hino K, Imai KS, Inaba K, Kano S, Kobayashi K, Kobayashi M, Lee BI, Makabe KW, Manohar C, Matassi G, Medina M, Mochizuki Y, Mount S, Morishita T, Miura S, Nakayama A, Nishizaka S, Nomoto H, Ohta F, Oishi K, Rigoutsos I, Sano M, Sasaki A, Sasakura Y, Shoguchi E, Shin-i T, Spagnuolo A, Stainier D, Suzuki MM, Tassy O, Takatori N, Tokuoka M, Yagi K, Yoshizaki F, Wada S, Zhang C, Hyatt PD, Larimer F, Detter C, Doggett N, Glavina T, Hawkins T, Richardson P, Lucas S, Kohara Y, Levine M, Satoh N, Rokhsar DS (2002) The draft genome of Ciona intestinalis: insights into chordate and vertebrate origins. Science 298:2157–2167

Dorus S, Anderson JR, Vallender EJ, Gilbert SL, Zhang L, Chemnick LG, Ryder OA, Li W, Lahn BT (2006) Sonic Hedgehog, a key development gene, experienced intensified molecular evolution in primates. Hum Mol Genet 15:2031–2037

Faraco J, Bashir M, Rosenbloom J, Francke U (1995) Characterization of the human gene for microfibril-associated glycoprotein (MFAP2), assignment to chromosome 1p36.1-p35, and linkage to D1S170. Genomics 25:630–637

Fares MA, Bezemer D, Moya A, Marin I (2003) Selection on coding regions determined Hox7 genes evolution. Mol Biol Evol 20:2104–2112

Flicek P, Amode MR, Barrell D, Beal K, Brent S, Carvalho-Silva D, Clapham P, Coates G, Fairley S, Fitzgerald S, Gil L, Gordon L, Hendrix M, Hourlier T, Johnson N, Kahari AK, Keefe D, Keenan S, Kinsella R, Komorowska M, Koscielny G, Kulesha E, Larsson P, Longden I, McLaren W, Muffato M, Overduin B, Pignatelli M, Pritchard B, Riat HS, Ritchie GR, Ruffier M, Schuster M, Sobral D, Tang YA, Taylor K, Trevanion S, Vandrovcova J, White S, Wilson M, Wilder SP, Aken BL, Birney E, Cunningham F, Dunham I, Durbin R, Fernandez-Suarez XM, Harrow J, Herrero J, Hubbard TJ, Parker A, Proctor G, Spudich G, Vogel J, Yates A, Zadissa A, Searle SM (2012) Ensembl 2012. Nucleic Acids Res 40:D84–D90.

Frankfater C, Maus E, Gaal K, Segade F, Copeland NG, Gilbert DJ, Jenkins NA, Shipley JM (2000) Organization of the mouse microfibril-associated glycoprotein-2 (MAGP-2) gene. Mamm Genome 11:191–195

Fraser HB (2006) Coevolution, modularity and human disease. Curr Opin Genet Dev 16:637–644

Fraser HB, Hirsh AE, Wall DP, Eisen MB (2004) Coevolution of gene expression among interacting proteins. Proc Natl Acad Sci U S A 101:9033–9038

Gans C (1989) Stages in the origin of vertebrates: analysis by means of scenarios. Biol Rev Camb Philos Soc 64:221–268

Gibson MA, Hughes JL, Fanning JC, Cleary EG (1986) The major antigen of elastin-associated microfibrils is a 31-kDa glycoprotein. J Biol Chem 261:11429–11436

Gibson MA, Sandberg LB, Grosso LE, Cleary EG (1991) Complementary DNA cloning establishes microfibril-associated glycoprotein (MAGP) to be a discrete component of the elastin-associated microfibrils. J Biol Chem 266:7596–7601

Gibson MA, Hatzinikolas G, Kumaratilake JS, Sandberg LB, Nicholl JK, Sutherland GR, Cleary EG (1996) Further characterization of proteins associated with elastic fiber microfibrils including the molecular cloning of MAGP-2 (MP25). J Biol Chem 271:1096–1103

Gibson MA, Finnis ML, Kumaratilake JS, Cleary EG (1998) Microfibril-associated glycoprotein-2 (MAGP-2) is specifically associated with fibrillin-containing microfibrils but exhibits more restricted patterns of tissue localization and developmental expression than its structural relative MAGP-1. J Histochem Cytochem 46:871–886

Gibson MA, Leavesley DI, Ashman LK (1999) Microfibril-associated glycoprotein-2 specifically interacts with a range of bovine and human cell types via alphaVbeta3 integrin. J Biol Chem 274:13060–13065

Giltay R, Timpl R, Kostka G (1999) Sequence, recombinant expression and tissue localization of two novel extracellular matrix proteins, fibulin-3 and fibulin-4. Matrix Biol 18:469–480

Gu X (1999) Statistical methods for testing functional divergence after gene duplication. Mol Biol Evol 16:1664–1674

Gu X (2001) Maximum-likelihood approach for gene family evolution under functional divergence. Mol Biol Evol 18:453–464

Gu X (2006) A simple statistical method for estimating type-II (cluster-specific) functional divergence of protein sequences. Mol Biol Evol 23:1937–1945

Hanada K, Vermeij M, Garinis GA, de Waard MC, Kunen MG, Myers L, Maas A, Duncker DJ, Meijers C, Dietz HC, Kanaar R, Essers J (2007) Perturbations of vascular homeostasis and aortic valve abnormalities in fibulin-4 deficient mice. Circ Res 100:738–746

Hartwell LH, Hopfield JJ, Leibler S, Murray AW (1999) From molecular to modular cell biology. Nature 402:C47–C52

Hatzinikolas G, Gibson MA (1998) The exon structure of the human MAGP-2 gene. Similarity with the MAGP-1 gene is confined to two exons encoding a cysteine-rich region. J Biol Chem 273:29309–29314

Hedges SB (2009) Vertebrates (Vertebrata). In: Hedges SB, Kumar S (eds) The timetree of life. Oxford University Press, Oxford, pp 309–314

Henderson M, Polewski R, Fanning JC, Gibson MA (1996) Microfibril-associated glycoprotein-1 (MAGP-1) is specifically located on the beads of the beaded-filament structure for fibrillin-containing microfibrils as visualized by the rotary shadowing technique. J Histochem Cytochem 44:1389–1397

Hesselson D, Newman C, Kim KW, Kimble J (2004) GON-1 and fibulin have antagonistic roles in control of organ shape. Curr Biol 14:2005–2010

Hoekstra HE, Coyne JA (2007) The locus of evolution: evo devo and the genetics of adaptation. Evolution 61:995–1016

Horiguchi M, Inoue T, Ohbayashi T, Hirai M, Noda K, Marmorstein LY, Yabe D, Takagi K, Akama TO, Kita T, Kimura T, Nakamura T (2009) Fibulin-4 conducts proper elastogenesis via interaction with cross-linking enzyme lysyl oxidase. Proc Natl Acad Sci U S A 106:19029–19034

Hubmacher D, Reinhardt DP (2011) Microfibrils and fibrillin. In: Mecham RP (ed) The extracellular matrix: an overview. Springer, Berlin, pp 233–266

Jaillon O, Aury JM, Brunet F, Petit JL, Stange-Thomann N, Mauceli E, Bouneau L, Fischer C, Ozouf-Costaz C, Bernot A, Nicaud S, Jaffe D, Fisher S, Lutfalla G, Dossat C, Segurens B, Dasilva C, Salanoubat M, Levy M, Boudet N, Castellano S, Anthouard V, Jubin C, Castelli V, Katinka M, Vacherie B, Biemont C, Skalli Z, Cattolico L, Poulain J, De Berardinis V, Cruaud C, Duprat S, Brottier P, Coutanceau JP, Gouzy J, Parra G, Lardier G, Chapple C, McKernan KJ, McEwan P, Bosak S, Kellis M, Volff JN, Guigo R, Zody MC, Mesirov J, Lindblad-Toh K, Birren B, Nusbaum C, Kahn D, Robinson-Rechavi M, Laudet V, Schachter V, Quetier F, Saurin W, Scarpelli C, Wincker P, Lander ES, Weissenbach J, Roest Crollius H (2004) Genome duplication in the teleost fish Tetraodon nigroviridis reveals the early vertebrate proto-karyotype. Nature 431:946–957

Kasahara M (2007) The 2R hypothesis: an update. Curr Opin Immunol 19:547–552

Kawasaki K (2011) The SCPP gene family and the complexity of hard tissues in vertebrates. Cells Tissues Organs 194:108–112

Kielty CM, Sherratt MJ, Shuttleworth CA (2002) Elastic fibres. J Cell Sci 115:2817–2828

Kielty CM, Sherratt MJ, Marson A, Baldock C (2005) Fibrillin microfibrils. Adv Protein Chem 70:405–436

King N, Hittinger CT, Carroll SB (2003) Evolution of key cell signaling and adhesion protein families predates animal origins. Science 301:361–363

Kobayashi N, Kostka G, Garbe JH, Keene DR, Bachinger HP, Hanisch FG, Markova D, Tsuda T, Timpl R, Chu ML, Sasaki T (2007) A comparative analysis of the fibulin protein family. Biochemical characterization, binding interactions, and tissue localization. J Biol Chem 282:11805–11816

Kortschak RD, Tamme R, Lardelli M (2001) Evolutionary analysis of vertebrate Notch genes. Dev Genes Evol 211:350–354

Kozel BA, Mecham RP, Rosenbloom J (2011) Elastin. In: Mecham RP (ed) Extracellular matrix: an overview. Springer, Berlin, pp 267–301

Kubota Y, Kuroki R, Nishiwaki K (2004) A fibulin-1 homolog interacts with an ADAM protease that controls cell migration in C. elegans. Curr Biol 14:2011–2018

Kumaratilake JS, Gibson MA, Fanning JC, Cleary EG (1989) The tissue distribution of microfibrils reacting with a monospecific antibody to MAGP, the major glycoprotein antigen of elastin-associated microfibrils. Eur J Cell Biol 50:117–127

Lemaire R, Bayle J, Mecham RP, Lafyatis R (2007) Microfibril-associated MAGP-2 stimulates elastic fiber assembly. J Biol Chem 282(1):800–808

McConnell CJ, Wright GM, DeMont ME (1996) The modulus of elasticity of lobster aorta microfibrils. Experientia 52:918–921

McConnell CJ, DeMont ME, Wright GM (1997) Microfibrils provide non-linear elastic behaviour in the abdominal artery of the lobster Homarus americanus. J Physiol 499(pt 2):513–526

McLaughlin PJ, Chen Q, Horiguchi M, Starcher BC, Stanton JB, Broekelmann TJ, Marmorstein AD, McKay B, Mecham R, Nakamura T, Marmorstein LY (2006) Targeted disruption of fibulin-4 abolishes elastogenesis and causes perinatal lethality in mice. Mol Cell Biol 26:1700–1709

McLaughlin PJ, Bakall B, Choi J, Liu Z, Sasaki T, Davis EC, Marmorstein AD, Marmorstein LY (2007) Lack of fibulin-3 causes early aging and herniation, but not macular degeneration in mice. Hum Mol Genet 16:3059–3070

McLysaght A, Hokamp K, Wolfe KH (2002) Extensive genomic duplication during early chordate evolution. Nat Genet 31:200–204

Mecham RP, Davis EC (1994) In: Yurchenko PD, Mecham RP (eds) Extracellular matrix assembly and structure. Academic, San Diego

Meyer A, Van de Peer Y (2005) From 2R to 3R: evidence for a fish-specific genome duplication (FSGD). Bioessays 27:937–945

Miyamoto A, Lau R, Hein PW, Shipley JM, Weinmaster G (2006) Microfibrillar proteins MAGP-1 and MAGP-2 induce Notch1 extracellular domain dissociation and receptor activation. J Biol Chem 281:10089–10097

Miyata T, Suga H (2001) Divergence pattern of animal gene families and relationship with the Cambrian explosion. Bioessays 23:1018–1027

Moltzer E, te Riet L, Swagemakers SM, van Heijningen PM, Vermeij M, van Veghel R, Bouhuizen AM, van Esch JH, Lankhorst S, Ramnath NW, de Waard MC, Duncker DJ, van der Spek PJ, Rouwet EV, Danser AH, Essers J (2011) Impaired vascular contractility and aortic wall degeneration in fibulin-4 deficient mice: effect of angiotensin II type 1 (AT1) receptor blockade. PLoS One 6:e23411

Morris PJ (1993) The developmental role of the extracellular-matrix suggests a monophyletic origin of the kingdom Animalia. Evolution 47:152–165

Muriel JM, Dong C, Hutter H, Vogel BE (2005) Fibulin-1C and Fibulin-1D splice variants have distinct functions and assemble in a hemicentin-dependent manner. Development 132:4223–4234

Muriel JM, Xu X, Kramer JM, Vogel BE (2006) Selective assembly of fibulin-1 splice variants reveals distinct extracellular matrix networks and novel functions for perlecan/UNC-52 splice variants. Dev Dyn 235:2632–2640

Nakamura T, Lozano PR, Ikeda Y, Iwanaga Y, Hinek A, Minamisawa S, Cheng CF, Kobuke K, Dalton N, Takada Y, Tashiro K, Ross J Jr, Honjo T, Chien KR (2002) Fibulin-5/DANCE is essential for elastogenesis in vivo. Nature 415:171–175

Nehring LC, Miyamoto A, Hein PW, Weinmaster G, Shipley JM (2005) The extracellular matrix protein MAGP-2 interacts with Jagged1 and induces its shedding from the cell surface. J Biol Chem 280:20349–20355

Pan TC, Sasaki T, Zhang RZ, Fassler R, Timpl R, Chu ML (1993) Structure and expression of fibulin-2, a novel extracellular matrix protein with multiple EGF-like repeats and consensus motifs for calcium binding. J Cell Biol 123:1269–1277

Pazos F, Juan D, Izarzugaza JM, Leon E, Valencia A (2008) Prediction of protein interaction based on similarity of phylogenetic trees. Methods Mol Biol 484:523–535

Putnam NH, Srivastava M, Hellsten U, Dirks B, Chapman J, Salamov A, Terry A, Shapiro H, Lindquist E, Kapitonov VV, Jurka J, Genikhovich G, Grigoriev IV, Lucas SM, Steele RE, Finnerty JR, Technau U, Martindale MQ, Rokhsar DS (2007) Sea anemone genome reveals ancestral eumetazoan gene repertoire and genomic organization. Science 317:86–94

Putnam NH, Butts T, Ferrier DE, Furlong RF, Hellsten U, Kawashima T, Robinson-Rechavi M, Shoguchi E, Terry A, Yu JK, Benito-Gutierrez EL, Dubchak I, Garcia-Fernandez J, Gibson-Brown JJ, Grigoriev IV, Horton AC, de Jong PJ, Jurka J, Kapitonov VV, Kohara Y, Kuroki Y, Lindquist E, Lucas S, Osoegawa K, Pennacchio LA, Salamov AA, Satou Y, Sauka-Spengler T, Schmutz J, Shin IT, Toyoda A, Bronner-Fraser M, Fujiyama A, Holland LZ, Holland PW, Satoh N, Rokhsar DS (2008) The amphioxus genome and the evolution of the chordate karyotype. Nature 453:1064–1071

Rahn DD, Acevedo JF, Roshanravan S, Keller PW, Davis EC, Marmorstein LY, Word RA (2009) Failure of pelvic organ support in mice deficient in fibulin-3. Am J Pathol 174:206–215

Ramirez F, Dietz HC (2007) Fibrillin-rich microfibrils: structural determinants of morphogenetic and homeostatic events. J Cell Physiol 213:326–330

Ramirez F, Sakai LY, Rifkin DB, Dietz HC (2007) Extracellular microfibrils in development and disease. Cell Mol Life Sci 64:2437–2446

Reber-Muller S, Spissinger T, Schuchert P, Spring J, Schmid V (1995) An extracellular matrix protein of jellyfish homologous to mammalian fibrillins forms different fibrils depending on the life stage of the animal. Dev Biol 169:662–672

Reinhardt DP, Sasaki T, Dzamba BJ, Keene DR, Chu ML, Gohring W, Timpl R, Sakai LY (1996) Fibrillin-1 and fibulin-2 interact and are colocalized in some tissues. J Biol Chem 271:19489–19496

Roark EF, Keene DR, Haudenschild CC, Godyna S, Little CD, Argraves WS (1995) The association of human fibulin-1 with elastic fibers: an immunohistological, ultrastructural, and RNA study. J Histochem Cytochem 43:401–411

Robertson I, Jensen S, Handford P (2011) TB domain proteins: evolutionary insights into the multifaceted roles of fibrillins and LTBPs. Biochem J 433:263–276

Robson P, Wright GM, Youson JH, Keeley FW (2000) The structure and organization of lamprin genes: multiple-copy genes with alternative splicing and convergent evolution with insect structural proteins. Mol Biol Evol 17:1739–1752

Rychel AL, Smith SE, Shimamoto HT, Swalla BJ (2006) Evolution and development of the chordates: collagen and pharyngeal cartilage. Mol Biol Evol 23:541–549

Santini F, Harmon LJ, Carnevale G, Alfaro ME (2009) Did genome duplication drive the origin of teleosts? A comparative study of diversification in ray-finned fishes. BMC Evol Biol 9:194

Satou Y, Kawashima T, Shoguchi E, Nakayama A, Satoh N (2005) An integrated database of the ascidian, Ciona intestinalis: towards functional genomics. Zoolog Sci 22:837–843

Segade F (2009) Functional evolution of the microfibril-associated glycoproteins. Gene 439:43–54

Segade F (2010) Molecular evolution of the fibulins: implications on the functionality of the elastic fibulins. Gene 464:17–31

Segade F, Mecham RP (2005) Regulatory elements of microfibril-associated glycoprotein-1 gene expression in muscle cells. Biochim Biophys Acta 1731:215–224

Segade F, Broekelmann TJ, Pierce RA, Mecham RP (2000) Revised genomic structure of the human MAGP1 gene and identification of alternate transcripts in human and mouse tissues. Matrix Biol 19:671–682

Segade F, Trask BC, Broekelmann TJ, Pierce RA, Mecham RP (2002) Identification of a matrix-binding domain in MAGP1 and MAGP2 and intracellular localization of alternative splice forms. J Biol Chem 277:11050–11057

Segade F, Suganuma N, Mychaleckyj JC, Mecham RP (2007) The intracellular form of human MAGP1 elicits a complex and specific transcriptional response. Int J Biochem Cell Biol 39:2303–2313

Shimeld SM, Holland PWH (2000) Vertebrate innovations. Proc Natl Acad Sci U S A 97:4449–4452

Spencer JA, Hacker SL, Davis EC, Mecham RP, Knutsen RH, Li DY, Gerard RD, Richardson JA, Olson EN, Yanagisawa H (2005) Altered vascular remodeling in fibulin-5-deficient mice reveals a role of fibulin-5 in smooth muscle cell proliferation and migration. Proc Natl Acad Sci U S A 102:2946–2951

Tamura K (1992) Estimation of the number of nucleotide substitutions when there are strong transition-transversion and G+C-content biases. Mol Biol Evol 9:678–687

Timpl R, Sasaki T, Kostka G, Chu ML (2003) Fibulins: a versatile family of extracellular matrix proteins. Nat Rev Mol Cell Biol 4:479–489

Trask BC, Broekelmann T, Ritty TM, Trask TM, Tisdale C, Mecham RP (2001) Posttranslational modifications of microfibril associated glycoprotein-1 (MAGP-1). Biochemistry 40:4372–4380

Vandepoele K, De Vos W, Taylor JS, Meyer A, Van de Peer Y (2004) Major events in the genome evolution of vertebrates: paranome age and size differ considerably between ray-finned fishes and land vertebrates. Proc Natl Acad Sci U S A 101:1638–1643

Vogel BE, Hedgecock EM (2001) Hemicentin, a conserved extracellular member of the immunoglobulin superfamily, organizes epithelial and other cell attachments into oriented line-shaped junctions. Development 128:883–894

Vogel BE, Muriel JM, Dong C, Xu X (2006) Hemicentins: what have we learned from worms? Cell Res 16:872–878

Wagenseil JE, Mecham RP (2007) New insights into elastic fiber assembly. Birth Defects Res C Embryo Today 81:229–240

Wagenseil JE, Mecham RP (2009) Vascular extracellular matrix and arterial mechanics. Physiol Rev 89:957–989

Weinbaum JS, Broekelmann TJ, Pierce RA, Werneck CC, Segade F, Craft CS, Knutsen RH, Mecham RP (2008) Deficiency in microfibril-associated glycoprotein-1 leads to complex phenotypes in multiple organ systems. J Biol Chem 283:25533–25543

Werneck CC, Trask BC, Broekelmann TJ, Trask TM, Ritty TM, Segade F, Mecham RP (2004) Identification of a major microfibril-associated glycoprotein-1-binding domain in fibrillin-2. J Biol Chem 279:23045–23051

Werneck CC, Vicente CP, Weinberg JS, Shifren A, Pierce RA, Broekelmann TJ, Tollefsen DM, Mecham RP (2008) Mice lacking the extracellular matrix protein MAGP1 display delayed thrombotic occlusion following vessel injury. Blood 111:4137–4144

Williamson MP (1994) The structure and function of proline-rich regions in proteins. Biochem J 297(pt 2):249–260

Yan Q, Sage EH (1999) SPARC, a matricellular glycoprotein with important biological functions. J Histochem Cytochem 47:1495–1506
Yanagisawa H, Davis EC (2010) Unraveling the mechanism of elastic fiber assembly: the roles of short fibulins. Int J Biochem Cell Biol 42:1084–1093
Yanagisawa H, Davis EC, Starcher BC, Ouchi T, Yanagisawa M, Richardson JA, Olson EN (2002) Fibulin-5 is an elastin-binding protein essential for elastic fibre development in vivo. Nature 415:168–171
Zheng Q, Davis EC, Richardson JA, Starcher BC, Li T, Gerard RD, Yanagisawa H (2007) Molecular analysis of fibulin-5 function during de novo synthesis of elastic fibers. Mol Cell Biol 27:1083–1095

Chapter 7
Evolution and Function of SPARC and Tenascins: Matricellular Counter-Adhesive Glycoproteins with Pleiotropic Effects on Angiogenesis and Tissue Fibrosis

Cristina A. Baratta, Theodore J. Brown, Fahd Al-Dhalaan, and Maurice J. Ringuette

Abstract The evolution of multicellular organisms composed of tissues with distinct morphologies and physiological functions has involved the appearance of novel genes coding for extracellular matrix (ECM) molecules. The ECM consists of four major classes of proteins: collagens, proteoglycans, non-collagenous structural proteins, and glycoproteins. Our knowledge of how these molecules act in concert to assemble intricate networks that provide tissues with their unique design, biomechanical properties, and stability has advanced rapidly over the last three decades. Although the ECM was first perceived to be an inert scaffold, it is now common knowledge that ECM components have dynamic and complex activities that affect all cellular activities. The focus of this chapter is on the tenascin family and Secreted Protein, Acidic, Rich in Cysteine, which are members of the "matricellular" subset of ECM glycoproteins. In contrast to polymer-forming pro-adhesive glycoproteins, such as fibronectin and laminins, these glycoproteins are not integrated into ECM scaffolds but, rather, exist as transient, diffusible, components of interstitial matrices and basement membranes.

7.1 Introduction

Tenascin family members and Secreted Protein, Acidic, Rich in Cysteine (SPARC) exhibit overlapping expression during normal and pathological development, tissue remodeling and repair. While the tenascins and SPARC have distinct modular organization and functions, they share counter-adhesive activity on cell–matrix

C.A. Baratta • F. Al-Dhalaan • M.J. Ringuette (✉)
Department of Cell and Systems Biology, University of Toronto, 25 Harbord Street, Toronto, ON, Canada M5S3G5
e-mail: maurice.ringuette@utoronto.ca

T.J. Brown
Samuel Lunenfeld Research Institute, 600 University Avenue, Toronto, ON, Canada M5G 1X5

F.W. Keeley and R.P. Mecham (eds.), *Evolution of Extracellular Matrix*, Biology of Extracellular Matrix, DOI 10.1007/978-3-642-36002-2_7,

interactions and regulate similar signaling pathways, including VEGF and transforming growth factor β (TGF-β). Tenascin-C, the first and most extensively investigated tenascin, plays a prominent role in promoting VEGF-mediated neovascularization, a process fundamental to organogenesis, tissue remodeling, wound repair, and tumor formation. Tenascin-C also plays an essential role in fibrosis by modulating TGF-β signaling. These effects are shared with SPARC, which contributes to angiogenesis and fibrosis by modulating VEGF and TGF-β signaling. This chapter presents an overview of the evolution of the tenascin family and SPARC and discusses their complementary roles in angiogenesis and tissue fibrosis.

7.2 Modular Organization and Evolution

7.2.1 SPARC

SPARC, the prototypical member of the SPARC family, which includes SPARC-like/hevin, SMOC-1 (SPARC-related modular calcium binding 1) and SMOC-2, and the testicans, is expressed in early metazoans (Fig. 7.1, Koehler et al. 2009). While SPARC orthologs have not been identified in sponges, they appear in cnidarians, the first metazoans to form true tissues. Four distinct SPARC genes are found in the genome of the starlet sea anemone *Nematostella vectensis* whereas only a single gene is present in other cnidarians such as the freshwater hydrozoan *Hydra magnipapillata*. In triploblastic organisms ranging from nematodes to mammals, SPARC is a 33-kDa glycoprotein comprised of three distinct structural domains. The glutamic acid-rich N-terminal domain (I) binds up to eight Ca^{2+} ions with low affinity ($Kd = 10^{-3}$–10^{-5} M), rendering SPARC sensitive to changes in extracellular Ca^{2+} levels. Domain I is absent from SPARC in diploblastic cnidarians, suggesting acquisition of additional or altered roles for SPARC in triploblastic organisms. Domain II contains a follistatin-like module with 11 conserved cysteine residues. A conserved N-glycosylation site is found in all animal phyla at the amino terminus of this domain with the exception of two of the four *N. vectensis* orthologs: one ortholog lacks N-glycosylation sites whereas the other has two putative sites. The presence of four orthologs with differences in glycosylation sites in *N. vectensis* may reflect distinct and overlapping functions for these SPARC variants in this organism. This is supported by the observations that differences in the type of N-glycosylation of mammalian SPARC affect its binding to extracellular matrix (ECM) components (Maillard et al. 1992; Kaufmann et al. 2004). With the exception of *N. vectensis*, a single gene encodes SPARC in all other species where *SPARC* has been detected. SPARC orthologs have also been identified in rotifers (Clark et al. 2012) and mollusks (Boutet et al. 2009). Additionally, SPARC has recently been identified in the flatworms *Clonorchis sinesis* (Wang et al. 2011), *Schistosoma japonicum* (Accession #: ACM78480),

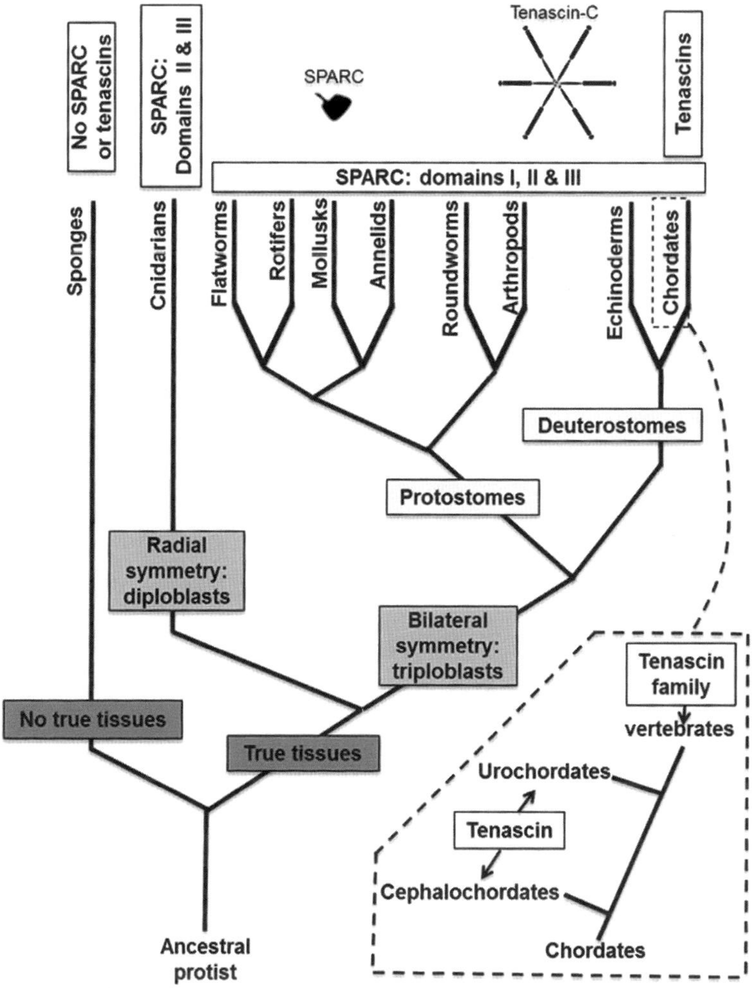

Fig. 7.1 Appearance of SPARC and tenascins during animal evolution. Matricellular glycoproteins have not been identified in organisms that do not form true tissues. SPARC from radially symmetrical diploblasts (Cnidarians) contains only the FC-EC domain pair (domains II and III). SPARC orthologs from bilaterally symmetrical triploblasts (protostomes and deuterostomes) also contain an N-terminal, Ca^{2+}-binding acidic domain (I). Tenascins are restricted to the chordate lineage. A single gene, coding for tenascin, has been described in cephalochordates, the closest animal phyla to invertebrates and, in urochordates, the closest animal phyla to vertebrates. In contrast, four tenascin paralogs are found in the vertebrate subphylum (*dashed box*). Schematic of tenascin-C derived from rotary shadow transmission electron photomicrographs (Erickson and Inglesias 1984)

and *Dugesia japonica* (Zhou 2009) but has not yet been fully characterized in these organisms.

The C-terminal half of SPARC (Domain III) is an α-helical-rich region containing two high-affinity Ca^{2+}-binding EF-hands: EF-hand 1 and 2, with Kds of 470 nM and 57 nM, respectively for mammalian SPARC (Fig. 7.2) (Busch et al. 2000). Because of the unique presence of Ca^{2+}-binding EF-hands on an extracellular molecule, Domain III is also referred to as the extracellular Ca^{2+} domain (EC domain). The high-affinity binding of Ca^{2+} is dependent on cooperative interactions between the EF-hands as well as between the EF-hands and Domain II (Busch et al. 2000). The presence of a disulfide-bridge in EF-hand 2, which is a unique feature of SPARC family members, is evolutionarily conserved in all animal phyla whose genomes have been annotated and constitutes the most conserved region of SPARC (Fig. 7.3a). A consensus sequence of SPARC EF-hand 2 across metazoan phyla highlights the evolutionary conservation of the cysteine residues. There is 70 % sequence identity of all the amino acids among vertebrates in this region (Fig. 7.3a, b). The amino acids critical for the high-affinity binding of a Ca^{2+} ion by EF-hand 2 are conserved among vertebrates, with many of these residues conserved in non-vertebrates as well (Fig. 7.3a, b). Domain III also contains two epitopes for fibrillar and network-forming collagen IV binding sites that are conserved from cnidarians to mammals (Koehler et al. 2009). The striking conservation of SPARC during metazoan evolution suggests that it plays a fundamental role in the development and function of these organisms.

Domains II and III are shared by all members of the SPARC family. SPARC-like, the most closely related ortholog, has the same modular organization as SPARC, except that domain I of SPARC-like is larger and more acidic. Phylogenetic analysis indicates that SPARC-like may have arisen from a duplication of the ancestral SPARC gene approximately 500 million years ago, after the emergence of teleosts (Kawasaki and Weiss 2004). SMOC-1 and SMOC-2 are characterized by a SMOC module flanked on either side by a thyroglobulin motif inserted between the FS-like and EC domains. Testicans, on the other hand, lack a SMOC module, but have a single thyroglobulin motif at the C-terminus. The SMOC module is restricted to vertebrate SMOC-1 and 2. However, SMOC-like and testican-like proteins are found in all animal phyla that form true tissues (Novinec et al. 2006). SPARC-like and testicans, similar to SPARC, are predominantly expressed in neural tissues, whereas SMOC-1 and SMOC-2 have broad tissue distributions during embryogenesis. However, unlike SPARC, they are expressed at high levels in adult tissues that are not undergoing remodeling. SPARC remains the most extensively investigated member of the SPARC family, particularly with respect to angiogenesis and fibrosis.

7.2.2 Tenascins

The signature features of tenascins are the presence of an N-terminal globular domain and heptad repeats, followed by a variable number of epidermal growth factor (EGF)-like repeats and fibronectin type III domains (FN3) and a C-terminal fibrinogen-related domain (FReD) (Fig. 7.2) (Tucker et al. 2006; Tucker and Chiquet-Ehrismann 2009). Four tenascin family members have been identified in

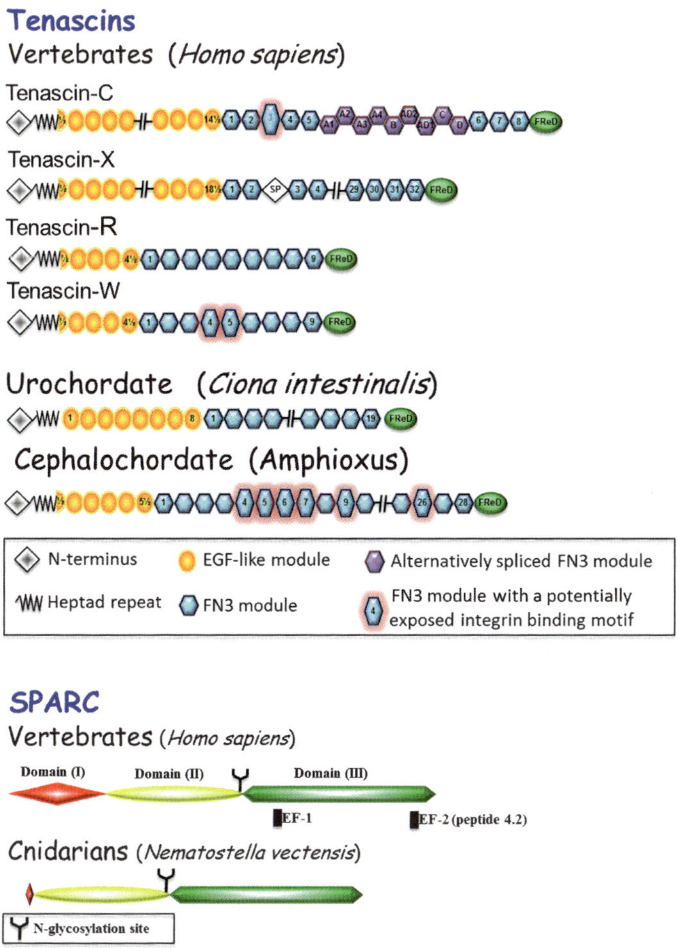

Fig. 7.2 Modular organization of tenascins and SPARC. The modular organization that defines tenascins includes an N-terminal globular domain, followed by heptad repeats, a variable number of EGF-like (*orange circles*) and FN-3 repeats (*blue hexagon*), and a C-terminal FReD domain. For simplicity, only representative tenascins from each Chordate subphyla are illustrated: vertebrates (*Homo sapiens*—tenascin-C,-X,-R,-W), urochordates (*Ciona intestinalis*), and cephalochordates (amphioxus). Alternatively spliced FN-3 modules are *numbered* and highlighted as *purple hexagons* for tenascin-C. FN-3 modules with predicted integrin-binding sites exposed are shown as larger *blue hexagons with pink*. A serine-proline motif (SP, *open rectangle*) is found in tenascin-X from vertebrates. Note that not all EGF-like and FN-3 repeats of the larger tenascins are shown (as indicated by *double slashes*). The schematics of tenascins have been modified from Chiquet-Ehrismann and Tucker (2011). Schematic of vertebrate (*Homo sapiens*) and Cnidarian (*Nematostella vectensis*) SPARC highlights that the acidic N-terminal domain I of SPARC is nearly absent in Cnidarians. The location of the two EF-hands in domain III are shown only for *Homo sapiens* SPARC. The N-glycosylation site *Homo sapiens* is conserved in all triploblastic organisms. Two of the four *N. vectensis* SPARC orthologs also contain putative N-glycosylation sites at that position. A synthetic peptide corresponding to EF-hand 2 is designated as peptide 4.2

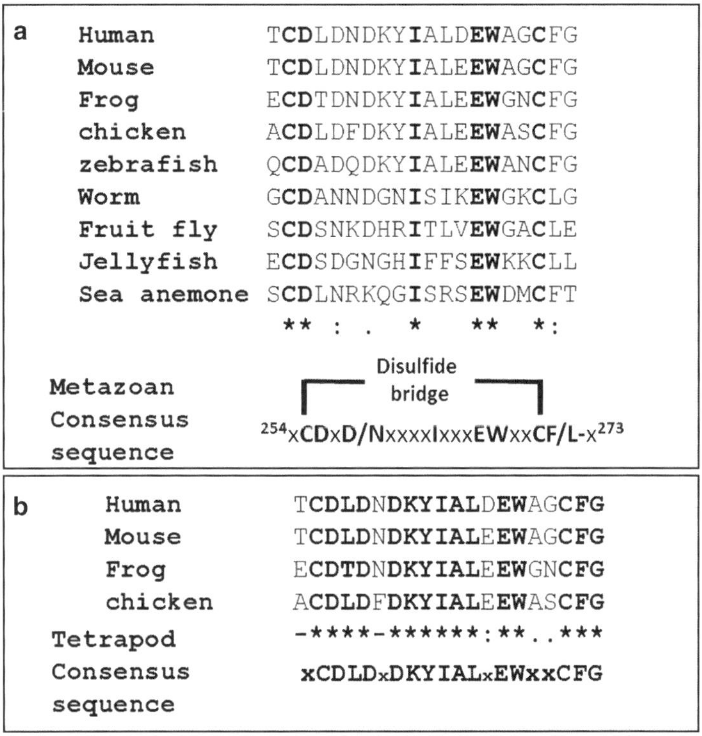

Fig. 7.3 Evolutionary conservation of SPARC EF-hand 2 (peptide 4.2). The most conserved domain of SPARC corresponds to the disulfide bridged EF-hand 2, a signature feature of the SPARC family. (**a**) The cysteine residues that form the disulfide bridge are conserved in all animal phyla whose genomes have been annotated. A consensus sequence (amino acids 254–273 from human SPARC) underscores the high degree of sequence conservation over this region. (**b**) A consensus sequence from four representative tetrapods indicates 70 % sequence identity in EF-hand 2

vertebrates: tenascin-C, tenascin-R, tenascin-W, and tenascin-X. Tenascin-R was originally identified in chicken and is significantly smaller than tenascin-C, owing to a fewer number of EGF-like domains (Rathjen et al. 1991). Tenascin-X was identified as a component of the major histocompatibility complex (MHC) class II in mouse and human (Matsumoto et al. 1992; Bristow et al. 1993). Structurally, this tenascin differs from tenascin-C and -R. While tenascin-X contains the standard sequence of heptad repeats and EGF-like domains, the FN3 domains are disrupted by a stretch of approximately 100 amino acids that is rich in proline (Tucker et al. 2006). Although the significance of this stretch is unknown, proline-rich regions have been shown to mediate protein–protein interactions. Tenascin-W was initially discovered in zebrafish by Weber et al. (1998) and has subsequently been found to overlap with the expression of tenascin-C in many tissues during mammalian development (Scherberich et al. 2004; Meloty-Kapella et al. 2006; Midwood et al. 2011).

Unlike SPARC, tenascin family members have only been identified in chordates. Tenascin-C was the first tenascin discovered and is highly expressed in embryonic neural, vascular, and skeletal tissues during embryonic development (Chiquet-Ehrismann and Tucker 2011). Minimal expression is evident in adult tissues except at myotendenous junctions (Järvinen et al. 2003). Similar to SPARC, tenascin-C is upregulated in response to trauma, wound repair, and during disease progression. Evidence for the existence of tenascins preceded the discovery of tenascin-C by nearly a decade when, in 1975, Yamada et al. reported the agglutinating activity of an embryonic chick fibroblast cell-surface protein. Nine years later, this protein was found by rotary shadowing in transmission electron microscopy preparations to form a six-armed structure, referred to as a "hexabrachion" (Erickson and Inglesias 1984; Fischer et al. 1997) (Fig. 7.1-schematic). While initially referred to by various names, including hexabrachion protein, cytotactin, and myotendinous antigen, it was renamed tenascin in 1986 (Chiquet-Ehrismann et al. 1986). With the discovery of other family members, this tenascin was designated as tenascin-C (C for cytotactin).

The ability to form a hexabrachion structure enables tenascin-C to function potently as a hub for the aggregation of interacting proteins. Hexabrachion formation begins with the trimerization of tenascin-C monomers into a coiled-coil structure that is stabilized by disulfide bonding between the amino terminus and the heptad repeats. Homophilic interactions between trimers and the formation of disulfide bonds result in the establishment of a complete hexamer (Kammerer et al. 1998). Tenascin-C (Jones and Jones 2000) and tenascin-W (Scherberich et al. 2004) are the only vertebrate tenascins that have been demonstrated to form hexabrachion structures; however, tenascin-R is also likely to form hexamers as it contains the necessary cysteine residue in the N-terminal region of the monomer required for association of tenascin trimers (Jones and Jones 2000). The EGF-like motif-containing regions within the hexabrachion arms appear narrower than the more distal regions of the arms, which contain the FN3 repeats (Chiquet-Ehrismann 1990). In contrast to the inelastic nature of the EGF-like repeats, FN3 repeats are more elastic owing to an absence of cysteine residues (Hynes 1999). This elasticity within the distal arms of the hexabrachion imparts flexibility, likely enhancing tenascin's ability to form molecular interactions. Tenascin-C also forms nonameric structures (Schenk and Chiquet-Ehrishmann 1994), suggesting that tenascin-C monomers might preferably arrange in multiples of three (Jones and Jones 2000). Heterotypic hexamers of tenascin-C have also been observed where two trimers of varying components associate covalently (Fischer et al. 1995, 1997; Kammerer et al. 1998). These heterotypic trimers could consist of different tenascin-C splice variants or, possibly, different tenascin family members altogether (Jones and Jones 2000). Thus, the ability to form heterotypic structures represents a viable method for increasing the functional diversity of this family. However, it is not abundantly clear how frequently tenascin heteroligomers are found in the ECM and what influences their prevalence (Jones and Jones 2000).

FN3 domains facilitate binding to other ECM proteins and to cell-surface receptors. These motifs contain 7 antiparallel β-strands organized into 2 sheets

(Patthy 1990; Leahy et al. 1992). While tenascin-C contains 8 constitutively expressed FN3 domains (Joester and Faissner 2001), 9 FN3 domains have been identified that are subject to alternative splicing (Jones et al. 1988, 1989; Spring et al. 1989; Gulcher et al. 1989, 1991; Nishi et al. 1991; Weller et al. 1991; Sriramarao and Bourdon 1993; Dorries and Schachner 1994; Derr et al. 1997) possibly generating up to 512 potential alternative monomers (Joester and Faissner 1999). For example, 27 different splice variants have been identified in the developing mouse brain (Joester and Faissner 1999). While the process of splice site selection has yet to be elucidated, the factors determining which splice variants are present in a given tissue have been identified. These include changes in the extracellular pH (Borsi et al. 1995), presence of specific growth factors (i.e., TGFβ) (Zhao and Young 1995), and the proliferative state of the composite cells (Borsi et al. 1994).

The C-terminal FReD contains a high-affinity Ca^{2+}-binding EF-hand that promotes interactions with other ECM components (Tufty and Kretsinger 1975). This globular domain is formed from protein loops held covalently by two intrachain disulfide bonds (Doolittle 1984).

Tenascins are particularly susceptible to cleavage by matrix metalloproteinases (MMPs), specifically MMP-1, -2, -3, and -7 (Imai et al. 1994; Siri et al. 1995), which could potentially expose neo-epitopes. Additionally, serine proteases, such as cathepsin G and leukocyte elastase (Siri et al. 1995) and plasmin (Gundersen et al. 1997), have been demonstrated to cleave tenascin-C. Interestingly, plasmin proteolysis converts tenascin-C into an adhesive substrate for T lymphocytes and potentially regulates the effects tenascin-C has on T-lymphocyte function (Gundersen et al. 1997).

Four members of the tenascin family have been identified in the vertebrate subphylum. However, *Tetradon nigroviridis* and *Takifugu rubripes*, the fresh water and Japanese pufferfish, respectively, appear to be an exception by possessing five tenascins; the fifth representing a gene duplication of tenascin-C (Tucker et al. 2006). The first invertebrate tenascin-like gene was identified in the sea squirt *Ciona intestinalis* (Tucker et al. 2006), a member of the chordate subphylum Urochordata, also referred to as Tunicata. This tenascin bears the hallmark of vertebrate tenascins (Tucker et al. 2006). More recently, a second tunicate, *Ciona savignyi*, was examined for the presence of a tenascin gene. The putative tenascin-like gene identified in this species lacks the N-terminal linker region and heptad repeats but encodes 8 EGF-like domains, 14 FN3 domains, and a FReD (Tucker and Chiquet-Ehrismann 2009). Closer examination of the tenascin from *C. intestinalis* predicts a possible capacity for integrin binding. This tenascin contains an RGE motif in its third FN3 domain, a position that coincides with the RGD motif of vertebrate tenascin-C. This RGE motif is absent from *C. savignyi* in spite of a high degree of similarity with the rest of this domain (Tucker and Chiquet-Ehrismann 2009). Nonetheless, this tenascin contains a putative integrin recognition domain (IDPPS) in its eighth FN3 domain, which is similar to the IDPPT found in the ninth FN3 domain of

C. intestinalis tenascin (Tucker and Chiquet-Ehrismann 2009; Tucker et al. 2006). Thus, the ability to interact with integrin is a conserved feature of the tenascins.

A tenascin is also encoded by the cephalochordate *Branchiostoma floridae* (amphioxus) genome (Tucker and Chiquet-Ehrismann 2009). This tenascin contains the same set of domains as those found in tunicates: a secretion signal containing several cysteine residues that could facilitate oligomerization, heptad repeats, 5 EGF-like domains, 38 FN3 domains, and a FReD at the C-terminus (Tucker and Chiquet-Ehrismann 2009). However, unlike the tenascin from *C. intestinalis* which contains an RGE motif, amphioxus tenascin contains a remarkable seven RGD motifs (Tucker and Chiquet-Ehrismann 2009), suggesting enhanced integrin interaction for this tenascin. Further analysis of amphioxus tenascin revealed that while its EGF-like repeats and FReD are most similar to those of vertebrate tenascins, the FN3 domains show similarity to proteins from organisms ranging from anthozoans to echinoderms but no similarity to either a tenascin or fibronectin (Tucker and Chiquet-Ehrismann 2009). More specifically, amphioxus FN3 domains show similarity to regions from, DCC (netrin receptor deleted in colorectal carcinoma/frazzled), (Dscam) Down syndrome cell adhesion molecule, and sidekick (Dscam-related adhesion molecule) (Tucker and Chiquet-Ehrismann 2009). As such, an interesting possibility raised by Tucker and Chiquet-Ehrismann (2009) is that these FN3 domains originated prior to the divergence of cephalochordates from the remainder of the chordate lineage.

As might be expected from the dynamic functional interactions between vertebrate tenascins and fibronectin (FN), a FN gene first appears in tunicates. Huxley-Jones et al. (2007) isolated a partial sequence of a FN-like gene from the *C. intestinalis* genome while an entire predicted fibronectin protein was identified from *C. savignyi* containing the signature FN1, FN2, and FN3 motifs (Tucker, and Chiquet-Ehrismann 2009). Interestingly, this predicted FN gene contains atypical immunoglobulin-like domains that lack potential RGD integrin-binding motifs, which are present in vertebrate fibronectin. The lack of these motifs raises the possibility that tunicate FN functions differently from that of vertebrates. In contrast to tenascin, a FN-like gene has not been identified in cephalochordates, leading Tucker and Chiquet-Ehrismann (2009) to propose that tenascin compensates for FN in the more ancient chordates. This is consistent with the increased integrin interaction sites in amphioxus tenascin. They further outline two scenarios to account for the absence of a FN-like gene in cephalochordates. One is that tenascin and FN coevolved after the divergence of chordates from echinoderms, but that FN was subsequently lost in cephalochordates. Alternatively, tenascin may have first appeared in an ancestral chordate, whereas FN did not appear until after the divergence of cephalochordates (Tucker and Chiquet-Ehrismann 2009).

7.3 Activities During Normal and Pathological Angiogenesis

7.3.1 Tenascin-C

Understanding the cellular and signaling events associated with the formation of new blood vessels through the process of angiogenesis has been the subject of intense investigation for many reasons. Principal among them is the critical role angiogenesis plays in ensuring tissues have sufficient oxygen and nutrients for their survival throughout development and following injury. This vital process is also important for the progression and survival of solid tumors. A growing body of evidence indicates that tenascin-C and SPARC make overlapping and distinct contributions to normal and pathological angiogenesis.

In vertebrates, endothelial cells line the luminal surface of blood vessels and play a central role in the angiogenic process. The endothelial cells are overlaid with a basement membrane with elastic fibers and contractile smooth muscle cells that surround the basement membrane of veins and arteries but not capillaries. The basement membranes of capillary networks are overlaid with pericytes, multifunctional cells that can differentiate into smooth muscle cells and other cell types to promote blood vessel maturation and remodeling.

During mammalian development, tenascin-C is transiently expressed in several tissues, including the developing heart and vasculature (Willems et al. 1996; Laky and Parascan 2007; van Obberghen et al. 2011). It was therefore somewhat unexpected that *tenascin-C*-null mice develop normally, suggesting that tenascin is not essential for blood vessel formation (Mitrovic and Schachner 1995). However, an impaired angiogenic phenotype emerges in response to environmental stress, wound repair, and disease progression (van Obberghen et al. 2011). Ballard et al. (2006) used a syngeneic cardiac allograph model to demonstrate that tenascin-C contributes to cardiac angiogenesis. This model involves transplantation of a neonatal heart from a donor strain into a host mouse. A benefit of this in vivo model is that a fibrin clot is formed around the cardiac allograph and that the allograft becomes vascularized by a combination of injected bone-marrow-derived progenitor cells from a donor and from host endothelial cells. Their data demonstrate that compared to wild-type mice, cardiac allographs implanted into *tenascin-C*-null mice fail to vascularize. They further demonstrated that cardiac endothelial cells to roundup and migrate towards VEGF and PDGF in the presence of tenascin-C. This, combined with their immunohistological data, demonstrated that tenascin-C co-localizes with the heart vasculature and transplanted endothelial progenitor cell and provides initial compelling evidence that tenascin-C plays a fundamental role in promoting early stages of angiogenesis in vivo. Tenascin-C has since been implicated in promoting angiogenesis in a broad spectrum of pathologies, including inflammatory diseases, diabetes, and cancer (as reviewed by Golledge et al. 2011; Obbergghen et al. 2011). The following discussion summarizes the role of tenascin-C during dermal wound repair, as the cellular and molecular events regulating angiogenesis during this process are shared with other remodeling events and pathologies.

Hypoxic conditions, generated by tissue injury, activate VEGF expression by keratinocytes and macrophages during the proliferative/granulation phase of wound repair. Other pro-angiogenic factors such as basic fibroblast growth factors (bFGFs), placenta growth factor (PIGF), and transforming growth factor β and α (TGFβ and TGFα) are activated by hypoxia-inducible factors (HIFs) (Kaur et al. 2005). VEGF is considered the most potent angiogenic factor under normal and most pathophysiological conditions. The paracrine activation of endothelial cells by VEGF induces the expression of matrix remodeling metalloproteases (MMPs) required to degrade the capillary basement membranes, setting the stage for the formation of branch points within existing blood vessels and sprouting behavior by the endothelial cells. Schenk et al. (1999) demonstrated that tenascin-C is expressed by sprouting but not by non-sprouting endothelial cells. Their in vitro data demonstrate that tenascin-C and bFGF act in concert to promote the reorganization of the actin cytoskeleton in endothelial cells exhibiting a spouting behavior. While a functional relationship between fibronectin and tenascin-C is well established, sprouting behavior was not induced when fibronectin was used as a substratum. Their data also demonstrated that the C-terminal FReD is primarily responsible for the counter-adhesive effect of tenascin-C on endothelial cells and on the induction of a sprouting phenotype. Based on studies demonstrating that bFGF also induces endothelial cell proliferation and migration, the authors hypothesize that tenascin-C acts in concert with bFGF to induce sprouting behavior by endothelial cells.

During wound repair, a complex mixture of matrix MMPs and other proteases are expressed by a variety of cell types, such as keratinocytes at the wound edge, inflammatory cells, macrophages, fibroblasts, myofibroblasts, and migrating endothelial cells. Their pleotropic effects on wound repair include the liberation of growth factors (reviewed by Parks 1999). MMP2, the predominant MMP during wound repair, is expressed by dermal fibroblasts and endothelial cells throughout the repair process (Salo et al. 1994; Siri et al. 1995) demonstrated that an FN3 repeat within the alternative splice domain of the large isoform of tenascin-C (300 kDa) is highly susceptible to MMP2 degradation. As the large tenascin-C isoform is widely distributed within the provisional matrix of the granulation tissue, it is conceivable that cleavage of tenascin-C by MMP2 at the leading edge of migrating endothelial cells enhances both their proliferation and migration.

VEGF increases vascular permeability, promoting the leakage of plasma proteins such as fibrinogen into the provisional matrix, which promotes the migration of several cell types, including endothelial cells. Brellier et al. (2011) recently reported that tenascin-C and fibrin co-localize in head and neck squamous cell carcinoma and that tissue plasmin activator expression is dramatically upregulated in *tenascin-C* deficient mouse embryonic fibroblasts. It remains to be determined if this correlation reflects a functional relationship between tenascin-C and the remodeling of the fibrin-rich (and collagen-rich) provisional matrix to promote angiogenesis during wound repair.

During the early phase of wound repair, peripheral blood fibrocytes migrate with inflammatory cells into the wound site. TGF-β, secreted by endothelial cells, induces the fibrocytes to differentiate into pericytes (Xueyong et al. 2008). In the final stages of angiogenesis PDGF, secreted by endothelial cells, recruits the differentiated pericytes to the growing vessels (Bjarnegård et al. 2004). Since tenascin-C is secreted by pericytes (Herold-Mende et al. 2002; Martina et al. 2010a), it is likely that they contribute to the broad distribution of tenascin-C within the provisional matrix and blood vessel basement membranes, promoting blood vessel maturation during the remodeling phase of angiogenesis.

FN, like tenascin-C, has pleotropic effects on angiogenesis. The adhesive properties of fibronectin and its interactions with growth factors that affect angiogenesis are modulated by interactions with tenascin-C and cell-surface heparin-sulfate proteoglycans. How these molecules are co-assembled into matrices and how they act in concert to promote angiogenesis and tumor vascular mimicry, which is beyond the scope of this chapter, has been comprehensively reviewed by Van Obberghen-Schilling et al. (2011).

In contrast to vertebrates, the circulatory vessels of invertebrate organisms of the cephalochordata and urochordata subphyla lack endothelial cells. Instead, the lumen of their vessels is lined with a laminin-containing basement membrane (Hartenstein and Mandal 2006). While it is unknown if tenascin associates with the vessel basement membranes in cephalochordates, tenascin is detected in the basement membrane of *C. intestinalis* vessels (Tucker et al. 2006); raising the possibility that tenascin has function in role the formation and/or stability of urochordate circulatory vessels.

7.3.2 SPARC

In vitro and in vivo studies indicate that SPARC has pleotropic effects on angiogenesis during wound repair and tissue remodeling. SPARC, like tenascins, is expressed at high levels by endothelial cells and plays a role in the recruitment of pericytes to nascent microvessels. As with other ECM components, studies on the effects of SPARC on angiogenesis have generated conflicting results that may be context-dependent and likely involving multiple mechanisms. Early in vitro studies by Sage and coworkers (Kupprion et al. 1998) indicated that SPARC exerts pleiotropic effects on endothelial cells by inducing G1 cell cycle arrest (Funk and Sage 1991) inhibiting proliferation (Sage et al. 1995) and migration (Hasselaar and Sage 1992), inducing cell rounding and increasing endothelial barrier permeability (Goldblum et al. 1994). They further showed that SPARC, as well as a small peptide corresponding to EF-hand 2 (designated peptide 4.2), binds to VEGF-A, preventing its binding to VEGFR1 on endothelial cells. Consistent with these studies, docking simulation modeling indicates that the SPARC amino acids sequences that interact with VEGF-A overlap with the VEGFR1 interaction site. Interestingly, while binding of SPARC to VEGF-A inhibited VEGFR1 phosphorylation, no effect on

VEGFR2 phosphorylation was observed (Kupprion et al. 1998). The pro-angiogenic effects of VEGF-A in cultured endothelial cells are largely mediated by VEGFR2 while VEGFR1 has been shown to positively or negatively regulate angiogenesis in vivo through a number of distinct mechanisms (Cao 2009). Thus SPARC's ability to regulate VEGF-mediated angiogenesis may be highly context-specific.

Studies have revealed that SPARC modulates VEGF activity in the microenvironment. Using an ocular neovascularization mouse model, Nozaki et al. (2006) observed that VEGFA exerted an anti-angiogenic effect on the choroid following injury, which they attributed to a transient decrease in the presence of SPARC. The authors hypothesized that with lower levels of SPARC, VEGF-A was able to activate VEGFR1, thereby inhibiting VEGFR2-mediated neovascularization. Consistent with these results, injection of VEGF-A into the vitreous humor of *Sparc*-null mice had no effect on neovascularization. Moreover, the authors demonstrated that VEGFR1 activates Src homology domain-2 containing tyrosine phosphatase-1 (SHP-1), resulting in the dephosphorylation and inactivation of VEGFR2, thereby inhibiting angiogenesis. Therefore, in the presence of SPARC, VEGFR1, and SHP-1 activation are diminished, enhancing pro-angiogenic VEGFR2 signaling. In contrast, SPARC inhibits the vascularization of gliomas by decreasing the expression and secretion of VEGFA (Yunker et al. 2008), adding further to the complexity of the effect of SPARC on angiogenesis. Consistent with this, dermal fibroblasts from *Sparc*-null mice express higher levels of VEGFA relative to cells from wild-type mice (Bradshaw et al. 2001). These complex and opposing actions on angiogenesis may reflect divergent contributions by the different domains of SPARC and their distinct molecular interactions.

A synthetic peptide with potent anti-angiogenic activity in vitro and in vivo was recently identified corresponding to the N-terminal region of the follistatin-like domain (CQNHHCKHGKV<u>CELDENNTPMC</u>), designated FS-E of SPARC. Moreover, the C-terminal region of this peptide (underlined sequence) had significantly higher anti-angiogenic activity than FS-E. Like tenascin-C, both peptides promoted the normalization of blood vessels in a neuroblastoma xenograft model (Chlenski et al. 2010). As the normalization of blood vessels enhances the delivery of chemotherapeutic agents, the normalization of blood vessels by SPARC peptides holds promise as an adjuvant in the treatment of cancers.

SPARC also inhibits angiogenesis by binding directly to platelet-derived growth factor (PDGF)-BB and PDGF-AB (Raines et al. 1992; Kupprion et al. 1998) and by decreasing fibroblast growth factor 2 (FGF2) and TGF-β signaling (Motamed et al. 2003; Rivera and Brekken 2011). PDGF-BB promotes endothelial cell proliferation and cord/tube formation (Battegay et al. 1994), an activity that is also promoted by SPARC (Iruela-Arispe et al. 1991; Goldblum et al. 1994; Liu et al. 2009), presenting an apparent contradiction. However, it is conceivable that the proliferation and movement of endothelial cells promoted by PDGF-BB during cord/tube formation is dampened by SPARC to prevent excessive blood vessel formation. The promoting effects of SPARC may thus be independent of PDGF. FGF2 also plays an important role in promoting angiogenesis by increasing endothelial cell proliferation and cord/tube formation. While SPARC does not bind FGF2, both VEGF-A signaling via VEGFR1, and FGF2 signaling via FGFR1, affect endothelial cell proliferation by activating the MAPK/ERK1/2 pathway. This led Rivera and Brekken (2011) to propose that the inhibition of FGF2 signaling by SPARC is in part mediated by its binding to VEGF-A. Consistent with this hypothesis, inhibition of FGF-2 signaling was ascribed to SPARC EF-hand2 (peptide 4.2), the region of SPARC that binds VEGF-A (Motamed et al. 2003).

TGF-β affects a broad spectrum of cellular activities, including cell proliferation and differentiation, and ECM secretion. During angiogenesis, TGF-β inhibits pericyte migration and promotes the secretion of basal membrane components by these cells that are necessary for the maturation phase of angiogenesis (Rivera and Brekken 2011). Hence, inhibition of TGF-β signaling in migrating pericytes may be important for the coordination of composite events in angiogenesis by preventing their premature differentiation/maturation. While the precise mechanisms remain poorly understood, a functional relationship has long been known for TGF-β and SPARC (see section on SPARC and tissue fibrosis). A recent study by Rivera and Brekken (2011) indicates that SPARC inhibits the activation of TGF-β-mediated signaling in pericytes by its binding to endoglin, a member of the TGF-β receptor signaling complex. TGF-β is secreted as an inactive form coupled to latent-associated peptide (LAP) that binds to α_V integrins. Based on these interactions and the fact that TGF-β signaling is increased in *Sparc*-null pericytes, the authors proposed that SPARC binding to endoglin inhibits the association of the LAP-integrin complex to TGF-β receptor 2 (TβRII), thereby silencing TGF-β1 signaling in the migrating pericytes. The authors further propose that upon proteolytic cleavage of SPARC, endoglin is able to promote TβRII-LAP-integrin complex interaction to trigger the signaling cascade. As a result, the maturation phase of angiogenesis is able to proceed.

The activities of SPARC can be altered or unmasked through proteolytic cleavage by MMPs. For example, digestion of human SPARC with MMP-3 generates cleavage products that have a dose-dependent effect on endothelial cell proliferation and migration (Sage et al. 2003). These include a 3.9 kDa peptide fragment containing the Cu^{2+}-binding sequence KHGK that stimulated the growth of blood vessels in a chorioallantoic membrane (CAM) assay. In contrast, a 6.1 kDa peptide (containing EF-hand 1) and 2.2 kDa peptide fragment, which were also generated, had no effect on vascular growth in CAM assays, but inhibited endothelial cell migration in vitro (Sage et al. 2003).

SPARC induces the expression and/or activation of several MMPs, including the collagenases MMP1 and MT1-MMP, and the gelatinases MMP-2, MMP-9, and MMP-11 during tissue remodeling, repair, and pathogenesis. Expression of MMP-1 and MMP9 by monocytes is also induced by a synthetic peptide corresponding to a segment of SPARC domain III (Shankavaram et al. 1997) whereas activation of MMP2 is mimicked by a peptide corresponding to the N-terminal domain (Gilles et al. 1998). SPARC has also been proposed to decrease angiogenesis by inhibiting MMP-9 expression (Bhoopathi et al. 2010). Therefore, as with tenascin-C, SPARC may have opposing activities at different stages of angiogenesis during wound repair as a consequence of its proteolytic processing or its impact on the MMP expression profile. For example, while MMP-1 and MMP-9 do not cleave mammalian SPARC (Sage et al. 2003), the interaction of SPARC with matrix components may be altered by these MMPs.

The counter-adhesive activity of SPARC may play a role in promoting endothelial cell migration and vessel remodeling during angiogenesis. SPARC promotes rounding of endothelial cells and fibroblasts in culture by decreasing the number of focal adhesion sites (Brekken and Sage 2000). This effect can be reproduced in vitro by small SPARC mimetic peptides corresponding to regions within all three domains, including peptide 4.2 (Sage et al. 1995).

7.4 Tissue Fibrosis and Wound Repair

7.4.1 Tenascin-C

The biophysical properties of connective and other tissues are determined by collagen-rich ECM. While remodeling of the ECM is essential for the development and growth of tissues, it is also required for adaptive changes in adult tissues, wound healing, and tissue regeneration. Matrix remodeling involves a complex repertoire of carefully orchestrated events that balance ECM component synthesis and degradation. Fibrosis occurs when the rate of formation of collagen and associated ECM components exceeds the rate of their degradation, which is manifested in the formation of scar tissue (Leask et al. 2004; Leask and Abraham 2003). In fibrotic diseases, matrix remodeling is misregulated, resulting in an excessive production and accumulation of collagen and other ECM components such as SPARC, fibronectin, thrombospondin, and tenascins. This often occurs in association with abnormal wound healing and in inflammatory diseases (Youssef and Tavill 2002; Gonzales and Coleman 1999; Peterson 2005; Carey et al. 2010). Fibrosis also occurs in response to tumor formation and is considered a critical part of tumor growth and survival (Desmoulière et al. 2004). Although the etiology of fibrosis is poorly understood, TGF-β and connective tissue growth factor (CTGF; CCN2) are prominent promoters of ECM formation (Leask et al. 2003, 2004; Leask and Abraham 2003, 2004). TGF-β stimulates the expression of CTGF (Leask et al. 2003), SPARC (Wrana et al. 1988, 1991a, b) and tenascin (Pearson et al. 1988), which are associated with ECM remodeling (Wasi et al. 1984; Lussier et al. 2001; Salonen et al. 1990; Maschler et al. 2004) and are upregulated in fibrotic tissues (Masson et al. 1998; Frizell et al. 1995; Francki and Sage 2001; Sadlier et al. 2004; Carey et al. 2010). The pro-fibrotic effects of TGF-β involve the promotion of myofibroblast differentiation, characterized by the expression of α-smooth muscle actin (α-SMA), a fibronectin isoform (EDA-FN), plasminogen activator inhibitor-1 (PAI-1), CTGF and SPARC and tenascin-C (Schiemann et al. 2003; Garrett et al. 2004; Tahashi et al. 2002; Yoshimura et al. 2011). The induction of α-SMA markedly enhances the contractile ability of myofibroblasts while EDA-FN promotes cell anchoring via increased focal adhesions. The combined action of these factors promotes collagen contraction associated with both wound healing and fibrosis. Increased SPARC (Zhou et al. 2005; Bassuk et al. 2000; Fujita et al. 2002; Francki et al. 2004) and tenascin (Carey et al. 2010) expression is a consistent observation in both these processes and may act in a feed-forward manner to enhance TGF-β signaling to promote fibrosis

Acute lung injury is characterized by a transient inflammatory phase that disrupts the capillary-endothelial cell interface, leading to the formation of life threatening pulmonary edemas. This is followed by a long-term reparative chronic phase with compromised tissue compliance due to variable amounts of interstitial fibrosis. While TGF-β signaling promotes pulmonary edema, it plays an instrumental role in interstitial fibrosis during the chronic phase. TGF-β signaling is attenuated in the lung interstitium of *tenascin-C* null mice relative to wild-type mice. Inhalation of bleomycin resulted in fewer myofibroblasts and decreased collagen I in the lung interstitium during the fibrotic phase of the null mice (Carey et al. 2010), suggesting that tenascin-C may promote the pro-fibrotic activity of TGF-β. In wild-type mice, tenascin-C transcript (Zhao et al. 1998) and protein levels increase dramatically during the acute inflammatory phase, but decrease during the ensuing fibrotic phase, thereby limiting the actions of TGF-β and the amount of fibrosis induced. In contrast, tenascin-C expression by human fibroblasts persists, resulting in prolonged TGF-β action promoting tissue fibrosis (Carey et al. 2010). The authors proposed that prolonged expression of tenascin-C in humans during the chronic phase contributes to the transformation of fibroblasts into contractile, matrix secreting myofibroblasts; however, the mechanism remains to be resolved.

7.4.2 Contributions by Other Members of the Tenascin Family to Angiogenesis and Tissue Fibrosis

In addition to tenascin-C, the vertebrate tenascin family consists of tenascin-R, -X, and -W, all of which exhibit anti-adhesive properties in vitro. Tenascin-W and tenascin-C have overlapping expression patterns during mammalian and teleost development (Weber et al. 1998; Scherberich et al. 2004). Unlike tenascin-C, tenascin-W is not increased during wound healing (Brellier et al. 2012), but similar to tenascin-C it is upregulated in the reactive stroma of several cancers (Degen et al. 2007; van Obberghen-Schilling et al. 2011; Martina et al. 2010b). Tenascin-W expression in gliomas localizes to the perivascular area of malignant tissues. Staining for this tenascin is not observed around blood vessels of nonmalignant tissues indicating that tenascin-W may promote tumor neovascularization. Consistent with such a role, tenascin-W induces endothelial cell sprouting (Martina et al. 2010a).

Tenascin-X, the largest member of the tenascin family, is expressed in muscle, the connective tissues of the brain meninges and choroid plexus (Kawakami and Matsumoto 2011). *Tenascin-X*-null mice are viable but show several mild phenotypes including heightened anxiety levels and improved memory, mild muscle weakness and myopathy with decrease collagen fibril density along peripheral nerves. While tenascin-X is expressed within the skin, it is not upregulated in response to wounding in granulation tissue of *tenascin-C*-null mice. Similar to tenascin-W, tenascin-X is an unlikely candidate to functionally compensate for the absence of tenascin-C during wound repair (Brellier et al. 2012). In humans, tenascin-X is associated with an autosomal recessive form of Ehlers-Danlos syndrome (Schalkwijk et al. 2001), characterized by hypermobile joints and skin and fragility of the perivascular capillary networks. The expression of tenascin-R is restricted to the central nervous system where it is thought to be involved in maintaining neural stem cells (Hargus et al. 2008). A role for this tenascin in angiogenesis and tissue fibrosis has not been addressed.

7.4.3 SPARC

Our understanding of the role of SPARC in fibrosis is derived from studies with mammalian tissue culture cells, animal models, and human pathologies. A striking upregulation in the expression of SPARC has been a consistent finding in fibroblasts from scleroderma patients (Zhou et al. 2002, 2005; Unemori and Amento 1991), indicating it is an important regulator of fibrosis. Studies with *Sparc*-null mice show impaired bleomycin-induced lung fibrosis and wound healing, further supporting the role of SPARC (Strandjord et al. 1999). Moreover, inhibition of SPARC expression by siRNA restores normal collagen deposition in scleroderma fibroblasts (Zhou et al. 2006). Studies indicate that SPARC regulates the initial stages of collagen fibrillogenesis (Giudici et al. 2008; Bradshaw 2009), but how SPARC specifically exerts its extracellular effects to modulate ECM deposition and remodeling is poorly understood. SPARC may act in part by promoting excessive collagen I deposition induced by TGF-β (Unemori and Amento 1991; Zhou et al. 2006; Sodek et al. 2002). Several studies also indicate that SPARC has intracellular functions (Huynh et al. 2000, 2004; Yan et al. 2005; Gooden et al. 1999; Barker et al. 2005a), which may modulate secretion of ECM proteins and contribute to fibrosis.

Almost invariably, upregulation of SPARC and collagen is associated with an increased expression of TGF-β. Studies have demonstrated TGF-β-stimulated SPARC expression in keratinocytes (Ford et al. 1993), fibroblasts (Zhou et al. 2005; Reed et al. 1994), and periodontal ligament cells (Fujita et al. 2002), and in a rat model of glomerulonephritis (Bassuk et al. 2000). Conversely, SPARC regulates TGF-β-dependent expression and signaling in primary glomerular mesangial cells (Francki et al. 2004; Bassuk et al. 2000), suggesting the existence of a positive feed-back loop between SPARC and TGF-β that may underlie the pro-fibrotic cycle. Schellings et al. (2009) demonstrated that SPARC has a protective effect on the heart following injury. Their study demonstrated that wild-type mice had a fourfold better survival rate than *Sparc*-null mice following myocardial infarction induced by coronary occlusion. The ECM of *Sparc*-null infarcted hearts had disorganized granulation tissue and immature collagens; a phenotype that was rescued by the infusion of TGFβ. Ectopic overexpression of SPARC in the wild-type mice enhanced TGFβ signaling, which had a protective effect on the heart following myocardial injury. As surprising observation was that *Sparc*-null mice had a significantly higher number of myofibroblasts, despite decreased TGFβ signaling. As elaborated on by McCurdy et al. (2010) in a comprehensive review of the activities of SPARC on cardiac ECM remodeling following injury and pressure overload, a possible explanation for this paradoxical result is the impact of SPARC on TGFβ signaling in fibroblasts may be distinct from other cell type and dependent on the precise composition of the ECM microenvironment.

SPARC is expressed at high levels in tissues that maintain a high rate of collagen turnover, such as bone, periodontal tissues (Salonen et al. 1990; Tung et al. 1985) and intestine (Limeback and Sodek 1979; Lussier et al. 2001). SPARC is over-expressed together with collagen in fibroblasts obtained from the skin of patients with scleroderma and systemic sclerosis (Macko et al. 2002). SPARC is also highly expressed in other fibrotic disorders including cardiac (Masson et al. 1998), kidney (Pichler et al. 1996; Francki and Sage 2001; Sadlier et al. 2004), liver (Frizell et al. 1995; Blazejewski et al. 1997; Nakatani et al. 2002), lung (Kuhn and Mason 1995) fibrosis and radiation-induced fibrosis (Strup-Perrot et al. 2004). A prominent phenotype in partially backcrossed *Sparc*-null mice is impaired collagen fibrillogenesis leading to the formation of dermal tissue with low tensile strength (Basu et al. 2001; Bradshaw et al. 2003). Relative to wild-type mice, the collagen fibrils in dermal tissues of *Sparc*-null mice are smaller and more uniform in size. Fibrotic responses to implanted foreign materials in these mice are also impaired and are characterized by smaller ECM capsule size, in part due to reduced collagen deposition (Strandjord et al. 1999; Pichler et al. 1996; Puolakkainen et al. 2005; Taneda et al. 2003). However, the absence of SPARC has no apparent impact on the initial acute inflammatory response that precedes chronic fibrosis (Barker et al. 2005b). That the absence of Sparc-like has no effect on fibrous encapsulation but enhances the initial inflammatory response, places particular emphasis on a pivotal role for SPARC in collagen synthesis and deposition. Further highlighting the functional importance of the affinity of SPARC for collagen is the observation that SPARC is not retained in the ECM of collagen-1 deficient Mov-13 mice (Iruela-Arispe et al. 1996).

Sparc-null mice also exhibit amelioration of diabetic nephropathy, a disease associated with chronic fibrotic responses. The decreased fibrosis in this model has been attributed to diminished glomerular deposition of collagen IV and laminin, and decreased interstitial deposition of collagens I and IV (Taneda et al. 2003). Thus, in addition to fibrillar collagens, SPARC also enhances the deposition of network-forming collagen IV in basal laminae. Studies have also shown a decrease in collagen type IV deposition in the reactive stroma of lobular mammary tumors grown in SPARC-null mice (Sangaletti et al. 2003). Our studies with *Drosophila melanogaster* indicate that SPARC is expressed by hemocytes and the fat body, which are the main source of basement membrane components for embryos (Martinek et al. 2002) and larvae. Loss of SPARC from the fat body leads to precocious cell rounding and accumulation of collagen IV, laminin, perlecan, and nidogen at the surface of adipocytes. Interestingly, the basal lamina surrounding the fat body is remarkably intact in spite of this fibrotic-like aberrant accumulation of ECM (Baratta et al., unpublished data).

Inhibiting the re-epithelization of wounds has been correlated with increased fibrosis following tissue injury. Emerging evidence indicates that keratinocytes secrete factors that reduce tissue fibrosis by decreasing ECM secretion by dermal fibroblasts. A recent study indicates that the keratinocytes secrete 14-3-3σ (stratafin) that interacts with SPARC to decrease secretion of type I collagen by dermal fibroblasts (Chavez-Muñoz et al. 2012). The authors speculate that by sequestering SPARC, stratafin prevents the induction of type I collagen by SPARC. In this scenario stratafin would serve to limit the pro-fibrotic actions of SPARC in the vicinity of migrating keratinocytes during the reepithelialization of wound repair. However, if a role of stratafin is to provide fine control over the activity of SPARC released by dermal fibroblasts, it remains to be clarified why keratinocytes simultaneously secrete SPARC with stratafin.

7.4.4 Contributions by Other Members of the SPARC Family to Angiogenesis and Fibrosis

In contrast to SPARC, little is known of the role of other SPARC family members in angiogenesis and fibrosis. SPARC-like activities have best characterized in neural tissues, where it is highly expressed during brain development and pathogenesis. Recent studies have indicated SPARC-like expression in several cancers, but its specific action remains unexplored. In the brain, SPARC-like is thought to regulate radial glial processes by modulating focal adhesion through its follistatin-like domain, thereby supporting neuronal migration. SPARC-like is expressed in the endothelia of the developing blood brain barrier, with some expression retained in adulthood. It is thought that SPARC-like may influence blood brain barrier ECM to affect the permeability of the barrier. A similar role outside of the neural system has not been addressed. Testicans, initially discovered is seminal plasma, are found primarily in neural tissue where they may regulate protease activity and neurite outgrowth (Schnepp et al. 2005). While testican has been reported to be expressed by endothelial cells (Marr et al. 1997), its relationship to angiogenesis remains unknown. SMOC-1 and -2 appear to affect the development of the gonads and Wolffian ducts (Pazin and Albrecht 2009). While there is no direct evidence that SMOC-1 affects angiogenesis, recent studies indicate an interesting role for SMOC-2 (Rocnik et al. 2006). SMOC-2 is ubiquitously expressed with the highest levels of expression reported in heart, spleen, and muscle (Rocnik et al. 2006). Ectopic expression of SMOC-2 in human umbilical vein endothelial cells induces DNA synthesis and network formation in matrigel angiogenesis assays whereas SMOC-2 knock down expression reduced both proliferation and network forming ability of the Rocnik et al. (2006). Moreover, these studies also indicate that SMOC-2 may potentiate the effects of VEGF and FGF on endothelial cell proliferation.

7.5 Concluding Remarks

Insight into the complex biological activities of matricellular glycoproteins may be obtained by an examination of their evolutionary histories. For example, the appearance of SPARC is coincident with the evolution of cnidarians, the first multicellular organisms to form true tissues. It is intriguing that the acidic N-terminal domain capable of binding several Ca^{2+} ions with low affinity, found in all other animal phyla, is virtually absent in SPARC expressed by diploblastic organisms (Koehler et al. 2009). The FS-EC domain pair that is conserved from cnidarians to mammals indicates that they act in concert to make a fundamental contribution to the development and physiology of tissue-forming animals. Why the acquisition of the N-terminal acidic domain is coincident with the evolutionary appearance of triploblastic organisms is interesting in light of the diverse functions of Ca^{2+} ions. The reasons are likely to be multifactorial as the biological activity of proteins can be altered by several factors, including changes in tissue distribution during evolution.

In the basal cnidarian *N. vectensis*, the four SPARC orthologs are first expressed by the endoderm, the germ layer which gives rise to the mesoderm in triploblastic organisms. Zygotic expression in most organisms begins following gastrulation within the mesoderm, with expression observable shortly thereafter within all germ layers (Damjanovski et al. 1994; Kang et al. 2008), indicating that SPARC activities are coincident with the formation of all tissues only after major embryonic induction events have occurred. Once organogenesis is complete, SPARC expression decreases to near basal levels except in tissues undergoing remodeling. As with tenascins, only in response to tissue injury or disease progression is SPARC expression reactivated, perhaps to promote the reestablishment of tissue homeostasis. However, in some cases, it appears that both SPARC and tenascins contribute to abnormal tissue remodeling and disease progression. In nonmammalian organisms that do not express SPARC-like, ectopic expression and knockdown of SPARC leads to severe developmental defects (Huynh et al. 1999, 2011; Kang et al. 2008; Fitzgerald and Schwarzbauer 1998). At present, there is little evidence that SPARC-like may substitute for the absence of SPARC in higher vertebrates. It is therefore conceivable that structurally unrelated matricellular glycoproteins (1) with overlapping expression patterns and activities, such as the tenascin family, or (2) SPARC family members that, with common domains, such as the SMOC subfamily, may functionally compensate for the loss of SPARC.

In organisms with a relatively simple ECM, such as nematodes and insects, SPARC is a major component of basement membranes. In contrast, SPARC is often not detected immunohistochemically in the basement membranes of organisms with an abundance of interstitial matrices, perhaps due to epitope masking. The absence of SPARC from the basement membrane of *Ceanorhabditis elegans* (Fitzgerald and Schwarzbauer 1998) and *D. melanogaster* (Baratta et al., unpublished data) leads to embryonic and larval lethality respectively. It remains to be determined if unmasking the activities of SPARC in the basement membranes by utilizing the genetic and molecular tools afforded by these model organisms, will lead to the discovery of novel activities of SPARC in higher organisms.

As with SPARC, the tenascin family is composed of modules that are traceable to the evolution of multicellularity. However, the organization into a novel collective of modules that is the signature of tenascins is a fairly recent event, coincident with the evolution of cephalochordates and urochordates. The transition from a single gene to a family of genes in the chordate phylum occurred with the evolution of vertebrates, which are characterized by a number of major developments such as an endoskeleton, closed circulatory system, and highly developed sense organs, including the brain. The twice-repeated genome duplication that occurred in ancestral vertebrates also gave rise to evolution of other new ECM molecules, such as FN. While the focus of this review has been on the shared activities of SPARC and tenascin-C, it is clear from numerous studies that tenascin-C and FN have formed a dynamic reciprocal relationship from their genesis (van Obberghen et al. 2011, and references therein).

Given the complexity of the ECM and its influence upon all cellular activities, deciphering the precise mechanisms of action of the structurally diverse SPARC and tenascin family members remains a formidable challenge. Moreover, posttranslational modifications and proteolytic processing alter their activities. While SPARC and tenascin-C have distinct functions, they converge in regulating angiogenesis and fibrosis. Unfortunately, these glycoproteins have thus far been studied independent of one another; however, a recent study has demonstrated that SMOC-2 binds tenascin-C with high affinity to oppose the pro-migratory effects of tenascin on glioma cells. It is, therefore, highly probable that members of a matricellular family may functionally compensate for a member of another family. Despite having distinct structural organization, they may act in concert to promote a specific developmental event or counteract one another in an attempt to maintain tissue homeostasis.

References

Ballard VL, Sharma A, Duignan I, Holm JM, Chin A, Choi R, Hajjar KA, Wong S-C, Edelberg JM (2006) Vascular tenascin-C regulates cardiac endothelial phenotype and neovascularization. FASEB J 20(6):717–719

Barker TH, Baneyx G, Cardó-Vila M, Workman GA, Weaver M, Menon PM, Dedhar S, Rempel SA, Arap W, Pasqualini R, Vogel V, Sage EH (2005a) SPARC regulates extracellular matrix organization through its modulation of integrin-linked kinase activity. J Biol Chem 280(43): 36483–36493

Barker TH, Framson P, Puolakkainen PA, Reed M, Funk SE, Sage EH (2005b) Matricellular homologs in the foreign body response: hevin suppresses inflammation, but hevin and SPARC together diminish angiogenesis. Am J Pathol 166(3):923–933

Bassuk JA, Pichler R, Rothmier JD, Pippen J, Gordon K, Meek RL, Bradshaw AD, Lombardi D, Strandjord TP, Reed M, Sage EH, Couser WG, Johnson R (2000) Induction of TGF-[bgr]1 by the matricellular protein SPARC in a rat model of glomerulonephritis. Kidney Int 57(1): 117–128

Basu A, Kligman L, Samulewicz S, Howe C (2001) Impaired wound healing in mice deficient in a matricellular protein SPARC (osteonectin, BM-40). BMC Cell Biol 2(1):15

Battegay EJ, Rupp J, Iruela-Arispe L, Sage EH, Pech M (1994) PDGF-BB modulates endothelial proliferation and angiogenesis in vitro via PDGF beta-receptors. J Cell Biol 125(4): 917–928

Bhoopathi P, Chetty C, Gujrati M, Dinh DH, Rao JS, Lakka SS (2010) The role of MMP-9 in the anti-angiogenic effect of secreted protein acidic and rich in cysteine. Br J Cancer 102(3): 530–540

Bjarnegård M, Enge M, Norlin J, Gustafsdottir S, Fredriksson S, Abramsson A, Takemoto M, Gustafsson E, Fässler R, Betsholtz C (2004) Endothelium-specific ablation of PDGFB leads to pericyte loss and glomerular, cardiac and placental abnormalities. Development 131(8): 1847–1857

Blazejewski SLBB, Boussarie L, Blanc JF, Malaval L, Okubo K, Saric J, Bioulac-Sage P, Rosenbaum J (1997) Osteonectin (SPARC) expression in human liver and in cultured human liver myofibroblasts. Am J Pathol 151(3):651–657

Borsi L, Balza E, Castellani P, Carnemolla B, Ponassi M, Querzé G, Zardi L (1994) Cell-cycle dependent alternative splicing of the tenascin primary transcript. Cell Commun Adhes 1(4): 307–317

Borsi L, Balza E, Gaggero B, Allemanni G, Zardi L (1995) The alternative splicing pattern of the tenascin-C pre-mRNA is controlled by the extracellular pH. J Biol Chem 270(11): 6243–6245

Boutet I, Jollivet D, Shillito B, Moraga D, Tanguy A (2009) Molecular identification of differentially regulated genes in the hydrothermal-vent species Bathymodiolus thermophilus and Paralvinella pandorae in response to temperature. BMC Genomics 10(1):222

Bradshaw A (2009) The role of SPARC in extracellular matrix assembly. J Cell Commun Signal 3 (3–4):239–246

Bradshaw AD, Reed MJ, Carbon JG, Pinney E, Brekken RA, Sage EH (2001) Increased fibrovascular invasion of subcutaneous polyvinyl alcohol sponges in SPARC-null mice. Wound Repair Regen 9(6):522–530

Bradshaw AD, Puolakkainen P, Dasgupta J, Davidson JM, Wight TN, Sage EH (2003) SPARC-null mice display abnormalities in the dermis characterized by decreased collagen fibril diameter and reduced tensile strength. J Invest Dermatol 120(6):949–955

Brekken RA, Sage EH (2000) SPARC, a matricellular protein: at the crossroads of cell–matrix. Matrix Biol 19(7):569–580

Brellier F, Hostettler K, Hotz H-R, Ozcakir C, Çöloğlu SA, Togbe D, Ryffel B, Roth M, Chiquet-Ehrismann R (2011) Tenascin-C triggers fibrin accumulation by downregulation of tissue plasminogen activator. FEBS Lett 585(6):913–920

Brellier FME, Chiquet M, Ferralli J, van der Heyden M, Orend G, Schittny JC, Chiquet-Ehrismann R, Tucker RP (2012) The adhesion modulating properties of tenascin-W. Int J Biol Sci 8(2): 187–194

Bristow J, Tee M, Gitelman S, Mellon S, Miller W (1993) Tenascin-X: a novel extracellular matrix protein encoded by the human XB gene overlapping P450c21B. J Cell Biol 122(1):265–278

Busch E, Hohenester E, Timpl R, Paulsson M, Maurer P (2000) Calcium affinity, cooperativity, and domain interactions of extracellular EF-hands present in BM-40. J Biol Chem 275(33): 25508–25515

Cao Y (2009) Positive and negative modulation of angiogenesis by VEGFR1 ligands. Sci Signal 2 (59):re1

Carey WA, Taylor GD, Dean WB, Bristow JD (2010) Tenascin-C deficiency attenuates TGF-β-mediated fibrosis following murine lung injury. Am J Physiol Lung Cell Mol Physiol 299(6): L785–L793

Chavez-Muñoz C, Hartwell R, Jalili RB, Jafarnejad M, Lai A, Nabai L, Ghaffari A, Hojabrpour P, Kanaan N, Duronio V, Guns E, Cherkasov A, Ghahary A (2012) SPARC/SFN interaction, suppresses type I collagen in dermal fibroblasts. J Cell Biochem 113(8):2622–2632

Chiquet-Ehrismann R (1990) What distinguishes tenascin from fibronectin? FASEB J 4(9): 2598–2604

Chiquet-Ehrismann R, Tucker, RP (2011) Tenascins and the importance of adhesion modulation. Cold Spring Harb Perspect Biol 3(5):pii: a004960

Chiquet-Ehrismann R, Mackie EJ, Pearson CA, Sakakura T (1986) Tenascin: an extracellular matrix protein involved in tissue interactions during fetal development and oncogenesis. Cell 47(1):131–139

Chlenski A, Guerrero L, Peddinti R, Spitz J, Leonhardt P, Yang Q, Tian Y, Salwen H, Cohn S (2010) Anti-angiogenic SPARC peptides inhibit progression of neuroblastoma tumors. Mol Cancer 9(1):138

Clark MS, Thorne MAS, Reinhardt R, Drungowski M, Albrecht MW, Klages S, Beck A, Kube M, Lubzens E (2012) Long-term survival of hydrated resting eggs from *Brachionus plicatilis*. PLoS One 7(1):e29365

Damjanovski S, Malaval L, Ringuette MJ (1994) Transient expression of SPARC in the dorsal axis of early Xenopus embryos: correlation with calcium-dependent adhesion and electrical coupling. Int J Dev Biol 38(3):439–446

Degen M, Brellier F, Kain R, Ruiz C, Terracciano L, Orend G, Chiquet-Ehrismann R (2007) Tenascin-W is a novel marker for activated tumor stroma in low-grade human breast cancer and influences cell behavior. Cancer Res 67(19):9169–9179

Derr LB, Chiquet-Ehrismann R, Gandour-Edwards R, Spence J, Tucker RP (1997) The expression of tenascin-C with the AD1 variable repeat in embryonic tissues, cell lines and tumors in various vertebrate species. Differentiation 62(2):71–82

Desmoulière A, Guyot C, Gabbiani G (2004) The stroma reaction myofibroblast: a key player in the control of tumor cell behavior. Int J Dev Biol 48:509–517

Doolittle RF (1984) Fibrinogen and fibrin. Annu Rev Biochem 53(1):195–229

Dorries U, Schachner M (1994) Tenascin mRNA isoforms in the developing mouse brain. J Neurosci Res 37(3):336–347

Erickson H, Inglesias JL (1984) A six-armed oligomer isolated from cell surface fibronectin preparations. Nature 311(5983):267–269

Fischer D, Chiquet-Ehrismann R, Bernasconi C, Chiquet M (1995) A single heparin binding region within the fibrinogen-like domain is functional in chick tenascin-C. J Biol Chem 270(7): 3378–3384

Fischer D, Tucker RP, Chiquet-Ehrismann R, Adams JC (1997) Cell-adhesive responses to tenascin-C splice variants involve formation of fascin microspikes. Mol Biol Cell 8(10): 2055–2075

Fitzgerald MC, Schwarzbauer JE (1998) Importance of the basement membrane protein SPARC for viability and fertility in Caenorhabditis elegans. Curr Biol 8(23):1285–1288

Ford R, Wang G, Jannati P, Adler D, Racanelli P, Higgins PJ, Staiano-Coico L (1993) Modulation of SPARC expression during butyrate-induced terminal differentiation of cultured human keratinocytes: regulation via a TGF-β-dependent pathway. Exp Cell Res 206(2):261–275

Francki A, Sage EH (2001) SPARC and the kidney glomerulus: matricellular proteins exhibit diverse functions under normal and pathological conditions. Trends Cardiovasc Med 11(1): 32–37

Francki A, McClure TD, Brekken RA, Motamed K, Murri C, Wang T, Sage EH (2004) SPARC regulates TGF-beta1-dependent signaling in primary glomerular mesangial cells. J Cell Biochem 91(5):915–925

Frizell E, Shu-Ling L, Abraham A, Ozaki I, Eghbali M, Sage EH, Zern MA (1995) Expression of SPARC in normal and fibrotic livers. Hepatology 21(3):847–854

Fujita T, Shiba H, Sakata M, Uchida Y, Ogawa T, Kurihara H (2002) Effects of transforming growth factor-beta 1 and fibronectin on SPARC expression in cultures of human periodontal ligament cells. Cell Biol Int 26(12):1065–1072

Funk SE, Sage EH (1991) The Ca2(+)-binding glycoprotein SPARC modulates cell cycle progression in bovine aortic endothelial cells. Proc Natl Acad Sci U S A 88(7):2648–2652

Garrett Q, Khaw PT, Blalock TD, Schultz GS, Grotendorst GR, Daniels JT (2004) Involvement of CTGF in TGF-β1—stimulation of myofibroblast differentiation and collagen matrix contraction in the presence of mechanical stress. Invest Ophthalmol Vis Sci 45(4):1109–1116

Gilles C, Bassuk JA, Pulyaeva H, Sage EH, Foidart J-M, Thompson EW (1998) SPARC/osteonectin induces matrix metalloproteinase 2 activation in human breast cancer cell lines. Cancer Res 58(23):5529–5536

Giudici C, Raynal N, Wiedemann H, Cabral WA, Marini JC, Timpl R, Bächinger HP, Farndale RW, Sasaki T, Tenni R (2008) Mapping of SPARC/BM-40/osteonectin-binding sites on fibrillar collagens. J Biol Chem 283(28):19551–19560

Goldblum SE, Ding X, Funk SE, Sage EH (1994) SPARC (secreted protein acidic and rich in cysteine) regulates endothelial cell shape and barrier function. Proc Natl Acad Sci U S A 91(8): 3448–3452

Golledge J, Clancy P, Maguire J, Lincz L, Koblar S (2011) The role of tenascin C in cardiovascular disease. Cardiovasc Res 92(1):19–28

Gonzales TS, Coleman GC (1999) Periodontal manifestations of collagen vascular disorders. Periodontol 2000 21(1):94–105

Gooden MD, Vernon RB, Bassuk JA, Sage EH (1999) Cell cycle-dependent nuclear location of the matricellular protein SPARC: association with the nuclear matrix. J Cell Biochem 74(2): 152–167

Gulcher JR, Nies DE, Marton LS, Stefansson K (1989) An alternatively spliced region of the human hexabrachion contains a repeat of potential N-glycosylation sites. Proc Natl Acad Sci U S A 86(5):1588–1592

Gulcher JR, Nies DE, Alexakos MJ, Ravikant NA, Sturgill ME, Marton LS, Stefansson K (1991) Structure of the human hexabrachion (tenascin) gene. Proc Natl Acad Sci U S A 88(21): 9438–9442

Gundersen D, Tran-Than C, Sordat B, Mourali F, Ruegg C (1997) Plasmin-induced proteolysis of tenascin-C: modulation by T lymphocyte-derived urokinase-type plasminogen activator and effect on T lymphocyte adhesion, activation, and cell clustering. J Immunol 158(3):1051–1060

Hargus G, Cui Y, Schmid J-S, Xu J, Glatzel M, Schachner M, Bernreuther C (2008) Tenascin-R promotes neuronal differentiation of embryonic stem cells and recruitment of host-derived neural precursor cells after excitotoxic lesion of the mouse striatum. Stem Cells 26(8): 1973–1984

Hartenstein V, Mandal L (2006) The blood/vascular system in a phylogenetic perspective. Bioessays 28(12):1203–1210

Hasselaar P, Sage EH (1992) SPARC antagonizes the effect of basic fibroblast growth factor on the migration of bovine aortic endothelial cells. J Cell Biochem 49(3):272–283

Herold-Mende C, Mueller MM, Bonsanto MM, Schmitt HP, Kunze S, Steiner H-H (2002) Clinical impact and functional aspects of tenascin-C expression during glioma progression. Int J Cancer 98(3):362–369

Huxley-Jones J, Robertson DL, Boot-Handford RP (2007) On the origins of the extracellular matrix in vertebrates. Matrix Biol 26(1):2–11

Huynh M-H, Helene Sage E, Ringuette M (1999) A calcium-binding motif in SPARC/osteonectin inhibits chordomesoderm cell migration during Xenopus laevis gastrulation: evidence of counter-adhesive activity in vivo. Dev Growth Differ 41(4):407–418

Huynh M-H, Hong H, Delovitch S, Desser S, Ringuette M (2000) Association of SPARC (osteonectin, BM-40) with extracellular and intracellular components of the ciliated surface ectoderm of Xenopus embryos. Cell Motil Cytoskeleton 47(2):154–162

Huynh M-H, Sodek K, Lee H, Ringuette M (2004) Interaction between SPARC and tubulin in *Xenopus*. Cell Tissue Res 317(3):313–317

Huynh M-H, Zhu S, Kollara A, Brown T, Winklbauer R, Ringuette M (2011) Knockdown of SPARC leads to decreased cell–cell adhesion and lens cataracts during post-gastrula development in *Xenopus laevis*. Dev Genes Evol 220(11):315–327

Hynes RO (1999) The dynamic dialogue between cells and matrices: Implications of fibronectin's elasticity. Proc Natl Acad Sci U S A 96(6):2588–2590

Imai K, Kusakabe M, Sakakura T, Nakanishi I, Okada Y (1994) Susceptibility of tenascin to degradation by matrix metalloproteinases and serine proteinases. FEBS Lett 352(2):216–218

Iruela-Arispe M, Diglio C, Sage E (1991) Modulation of extracellular matrix proteins by endothelial cells undergoing angiogenesis in vitro. Arterioscler Thromb Vasc Biol 11(4):805–815

Iruela-Arispe ML, Vernon RB, Wu H, Jaenisch R, Sage EH (1996) Type I collagen-deficient Mov-13 mice do not retain SPARC in the extracellular matrix: implications for fibroblast function. Dev Dyn 207(2):171–183

Järvinen TAH, Józsa L, Kannus P, Järvinen TLN, Hurme T, Kvist M, Pelto-Huikko M, Kalimo H, Järvinen M (2003) Mechanical loading regulates the expression of tenascin-C in the myotendinous junction and tendon but does not induce de novo synthesis in the skeletal muscle. J Cell Sci 116(5):857–866

Joester A, Faissner A (1999) Evidence for combinatorial variability of tenascin-C isoforms and developmental regulation in the mouse central nervous system. J Biol Chem 274(24): 17144–17151

Joester A, Faissner A (2001) The structure and function of tenascins in the nervous system. Matrix Biol 20(1):13–22

Jones FS, Burgoon MP, Hoffman S, Crossin KL, Cunningham BA, Edelman GM (1988) A cDNA clone for cytotactin contains sequences similar to epidermal growth factor-like repeats and segments of fibronectin and fibrinogen. Proc Natl Acad Sci U S A 85(7):2186–2190

Jones FS, Hoffman S, Cunningham BA, Edelman GM (1989) A detailed structural model of cytotactin: protein homologies, alternative RNA splicing, and binding regions. Proc Natl Acad Sci U S A 86(6):1905–1909

Jones FS, Jones PL (2000) The tenascin family of ECM glycoproteins: Structure, function and regulation during embryonic development and tissue remodeling. Dev Dyn 218(2):235–259

Kammerer RA, Schulthess T, Landwehr R, Lustig A, Fischer D, Engel J (1998) Tenascin-C hexabrachion assembly is a sequential two-step process initiated by coiled-coil α-helices. J Biol Chem 273(17):10602–10608

Kang Y-J, Stevenson A, Yau P, Kollmar R (2008) Sparc protein is required for normal growth of zebrafish otoliths. J Assoc Res Otolaryngol 9(4):436–451

Kaufmann B, Müller S, Hanisch F-G, Hartmann U, Paulsson M, Maurer P, Zaucke F (2004) Structural variability of BM-40/SPARC/osteonectin glycosylation: implications for collagen affinity. Glycobiology 14(7):609–619

Kaur BKF, Severson EA, Matheny SL, Brat DJ, Van Meir EG (2005) Hypoxia and the hypoxia-inducible-factor pathway in glioma growth and angiogenesis. Neuro Oncol 7(2):134–153

Kawakami K, Matsumoto K-I (2011) Behavioral alterations in mice lacking the gene for tenascin-X. Biol Pharm Bull 34(4):590–593

Kawasaki KST, Weiss K (2004) Genetic basis for the evolution of vertebrate mineralized tissue. Proc Natl Acad Sci U S A 101(31):11356–11361

Koehler A, Desser S, Chang B, MacDonald J, Tepass U, Ringuette M (2009) Molecular evolution of SPARC: absence of the acidic module and expression in the endoderm of the starlet sea anemone, & *Nematostella vectensis*. Dev Genes Evol 219(9):509–521

Kuhn C, Mason RJ (1995) Immunolocalization of SPARC, tenascin, and thrombospondin in pulmonary fibrosis. Am J Pathol 147(6):1759–1769

Kupprion C, Motamed K, Sage EH (1998) SPARC (BM-40, osteonectin) inhibits the mitogenic effect of vascular endothelial growth factor on microvascular endothelial cells. J Biol Chem 273(45):29635–29640

Laky D, Parascan L (2007) Hibernating myocardium, morphological studies on intraoperatory myocardial biopsies and on chronic ischemia experimental model. Rom J Morphol Embryol 48(4):407–413

Leahy DJ, Hendrickson WA, Aukhil I, Erickson HP (1992) Structure of a fibronectin type III domain from tenascin phased by MAD analysis of the selenomethionyl protein. Science 258 (5084):987–991

Leask A, Abraham DJ (2003) The role of connective tissue growth factor, a multifunctional matricellular protein, in fibroblast biology. Biochem Cell Biol 81(6):355–363

Leask A, Abraham DJ (2004) TGF-ß signaling and the fibrotic response. FASEB J 18(7):816–827

Leask A, Holmes A, Black CM, Abraham DJ (2003) Connective tissue growth factor gene regulation. J Biol Chem 278(15):13008–13015

Leask A, Denton CP, Abraham DJ (2004) Insights into the molecular mechanism of chronic fibrosis: the role of connective tissue growth factor in scleroderma. J Invest Dermatol 122(1): 1–6

Limeback HF, Sodek J (1979) Procollagen synthesis and processing in periodontal ligament in vivo and in vitro. Eur J Biochem 100(2):541–550

Liu A, Mosher DF, Murphy-Ullrich JE, Goldblum SE (2009) The counteradhesive proteins, thrombospondin 1 and SPARC/osteonectin, open the tyrosine phosphorylation-responsive paracellular pathway in pulmonary vascular endothelia. Microvasc Res 77(1):13–20

Lussier C, Sodek J, Beaulieu J-F (2001) Expression of SPARC/osteonectin/BM4O in the human gut: predominance in the stroma of the remodeling distal intestine. J Cell Biochem 81(3): 463–476

Macko RF, Gelber AC, Young BA, Lowitt MH, White B, Wigley FM, Goldblum SE (2002) Increased circulating concentrations of the counteradhesive proteins SPARC and thrombospondin-1 in systemic sclerosis (scleroderma). Relationship to platelet and endothelial cell activation. J Rheumatol 29(12):2565–2570

Maillard C, Malaval L, Delmas PD (1992) Immunological screening of SPARC/osteonectin in nonmineralized tissues. Bone 13(3):257–264

Marr H, Basalamah MA, Edgell CJ (1997) Endothelial cell expression of testican mRNA. Endothelium 5(3):209–219

Martina E, Degen M, Rüegg C, Merlo A, Lino MM, Chiquet-Ehrismann R, Brellier F (2010a) Tenascin-W is a specific marker of glioma-associated blood vessels and stimulates angiogenesis in vitro. FASEB J 24(3):778–787

Martina E, Chiquet-Ehrismann R, Brellier F (2010b) Tenascin-W: an extracellular matrix protein associated with osteogenesis and cancer. Int J Biochem Cell Biol 42(9):1412–1415

Martinek N, Zou R, Berg M, Sodek J, Ringuette M (2002) Evolutionary conservation and association of SPARC with the basal lamina in *Drosophila*. Dev Genes Evol 212(3):124–133

Maschler S, Grunert S, Danielopol A, Beug H, Wirl G (2004) Enhanced tenascin-C expression and matrix deposition during Ras/TGF-[beta]-induced progression of mammary tumor cells. Oncogene 23(20):3622–3633

Masson S, Arosio B, Luvarà G, Gagliano N, Fiordaliso F, Santambrogio D, Vergani C, Latini R, Annoni G (1998) Remodelling of cardiac extracellular matrix during β-adrenergic stimulation: upregulation of SPARC in the myocardium of adult rats. J Mol Cell Cardiol 30(8):1505–1514

Matsumoto K-I, Arai M, Ishihara N, Ando A, Inoko H, Ikemura T (1992) Cluster of fibronectin type III repeats found in the human major histocompatibility complex class III region shows the highest homology with the repeats in an extracellular matrix protein, tenascin. Genomics 12 (3):485–491

McCurdy S, Baicu CF, Heymans S, Bradshaw AD (2010) Cardiac extracellular matrix remodeling: fibrillar collagens and Secreted Protein Acidic and Rich in Cysteine (SPARC). J Mol Cell Cardiol 48(3):544–549

Meloty-Kapella CV, Degen M, Chiquet-Ehrismann R, Tucker RP (2006) Avian tenascin-W: expression in smooth muscle and bone, and effects on calvarial cell spreading and adhesion in vitro. Dev Dyn 235(6):1532–1542

Midwood K, Hussenet T, Langlois B, Orend G (2011) Advances in tenascin-C biology. Cell Mol Life Sci 68(19):3175–3199

Mitrovic N, Schachner M (1995) Detection of tenascin-C in the nervous system of the tenascin-C mutant mouse. J Neurosci Res 42(5):710–717

Motamed K, Blake DJ, Angello JC, Allen BL, Rapraeger AC, Hauschka SD, Sage EH (2003) Fibroblast growth factor receptor-1 mediates the inhibition of endothelial cell proliferation and the promotion of skeletal myoblast differentiation by SPARC: a role for protein kinase A. J Cell Biochem 90(2):408–423

Nakatani K, Seki S, Kawada N, Kitada T, Yamada T, Sakaguchi H, Kadoya H, Ikeda K, Kaneda K (2002) Expression of SPARC by activated hepatic stellate cells and its correlation with the stages of fibrogenesis in human chronic hepatitis. Virchows Arch 441(5):466–474

Nishi T, Weinstein J, Gillespie WM, Paulson JC (1991) Complete primary structure of porcine tenascin. Eur J Biochem 202(2):643–648

Novinec M, Kordiš D, Turk V, Lenarčič B (2006) Diversity and evolution of the thyroglobulin type-1 domain superfamily. Mol Biol Evol 23(4):744–755

Nozaki M, Sakurai E, Raisler BJ, Baffi JZ, Witta J, Ogura Y, Brekken RA, Sage EH, Ambati BK, Ambati J (2006) Loss of SPARC-mediated VEGFR-1 suppression after injury reveals a novel antiangiogenic activity of VEGF-A. J Clin Invest 116(2):422–429

Parks WC (1999) Matrix metalloproteinases in repair. Wound Repair Regen 7(6):423–432

Patthy L (1990) Homology of a domain of the growth hormone/prolactin receptor family with type III modules of fibronectin. Cell 61(1):13–14

Pazin DE, Albrecht KH (2009) Developmental expression of Smoc1 and Smoc2 suggests potential roles in fetal gonad and reproductive tract differentiation. Dev Dyn 238(11):2877–2890

Pearson C, Pearson D, Shibahara S, Hofsteenge J, Chiquet-Ehrismann R (1988) Tenascin: cDNA cloning and induction by TGF-beta. EMBO J 7(10):2977–2982

Peterson MC (2005) Circulating transforming growth factor beta-1: a partial molecular explanation for associations between hypertension, diabetes, obesity, smoking and human disease involving fibrosis. Med Sci Monit 11(7):RA229–RA232

Pichler RH, Hugo C, Shankland SJ, Reed MJ, Bassuk JA, Andoh TF, Lombardi DM, Schwartz SM, Bennett WM, Alpers CE, Sage EH, Johnson RJ, Couser WG (1996) SPARC is expressed in renal interstitial fibrosis and in renal vascular injury. Kidney Int 50(6):1978–1989

Puolakkainen PA, Bradshaw AD, Brekken RA, Reed MJ, Kyriakides T, Funk SE, Gooden MD, Vernon RB, Wight TN, Bornstein P, Sage EH (2005) SPARC-thrombospondin-2-double-null mice exhibit enhanced cutaneous wound healing and increased fibrovascular invasion of subcutaneous polyvinyl alcohol sponges. J Histochem Cytochem 53(5):571–581

Raines EW, Lane TF, Iruela-Arispe ML, Ross R, Sage EH (1992) The extracellular glycoprotein SPARC interacts with platelet-derived growth factor (PDGF)-AB and -BB and inhibits the binding of PDGF to its receptors. Proc Natl Acad Sci U S A 89(4):1281–1285

Rathjen FG, Wolff JM, Chiquet-Ehrismann R (1991) Restrictin: a chick neural extracellular matrix protein involved in cell attachment co-purifies with the cell recognition molecule F11. Development 113(1):151–164

Reed MJ, Vernon RB, Abrass IB, Sage EH (1994) TGF-beta 1 induces the expression of type I collagen and SPARC, and enhances contraction of collagen gels, by fibroblasts from young and aged donors. J Cell Physiol 158(1):169–179

Rivera LB, Brekken RA (2011) SPARC promotes pericyte recruitment via inhibition of endoglin-dependent TGF-β1 activity. J Cell Biol 193(7):1305–1319

Rocnik EF, Liu P, Sato K, Walsh K, Vaziri C (2006) The novel SPARC family member SMOC-2 potentiates angiogenic growth factor activity. J Biol Chem 281(32):22855–22864
Sadlier DM, Connolly SB, Kieran NE, Roxburgh S, Brazil DP, Kairaitis L, Wang Y, Harris DCH, Doran P, Brady HR (2004) Sequential extracellular matrix-focused and baited-global cluster analysis of serial transcriptomic profiles identifies candidate modulators of renal tubulointerstitial fibrosis in murine adriamycin-induced nephropathy. J Biol Chem 279(28): 29670–29680
Sage E, Bassuk JA, Yost JC, Folkman MJ, Lane TF (1995) Inhibition of endothelial cell proliferation by SPARC is mediated through a Ca2+-binding EF-hand sequence. J Cell Biochem 57(1):127–140
Sage EH, Reed M, Funk SE, Truong T, Steadele M, Puolakkainen P, Maurice DH, Bassuk JA (2003) Cleavage of the matricellular protein SPARC by matrix metalloproteinase 3 produces polypeptides that influence angiogenesis. J Biol Chem 278(39):37849–37857
Salo T, Mäkelä M, Kylmäniemi M, Autio-Harmainen H, Larjava H (1994) Expression of matrix metalloproteinase-2 and -9 during early human wound healing. Lab Invest 70(2):176–182
Salonen J, Domenicucci C, Goldberg HA, Sodek J (1990) Immunohistochemical localization of SPARC (osteonectin) and denatured collagen and their relationship to remodelling in rat dental tissues. Arch Oral Biol 35(5):337–346
Sangaletti S, Stoppacciaro A, Guiducci C, Torrisi MR, Colombo MP (2003) Leukocyte, rather than tumor-produced SPARC, determines stroma and collagen type IV deposition in mammary carcinoma. J Exp Med 198(10):1475–1485
Schalkwijk J, Zweers MC, Steijlen PM, Dean WB, Taylor G, van Vlijmen IM, van Haren B, Miller WL, Bristow J (2001) A recessive form of the Ehlers–Danlos syndrome caused by tenascin-X deficiency. N Engl J Med 345(16):1167–1175
Schellings MW, Vanhoutte D, Swinnen M, Cleutjens JP, Debets J, van Leeuwen RE, d'Hooge J, Van de Werf F, Carmeliet P, Pinto YM, Sage EH, Heymans S (2009) Absence of SPARC results in increased cardiac rupture and dysfunction after acute myocardial infarction. J Exp Med 206(1):113–123
Schenk S, Chiquet-Ehrismann R (1994) Teanscins. In: Methods in Enzymology, Vol 245. New York Academic Press p 52–61
Schenk S, Chiquet-Ehrismann R, Battegay EJ (1999) The fibrinogen globe of tenascin-C promotes basic fibroblast growth factor-induced endothelial cell elongation. Mol Biol Cell 10(9): 2933–2943
Scherberich A, Tucker RP, Samandari E, Brown-Luedi M, Martin D, Chiquet-Ehrismann R (2004) Murine tenascin-W: a novel mammalian tenascin expressed in kidney and at sites of bone and smooth muscle development. J Cell Sci 117(4):571–581
Schiemann BJ, Neil JR, Schiemann WP (2003) SPARC inhibits epithelial cell proliferation in part through stimulation of the transforming growth factor-β-signaling system. Mol Biol Cell 14(10):3977–3988
Schnepp A, Lindgren PK, Hülsmann H, Kröger S, Paulsson M, Hartmann U (2005) Mouse testican-2. J Biol Chem 280(12):11274–11280
Shankavaram UT, DeWitt DL, Funk SE, Sage EH, Wahl LM (1997) Regulation of human monocyte matrix metalloproteinases by SPARC. J Cell Physiol 173(3):327–334
Siri A, Knäuper V, Veirana N, Caocci F, Murphy G, Zardi L (1995) Different susceptibility of small and large human tenascin-C isoforms to degradation by matrix metalloproteinases. J Biol Chem 270(15):8650–8654
Sodek J, Zhu B, Huynh MH, Brown TJ, Ringuette M (2002) Novel functions of the matricellular proteins osteopontin and osteonectin/SPARC. Connect Tissue Res 43(2–3):308–319
Spring J, Beck K, Chiquet-Ehrismann R (1989) Two contrary functions of tenascin: dissection of the active sites by recombinant tenascin fragments. Cell 59(2):325–334
Sriramarao P, Bourdon MA (1993) A novel tenascin type III repeat is part of a complex of tenascin mRNA alternative splices. Nucleic Acids Res 21(1):163–168
Strandjord TP, Madtes DK, Weiss DJ, Sage EH (1999) Collagen accumulation is decreased in SPARC-null mice with bleomycin-induced pulmonary fibrosis. Am J Physiol Lung Cell Mol Physiol 277(3):L628–L635

Strup-Perrot C, Mathé D, Linard C, Violot D, Milliat F, François A, Bourhis J, Vozenin-Brotons M-C (2004) Global gene expression profiles reveal an increase in mRNA levels of collagens, MMPs, and TIMPs in late radiation enteritis. Am J Physiol Gastrointest Liver Physiol 287(4): G875–G885

Tahashi Y, Matsuzaki K, Date M, Yoshida K, Furukawa F, Sugano Y, Matsushita M, Himeno Y, Inagaki Y, Inoue K (2002) Differential regulation of TGF-β signal in hepatic stellate cells between acute and chronic rat liver injury. Hepatology 35(1):49–61

Taneda S, Pippin JW, Sage EH, Hudkins KL, Takeuchi Y, Couser WG, Alpers CE (2003) Amelioration of diabetic nephropathy in SPARC-null mice. J Am Soc Nephrol 14(4):968–980

Tucker RP, Chiquet-Ehrismann R (2009) Evidence for the evolution of tenascin and fibronectin early in the chordate lineage. Int J Biochem Cell Biol 41(2):424–434

Tucker R, Drabikowski K, Hess J, Ferralli J, Chiquet-Ehrismann R, Adams J (2006) Phylogenetic analysis of the tenascin gene family: evidence of origin early in the chordate lineage. BMC Evol Biol 6(1):60

Tufty RM, Kretsinger RH (1975) Troponin and parvalbumin calcium binding regions predicted in myosin light chain and T4 lysozyme. Science 187(4172):167–169

Tung PS, Domenicucci C, Wasi S, Sodek J (1985) Specific immunohistochemical localization of osteonectin and collagen types I and III in fetal and adult porcine dental tissues. J Histochem Cytochem 33(6):531–540

Unemori E, Amento EP (1991) Connective tissue metabolism including cytokines in scleroderma. Curr Opin Rheumatol 3(6):953–959

Van Obberghen-Schilling E, Tucker RP, Saupe F, Gasser I, Cseh B, Orend G (2011) Fibronectin and tenascin-C: accomplices in vascular morphogenesis during development and tumor growth. Int J Dev Biol 55:511–525

Wang X, Chen W, Huang Y, Sun J, Men J, Liu H, Luo F, Guo L, Lv X, Deng C, Zhou C, Fan Y, Li X, Huang L, Hu Y, Liang C, Hu X, Xu J, Yu X (2011) The draft genome of the carcinogenic human liver fluke Clonorchis sinensis. Genome Biol 12(10):R107

Wasi S, Otsuka K, Yao KL, Tung PS, Aubin JE, Sodek J, Termine JD (1984) An osteonectinlike protein in porcine periodontal ligament and its synthesis by periodontal ligament fibroblasts. Can J Biochem Cell Biol 62(6):470–478

Weber P, Montag D, Schachner M, Bernhardt RR (1998) Zebrafish tenascin-W, a new member of the tenascin family. J Neurobiol 35(1):1–16

Weller A, Beck S, Ekblom P (1991) Amino acid sequence of mouse tenascin and differential expression of two tenascin isoforms during embryogenesis. J Cell Biol 112(2):355–362

Willems IE, Arends J, Daemen M (1996) Tenascin and fibronectin expression in healing human myocardial scars. J Pathol 179(3):321–325

Wrana JL, Maeno M, Hawrylyshyn B, Yao KL, Domenicucci C, Sodek J (1988) Differential effects of transforming growth factor-beta on the synthesis of extracellular matrix proteins by normal fetal rat calvarial bone cell populations. J Cell Biol 106(3):915–924

Wrana JL, Overall CM, Sodek J (1991a) Regulation of the expression of a secreted acidic protein rich in cysteine (SPARC) in human fibroblasts by transforming growth factor β. Eur J Biochem 197(2):519–528

Wrana JLKT, Zhang Q, Overall CM, Aubin JE, Butler WT, Sodek J (1991b) Regulation of transformation-sensitive secreted phosphoprotein (SPPI/osteopontin) expression by transforming growth factor-beta. Comparisons with expression of SPARC (secreted acidic cysteine-rich protein). Biochem J 1(273):523–531

Xueyong L, Shaozong C, Wangzhou L, Yuejun L, Xiaoxing L, Jing L, Yanli W, Jinqing L (2008) Differentiation of the pericyte in wound healing: the precursor, the process, and the role of the vascular endothelial cell. Wound Repair Regen 16(3):346–355

Yamada KM, Yamada SS, Pastan I (1975) The major cell surface glycoprotein of chick embryo fibroblasts is an agglutinin. Proc Natl Acad Sci U S A 72(8):3158–3162

Yan Q, Weaver M, Perdue N, Sage EH (2005) Matricellular protein SPARC is translocated to the nuclei of immortalized murine lens epithelial cells. J Cell Physiol 203(1):286–294

Yoshimura H, Michishita M, Ohkusu-Tsukada K, Takahasi K (2011) Appearance and distribution of stromal myofibroblasts and tenascin-C in feline mammary tumors. Histol Histopathol 26(3): 297–305

Youssef W, Tavill AS (2002) Connective tissue diseases and the liver. J Clin Gastroenterol 35(4): 345–349

Yunker CK, Golembieski W, Lemke N, Schultz CR, Cazacu S, Brodie C, Rempel SA (2008) SPARC-induced increase in glioma matrix and decrease in vascularity are associated with reduced VEGF expression and secretion. Int J Cancer 122(12):2735–2743

Zhao Y, Young SL (1995) TGF-beta regulates expression of tenascin alternative-splicing isoforms in fetal rat lung. Am J Physiol Lung Cell Mol Physiol 268(2):L173–L180

Zhao Y, Young SL, McIntosh JC (1998) Induction of tenascin in rat lungs undergoing bleomycin-induced pulmonary fibrosis. Am J Physiol Lung Cell Mol Physiol 274(6):L1049–L1057

Zhou Y (2009) The Schistosoma japonicum genome reveals features of host-parasite interplay. Nature 460(7253):345–351

Zhou X, Tan FK, Reveille JD, Wallis D, Milewicz DM, Ahn C, Wang A, Arnett FC (2002) Association of novel polymorphisms with the expression of SPARC in normal fibroblasts and with susceptibility to scleroderma. Arthritis Rheum 46(11):2990–2999

Zhou X, Tan FK, Guo X, Wallis D, Milewicz DM, Xue S, Arnett FC (2005) Small interfering RNA inhibition of SPARC attenuates the profibrotic effect of transforming growth factor β1 in cultured normal human fibroblasts. Arthritis Rheum 52(1):257–261

Zhou X, Tan FK, Guo X, Arnett FC (2006) Attenuation of collagen production with small interfering RNA of SPARC in cultured fibroblasts from the skin of patients with scleroderma. Arthritis Rheum 54(8):2626–2631

Chapter 8
Thrombospondins and Their Receptors: Evolving Functions

David D. Roberts and William A. Frazier

Abstract Since the discovery of thrombospondin-1 (TSP1) in the 1970s, and the gradual unveiling of four more members of the mammalian TSP family, the field has been in pursuit of two fundamental questions: what are the physiological roles of these large multidomain glycoproteins and which receptors and signaling pathways mediate these functions? This has been a complex task because TSPs interact with multiple receptors and extracellular binding partners via their several domains. The multidomain structure of the TSP subunits, their trimeric or pentameric quaternary structure, and the promiscuity of their receptors combine to present us with a veritable "Rubic's Cube" of possible functions and mechanisms of action. In this chapter we provide a current overview of progress in solving this puzzle. We present a brief overview of the structure and evolution of the TSP family members and the receptors that recognize various domains of TSPs. These advances have led to the identification of important pathophysiological roles for certain TSP family members and revealed opportunities to develop novel therapeutics.

8.1 Beginnings

Platelet thrombospondin, now termed thrombospondin-1 or TSP1, was first seen as a high-molecular-weight protein band on SDS gels of material secreted from activated platelets whose apparent molecular weight was reduced by thrombin

D.D. Roberts
Laboratory of Pathology, Center for Cancer Research, National Cancer Institute, National Institutes of Health, Bethesda, MD 20892, USA

W.A. Frazier (✉)
Departments of Biochemistry and Molecular Biophysics and Cell Biology, Washington University School of Medicine, 660 South Euclid Avenue, St. Louis, MO 63110, USA
e-mail: frazier@wustl.edu

F.W. Keeley and R.P. Mecham (eds.), *Evolution of Extracellular Matrix*, Biology of Extracellular Matrix, DOI 10.1007/978-3-642-36002-2_8,

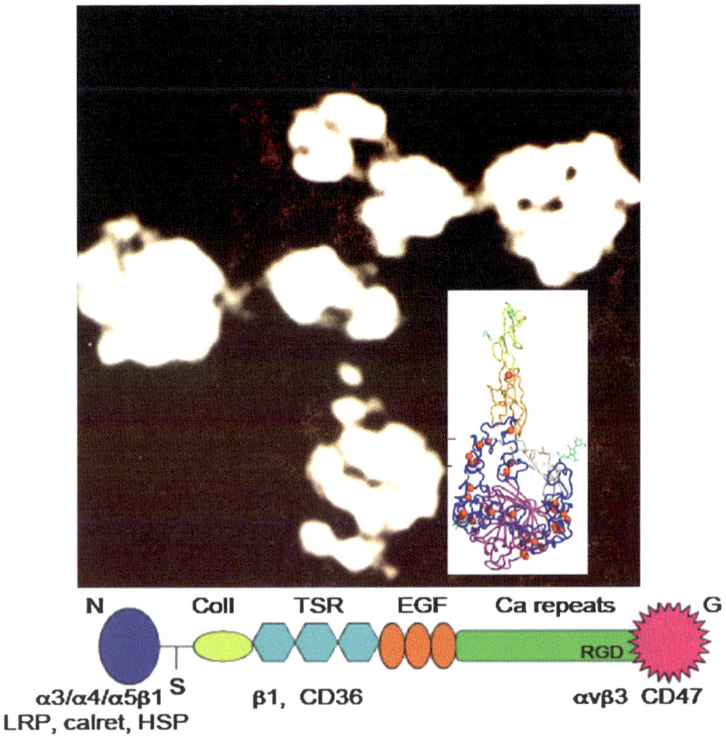

Fig. 8.1 The structure of TSP1 was first revealed by electron microscopy as seen here in an image of a human TSP1 trimer atomized onto a mice surface and rotary shadowed with tungsten (Galvin et al. 1985). As the complete cDNA sequence of TSP1 became available (Hennessy et al. 1989), it was possible to assign specific amino acid sequences to structural and functional domains of TSP1 leading to the model shown under the EM image with the domains labeled (above the model) as in the text and the receptors for the domains indicated below the model. The inset in the EM image is a ribbon diagram calculated from the crystal structure of the human TSP2 "signature domain" (Carlson et al. 2005). *Red spheres* represent calcium ions bound to the "calcium wire" composed of calcium-binding repeats

(Baenziger et al. 1971). Over the next several years, a number of groups reported purification of TSP1 (Baenziger et al. 1972; Lawler et al. 1978), its trimeric quaternary structure (Margossian et al. 1981; Dixit et al. 1984), and the fact that it could be proteolyzed into relatively protease-stable domains (Haverstick et al. 1984, 1985). These results along with the subsequent cloning and sequencing of full-length TSP1 cDNA (Hennessy et al. 1989), publication of rotary-shadowed EM images (Lawler et al. 1978; Galvin et al. 1985), and X-ray crystallographic analysis of several recombinant TSP1 domains (Tan et al. 2002, 2006; Kvansakul et al. 2004; Carlson et al. 2005) yielded a picture of TSP1 as a large (Mr ca 450,000) glycoprotein made up of three identical subunits with the domain structure shown in Fig. 8.1. Each subunit consists of an N-terminal globular laminin G-like domain that binds heparin and other GAGs followed by an extended stalk-like region composed of a coiled-coil region with the two cysteine residues that stabilize the trimer and three TSR repeats (related to properdins). These are followed by three EGF-like repeats and 12 type 3 calcium-binding loops stabilized by disulfide bonds. The C-terminal G or "cell-binding domain" is a globular β-jelly roll structure related to the L-type lectin family (Kvansakul et al. 2004).

Electron micrographs of TSP1 in the presence and absence of calcium ions revealed a massive conformational change. In calcium concentrations above about 0.1 mM, the EGF domains, the calcium-binding loops, and the C-terminal G domain form a folded unit that appears as a large globular domain in such images. At low calcium concentrations, Ca dissociates from the 12 type three loops and they unravel resulting in a structure that now appears as two nearly equal-sized globular domains connected by a much longer, more flexible stalk (Dixit et al. 1986). This extended conformational transition is reversible upon the re-addition of calcium. As yet we have not known the physiological function for this structural transition. Since the extracellular calcium concentration is nearly always higher than the level required to maintain the tightly folded state of the Ca-binding loops (also referred to as the "calcium wire" (Carlson et al. 2005)), it has been difficult to rationalize a function for this structural transition. It may allow for a gradual or sequential folding of TSP1 as it encounters different calcium concentrations as it transits through the secretory pathway of the cell. It may also have a role in regulating binding to other membrane proteins or TSP1 receptors during vesicular transit. Notably, several antibodies specific for the calcium-depleted conformation of TSP1 recognize the protein when it is present on the surface of cells such as activated platelets (Calzada et al. 2008), suggesting that physiological interactions of TSP1 can induce such conformation transitions in the presence of calcium.

8.2 The TSP Signature Domain

The EGF-like repeats, Ca-binding loops, and the G domain form a compact structural unit (in the calcium replete state) that has been referred to as the TSP "signature domain" since it defines the TSP family (Carlson et al. 2005). In the TSP A subgroup, which is specific to chordates, TSP1 and TSP2 have the same domain structure and both exist as trimers (Adams et al. 2003). The members of the TSP B group, TSP3, TSP4, and TSP5 (also called cartilage oligomeric matrix protein or COMP), have very divergent N-terminal regions and exist as pentamers rather than trimers (Anilkumar et al. 2002). Chickens and *Xenopus* have orthologs of all five TSPs while fish have 6–8 TSP genes, none of which appears to represent an ortholog of TSP5/COMP. Syntenic analysis revealed that TSP5 is an ancient gene, predating the divergence of fish and tetrapods about 450 million years ago (McKenzie et al. 2006) and was subsequently lost in fish. More primitive species such as shrimp and sea urchins express a single, monomeric TSP with an N-terminal chitin-binding domain and discoidin domain respectively (McKenzie et al. 2006). Suffice it to say that the TSP gene family is ancient and has undergone a number or changes to acquire new structural features such as different oligomerization or coiled-coil domains leading to monomeric, dimeric, trimeric, and pentameric structures, different N-terminal domains (including the near extinction of the N-terminal region of TSP5), and different numbers of TSR and EGF domains. The essential signature domain of three EGF-like repeats, ca 12 Ca-binding loops (some binding upwards of 30 Ca ions), and the C-terminal G domain is present in all TSPs identified thus far. Therefore, we expect that it embodies the set of functions that define the original, and still useful, raison d'etre of the TSP family.

In spite of the large body of knowledge that has accumulated about the structure and evolutionary history of the TSP family, elucidating the physiological functions of TSPs has been a difficult and protracted task. In the early days of TSP research, the focus was on defining the functions of the recognized TSP1 domains. Since TSP1 constitutes about 25 % of the platelet secretome, it was widely expected that TSP1 must have an important role in some platelet function such as activation, cross-linking platelets to one another, or spreading of platelets on thrombogenic surfaces such as subendothelial collagens or basement membrane. This search for TSP1 functions had about it the sense of a group of blind men describing an elephant. Each laboratory tended to focus on its favorite (or easily obtained) piece of TSP1 and tried to define its role. Part of the early characterization of TSP1 revealed interactions of its various domains with other extracellular proteins including fibrinogen, collagens, proteoglycans, and thrombin among them (Roberts et al. 1985; Galvin et al. 1987; Adams and Lawler 1993). Using mostly in vitro assays such as cell adhesion, spreading, migration, etc. it was shown that TSP1 could affect these cell activities, and the corresponding receptors involved in these effects and their mechanisms are increasingly being defined (Gao et al. 1996a, b; Krutzsch et al. 1999; Calzada et al. 2004a, b; Staniszewska et al. 2007).

8.3 What Is the Role of TSP Domain Structure and Oligomerization?

One of the persistent questions about TSP1 that remains unanswered is "what is the purpose of collecting together over the course of evolution the various domains that make up the TSPs?" It has been argued that since the "signature domain," consisting of the TSRs, the EGF-like repeats, the calcium "wire," and the G domain, is found in all TSPs, that it represented the first collection of functional domains to confer some selective advantage early in the evolution of metazoans (McKenzie et al. 2006). Further, the selection of the pair of cysteine residues that reside in the trimer- and pentamer-forming coiled-coil regions upstream of the TSR domains suggests that multivalency is also an important feature of TSPs, at least those of the A and B type.

What might be the role(s) of multivalency, the trimeric, or pentameric structure? One rather obvious effect of a multivalent structure is that it increases the avidity of an interaction by simultaneously binding to more than one low affinity receptor at a time. This is relevant to the cell adhesive activity of immobilized TSP1, which is mediated by its relatively low affinity binding to several integrins. Clustering of integrins is part and parcel of their signaling mechanism, and integrin clustering seems to be important for "outside in" signaling as seen in the amplification of platelet activation by inward signaling from clustered αIIbβ3 integrin (Derrick et al. 2001). Multivalency may facilitate incorporation of TSP1 into the extracellular matrix via its relatively low affinity binding to extracellular matrix components including proteoglycans, collagens, hyaladherins, fibrin, and fibronectin. TSP1 is not a structural matrix protein in healthy tissues, but it accumulates transiently during tissue remodeling and becomes a long-term component of ECM associated with several chronic diseases of aging. In this context, some subunits may engage the matrix while others interact with cell surface receptors. Thus, trimeric and pentameric TSPs may serve as bridging molecules between cells and matrix. This is analogous to the bridging function of laminins between cells and basement membranes (Yurchenco 2011).

Since TSPs consist of domains that bind many different receptors and proteins, a trimeric or pentameric TSP can assemble collections of different receptors into a supramolecular cluster on the plasma membrane of target cells. TSP1 may thus serve a function on the outside of the plasma membrane analogous to the function of scaffolding proteins that accumulate on the inner surface of the membrane to assemble signaling complexes for efficient and directed signal transduction. While we know what some of these receptors/players are, we do not yet know the complete composition or indeed the purpose of these putative signaling exocomplexes. One example of such receptor clustering that may be of functional importance occurs because TSP1 can bind to a number of integrins and also to CD47. This could have the effect of bringing CD47 close to an integrin in the plane of the membrane thus facilitating direct interactions between CD47 and the integrin which can then promote activation of the integrin (McDonald et al. 2004). Another potential case of bivalent receptor binding involves the N or heparin-binding domain of TSP1 and its C-terminal G1 domain. It was recently reported that CD47, a receptor for the G1 domain of TSP1, is expressed on some cells with glycosaminoglycan chains attached to serines 64 and 79 (Kaur et al. 2011). This would allow TSP1 to bind to the heparan sulfate chains of CD47 via its N-domain which would then increase the local concentration of the G1 domain and facilitate its interaction with CD47.

8.4 Thrombospondin Receptors

Let us first consider the list of receptors and binding partners of TSP1 as an example. From N-terminus to C-terminus of the mature TSP1 subunit these are:

N/HBD domain (TSP1 19–240): HSPGs and sulfated lipids (sulfatides) (Roberts et al. 1985), several link-domain containing proteins (Kuznetsova et al. 2006), soluble frizzled-related protein-1 (Martin-Manso et al. 2011), Calreticulin/LRP1 (Sweetwyne and Murphy-Ullrich 2012), α3β1 (Krutzsch et al. 1999), α4β1 (Calzada et al. 2004a, b), α6β1 (Calzada et al. 2003), and α9β1 integrins (Staniszewska et al. 2007).

Coiled-coil oligomerization domain (241–262): Other TSP subunits for oligomer formation which is then stabilized by disulfide bonds. The type A TSPs (1 and 2) have a coiled-coil domain that assembles trimmers while the coiled-coil domains of type B TSPs (3, 4, and 5) assemble into pentamers (Anilkumar et al. 2002). Both type A and type B domains contain two cysteines that link the subunits covalently to stabilize the trimeric or pentameric structures.

Procollagen-like or vWF type A-like Domain (263–350): collagens

TSR repeats (351–418, 419–475, 476–532): CD36 (Pearce et al. 1995), TGFβ1 and latency-associated peptide (Sweetwyne and Murphy-Ullrich 2012), β1 integrins (Calzada et al. 2004a, b), histidine-rich glycoprotein (Simantov et al. 2001), and fibrinogen (Panetti et al. 1999).

EGF-like repeats (533–920): Ca++, integrins (Calzada et al. 2004a, b).

Calcium-binding loops ("wire" domain): Calcium, cathepsin G (Hogg et al. 1993), FGF2 (Margosio et al. 2008), disulfide isomerization-dependent αvβ3/αIIbβ3 integrin binding to RGDA sequence in the last Ca loop (Hotchkiss et al. 1998).

G1 or cell-binding domain (921–1152): CD47 (Gao et al. 1996a, b; Frazier et al. 1999).

8.5 Promiscuity of TSP Receptors

One of the remarkable features of TSP1's interactions with receptors is that in no case does a TSP1-binding protein or receptor bind only TSP1. This raises some interesting and experimentally confounding possibilities:

1. TSP1 may in some cases act as a competitor of the binding of some other extracellular protein to a given receptor. For example, TSP1 binding to CD36 can compete for binding of collagens and oxLDL and uptake of free fatty acids via CD36 (Isenberg et al. 2007b). SIRPα, binding to CD47 with relatively high affinity, can compete (when present) with TSP1 binding to CD47 (Isenberg et al. 2009a).
2. In different physiological settings/scenarios one of the several ligands of a particular receptor may be more prominent or present in higher concentrations than the others. For example, when platelets are activating, TSP1 is secreted in large amounts and likely swamps out the binding of other ligands to CD36 and CD47 while acting as a linker to fibrinogen to aid in cross-linking of the platelet aggregate (Dixit et al. 1985).

3. Because of its trimeric nature, TSP1 may bind with high avidity to certain receptors even though the intrinsic affinity of each interaction may not be particularly high. It is this increased avidity conferred by multivalency that allows TSP1 to share receptors with other ligands and successfully compete for binding even though the measured affinity of the TSP1 domain for that receptor may not be high.
4. Defining phenotypes of TSP knockout mice has been difficult partially because TSP1 can bind to so many different receptors and proteins. Therefore in most studies it has been necessary to study the receptor knockout to learn the specific role of TSP1 domains. However as argued above, the aggregate phenotype may be a reflection of the ability of TSPs to cluster different receptors to accomplish outcomes that are unrecognizable as resulting from perturbation of a single receptor's pathway.

Nonetheless, the recent discoveries that TSP1 acting via CD47 plays a major regulatory role in cardiovascular regulation through modulating nitric oxide-cGMP signaling (Roberts et al. 2012), mitochondrial homeostasis (Frazier et al. 2011), and in regulating radioresistance (Isenberg et al. 2008a, b) are three cases in which there has been complete concordance between the phenotypes of ligand (TSP1) knockouts and receptor (CD47) knockouts. That is each phenotype revealed for the CD47 null mice is replicated in the TSP1 null animals. Conversely, some phenotypes of TSP1 null mice are not replicated in CD47 nulls, reflecting the fact that other domains of TSP1 and binding partners play significant physiological roles. A case in point is the finding reported in the paper first describing the TSP1 knockout mice (Lawler et al. 1998) that the mice were susceptible to pneumonia and died of lung inflammation when housed in the Harvard animal facility. This turned out to be due in part to a sensitivity of the TSP1 nulls to adventitious environmental pathogens that are normally cleared from the lungs of WT animals. When the mice were housed in the Washington University animal facility, they did not contract pneumonia until they were removed from the barrier facility for a procedure, after which many of the animals died of massive lung inflammation. This effect in the TSP1 null mice was exacerbated by the increased inflammatory milieu of the ApoE null background in TSP1XApoE double null animals (Frazier lab, unpublished). CD47 nulls or CD47XApoE double null mice do not display this lung inflammatory phenotype, indicating that some interaction of TSP1 besides that with CD47 is responsible. A likely explanation is that TSP1 can serve as an opsonin to link pathogens to receptors such as LRP1, integrins, and/or CD47 on phagocytes (Ren and Savill 1995; Gardai et al. 2003) thus facilitating their clearance. Impaired activation of latent TGFβ1, which has a potent anti-inflammatory activity in the immune system, also contributes to the lung inflammatory phenotype of TSP1 null mice (Crawford et al. 1998). However, this phenotype differs from the lethal inflammatory responses observed in TGFβ1 null mice. The mild phenotype of the TSP1 null results from the ability of several TSP1-independent pathways to maintain physiological latent TGFβ1 activation in the TSP1 null except when subjected to specific injury models (Annes et al. 2003; Daniel et al. 2004, 2007). Among the thrombospondins, activation of latent TGFβ1 is limited to TSP1 (Crawford et al. 1998), suggesting that this function was acquired by TSP1 after its divergence in evolution from TSP2.

8.6 TSP1 and CD47 Regulation of Cyclic Nucleotide Signaling

8.6.1 Cyclic AMP Signaling

Shortly after the identification of CD47 as a receptor for TSP1 (Gao et al. 1996a), studies performed by the Frazier lab revealed that *pertussis* toxin is a potent inhibitor of the effects of TSP1 on cellular functions such as cell spreading and migration (Gao et al. 1996b). This suggested that CD47 functions as a G protein-coupled receptor acting via heterotrimeric Gi. Further, Giα along with Gβ were detected in immunoprecipitates of CD47 and CD47 eluted from an affinity column of 4N1K peptide (but not control peptide 4NGG) with Giα and Gβ and β3 integrin (Chung et al. 1997) Further, GTPγS or $AlFl_4$ eluted Gi from the 4N1K column, suggesting that activation of the G protein released it from CD47. Interestingly, GTPγS and $AlFl_4$ (but not GDPβS or ATP) also eluted CD47 from 4N1K beads, suggesting a "negative heterotropic effect" of G protein activation on ligand binding to CD47 (Frazier et al. 1999), a phenomenon well documented for many GPCRs that results from the receptor being in a high affinity state when it is associated with inactive (GDP bound) G protein and a transition to a lower affinity state when the G protein dissociates from the receptor (Lefkowitz and Caron 1988).

Consistent with CD47-Gi activation, ligation of CD47 by TSP1 or peptides such as 4N1K and 7N3 dramatically reduced intracellular levels of cAMP in platelets (Chung et al. 1997), smooth muscle cells (Chung et al. 1999), and lymphocytes (Manna and Frazier 2003) and in all cases, this reduction in cAMP was prevented by pretreatment of the cells with *pertussis* toxin.

In many cell types, activation of integrins is suppressed by cAMP (Ginsberg et al. 2005), thus the ability of CD47 to rapidly depress cAMP levels could explain its stimulation of integrin-dependent functions. Given the fact that virtually all eukaryotic cells express one or more integrins and that CD47 can associate with and activate a variety of integrins, it has not so far been possible to demonstrate that CD47 in and of itself can activate Gi. It was proposed that the five transmembrane segment domain of CD47 could associate with two additional TM domains of an integrin to form a functional seven TM bundle that activates Gi and perhaps other G proteins. For example, expression of human CD47 in Sf9 insect cells resulted in TSP1-dependent increases in cAMP levels, possibly due to the activation of endogenous Gs, the only G protein expressed in Sf9 cells (Frazier, unpublished data).

The physiological relevance of CD47's coupling to Gi has been well demonstrated by the recent finding that mice lacking CD47 (CD47 KO mice) or TSP1 have significantly elevated levels of cAMP in muscle tissues and in VSMCs isolated from these null mice (Isenberg et al. 2009b; Yao et al. 2011).

8.7 Cyclic GMP Signaling

TSP1, a recombinant form of its signature domain, and a minimal recombinant form of the C-terminal G1 domain, at picomolar concentrations inhibit NO-stimulated activation of soluble guanylate cyclase (sGC) in endothelial cells, VSMC, platelets, and T lymphoma cells (Isenberg et al. 2005, 2006b, 2008c, d). This activity is absent in CD47 null but not in CD36 null cells and is mimicked by peptides derived from TSP1 that bind to CD47 (Isenberg et al. 2006a). Ligation of CD36 also inhibits sGC activation, but only at higher concentrations of TSP1 and only in cells that also express CD47 (Isenberg et al. 2006a). Tissue cGMP levels are elevated in TSP1 and CD47 null mice. Therefore, physiological levels of TSP1 significantly limit NO/cGMP signaling, and CD47 is required to mediate this inhibitory activity of TSP1. TSP1 also inhibits sGC activation by heme-independent activators including chemical agents such as YC1 and BAY 41–2272 (Miller et al. 2011). These inhibitory activities are independent of any effects on cGMP-targeting phosphodiesterases, and at least in T cells involve regulation of cytoplasmic calcium levels (Ramanathan et al. 2011). The proximal molecular mechanism by which sGC activation is inhibited by CD47 signaling remains to be identified, but its reversal by staurosporine suggests involvement of an inhibitory phosphorylation.

Inhibitory signaling through CD47 further regulates NO signaling both upstream and downstream of sGC. In platelets, endothelial cells, and VSMC TSP1 inhibits responses to the permeable cGMP analog 8Br-cGMP (Isenberg et al. 2008c, d). One proximal target of cGMP is cGMP-dependent protein kinase, and TSP1 signaling through CD47 was demonstrated to inhibit the ability of this enzyme to phosphorylate a defined peptide substrate (Isenberg et al. 2008c, d). Upstream of sGC, NO is produced in endothelial cells and platelets by eNOS. TSP1 binding to CD47 on endothelial cells was shown to inhibit the enzymatic conversion of arginine to citrulline plus NO (Bauer et al. 2010). This was accompanied by decreased phosphorylation of eNOS at the activating site Ser^{1177}. Physiologically, eNOS in endothelial cells becomes activated in response to signaling through VEGF receptor-2, and subsequent studies revealed that CD47 is a proximal binding partner of VEGFR2 and that TSP1 binding to CD47 disrupts this complex and prevents activation of the tyrosine kinase activity of VEGFR2 upon VEGF binding to the receptor (Kaur et al. 2010).

The ability of TSP1/CD47 signaling to redundantly inhibit NO signaling has important ramifications in understanding its physiological significance in the cardiovascular system. TSP1 circulates in healthy individuals at ~100 picomolar levels, which are sufficient to tonically limit synthesis of NO in endothelial cells by eNOS. Although circulating TSP1 in blood plasma is too large to cross the vascular basement membrane to engage CD47 on the underlying VSMC, its ability to regulate endothelial cell NO synthesis enables it to regulate vascular tone by controlling endothelial production of the diffusible mediator NO. Thus, intravenous injection of TSP1 or a CD47 agonist antibody acutely raises blood pressure, and both TSP1 and CD47 null mice show exaggerated cardiovascular responses to hypotensive agents (Isenberg et al. 2009b; Bauer et al. 2010).

NO signaling plays an important role in tissue responses to ischemic injuries by causing vasorelaxation and redirection of blood flow that restores perfusion of the injured tissues. TSP1 and CD47 null mice exhibit a profound resistance to fixed ischemic injuries relative to WT mice. Conversely, treatment of WT mice or pigs with CD47 or TSP1 blocking antibodies or with antisense morpholino oligonucleotides that temporarily decrease CD47 expression in the injured tissues recapitulates the resistance to ischemic injuries observed in the null mice (Isenberg et al. 2007c, 2008c, d). Therefore, endogenous TSP1 plays an important role in limiting recovery from fixed ischemic injuries, and agents targeting CD47 could be useful to treat such injuries.

Recovery from ischemic injuries is known to become impaired with increasing age, and based on the diminished levels of NO synthesis and tissue cGMP in aged animals, aging has come to be considered a condition of NO insufficiency. Efforts to restore tissue NO signaling in aged animals and patients have had limited success, and this may be explained in part by reports that tissue levels of TSP1 increase with age (van Almen et al. 2011; Isenberg et al. 2007a; Reissen et al. 1998). Correspondingly, we found that tissue levels of cGMP remained elevated in aged CD47 and TSP1 null mice (Isenberg et al. 2007a). Unlike WT mice, which showed diminished recovery from a soft tissue ischemic injury at 18–24 months age, aged CD47 and TSP1 null mice exhibited similar recovery from ischemic injuries as 3-month-old null mice. Conversely, aged apoE null mice fed a high-fat diet that model the impaired injury responses of aged humans showed excellent tissue survival of fixed ischemia when pretreated with CD47 blocking agents. Thus, the TSP1/CD47 pathway is a significant contributor to age-associated NO insufficiency.

Reperfusion incurs an additional inflammatory response accompanied by release of reactive oxygen species that further complicates recovery from ischemia/reperfusion (I/R) injuries. TSP1 null mice were resistant to a kidney I/R injury (Thakar et al. 2005), and both TSP1 and CD47 null mice exhibited resistance to a warm I/R injury of the liver (Isenberg et al. 2008a, b). Treatment of WT mice or rats with CD47 blocking agents correspondingly protected the animals from I/R injuries, and protection in the rat I/R injury model was remarkably observed even when treatment with a CD47-blocking antibody was delayed until 30 min after reperfusion (Maxhimer et al. 2009a).

CD47 and TSP1 null mice also displayed similar resistance to radiation injuries (Isenberg et al. 2008a, b). Although NO has been demonstrated to have some radioprotective activities, the ability of CD47 blockade to suppress radiation injuries in WT mice was not dependent on either NO synthesis or cGMP signaling (Maxhimer et al. 2009b). Thus, additional cytoprotective signaling pathways relevant to radioresistance are regulated by TSP1/CD47 signaling. Remarkably, this cytoprotection was restricted to healthy tissues as tumors showed increased sensitivity to radiation when subjected to CD47 blockade prior to irradiation. The mechanisms responsible for this selective radioprotection are under investigation.

In addition to being a regulator of NO/cGMP signaling, expression of TSP1 is controlled by this pathway and thereby creates a negative feedback loop to control

TSP1 expression (Ridnour et al. 2005). This was first demonstrated in endothelial cells, where low concentrations of NO inhibited TSP1 mRNA and protein expression. Feedback inhibition of TSP1 expression in endothelium may be important to facilitate angiogenic responses driven by NO. This feedback regulation was recently shown to have pathologic relevance in the context of glomerulonephritis (Hohenstein et al. 2008). TSP1 expression increases in glomerulonephritis and contributes to fibrosis through increased latent TGFβ activation. Pharmacological inhibition of the cGMP phosphodiesterase PDE5 using vardenafil, which elevates tissue cGMP levels, suppressed TSP1 expression by mesangial cells in injured kidney, thereby limiting fibrosis. The development of such negative feedback between TSP1 expression and NO/cGMP signaling provides further support for the importance of TSP1 as a regulator of this pathway and indicates the coevolution of TSP1 with other components of this pathway in higher vertebrates.

8.8 Evolution of TSP-Receptor Interactions

8.8.1 N Module Ligands and Receptors

Despite sequence divergence in their N modules, heparin binding is conserved between TSP1 and TSP2. These interactions are important for the heparan sulfate-dependent clearance of group A TSPs via the LDL receptor-related protein-1 (LRP1) scavenger receptor pathway. In addition, binding to HSPGs may be important to sequester these TSPs in provisional extracellular matrix during tissue remodeling. Heparin binding has also been reported for TSP3, TSP4, and COMP (Qabar et al. 1994; Lawler et al. 1995; Chen et al. 2007), so glycosaminoglycan binding is a highly conserved function of the TSP1 family.

The N module of TSP1 binds with high affinity to several proteins that contain hyaluronan-binding link domains. The first such interaction was identified for TSG6 (Kuznetsova et al. 2005). This interaction is conserved between TSP2 and TSP1 and is inhibited by heparin but not by hyaluronan. Thus, TSP1 and TSP2 can presumably bind to TSG6 while it is bound to hyaluronan. TSP1 binding to TSG6 was further shown to enhance its covalent association with the heavy chain of inter-α-trypsin inhibitor (IαI) and to enhance the covalent modification of IαI by TSG6 and the resulting transfer of its heavy chains to hyaluronan. IαI plays important roles in inflammation and expansion of the cumulus oocyte complex during fertilization. The fertility and inflammation phenotypes of TSG6 null mice show some parallels with phenotypes of TSP1 and TSP2 nulls, suggesting possible mechanistic bases. Subsequent studies extended link module recognition to versican and aggrecan (Kuznetsova et al. 2006). Notably, recognition of the link domain of versican was only weakly conserved in TSP2. This suggests another divergence between the functions of TSP1 and TSP2. Secretion of both versican and TSP1 was induced by TLR3 signaling in VSMC, and TSP1 decorated elastin fibrils on these cells mediated by its binding to versican.

Soluble frizzled-related proteins (sFRPs) are important regulators of Wnt signaling (Mii and Taira 2011). The N module of TSP1 binds with relatively high affinity to sFRP1 and sFRP2 but not to sFRP3 (Martin-Manso et al. 2011). This interaction interferes with the recognition of the N module by α3β1 integrin. It is unclear whether this interaction is conserved in other TSPs.

8.9 Integrin Recognition and Function

Integrins appear to be an evolutionarily ancient class of receptors for thrombospondins. The primitive TSP in *Drosophila* is recognized by the integrin alpha-PS2. This interaction is required for muscle attachment to tendon (Subramanian et al. 2007). In mammals, the number of integrins that recognize thrombospondins has expanded, and these recognize sites in several domains of the thrombospondins. There is evidence for integrin interactions with most of the TSPs, but the specificities have been most completely mapped for TSP1 and TSP2. α4β1, α9β1, and α6β1 binding is conserved among the type A TSPs; α3β1 recognition is unique to TSP1, so this may be the most recently evolved integrin interaction. Integrin α3β1 is highly expressed by endothelial cells, and this interaction with TSP1 results in pro-angiogenic responses (Chandrasekaran et al. 2000). Likewise, α4β1 and α9β1 interactions with the N-domains of TSP1 and TSP2 stimulate angiogenic responses in vitro and in vivo (Staniszewska et al. 2007; Calzada et al. 2004a, b). Thus, although intact TSP1 and TSP2 are primarily anti-angiogenic proteins, their N-terminal domains can clearly stimulate angiogenesis. Because as noted above TSP1 is sensitive to a number of proteases that cleave the N-terminal domain, one can speculate that evolution of multiple pro-angiogenic integrin-binding sites in the N-domain could permit a net anti-angiogenic activity of TSP1 and TSP2 to be converted to a net pro-angiogenic activities in a protease-rich environment such as occurs during wound repair. In this regard, it is interesting that TSP1 null mice displayed a delay in repair of an excisional skin wound (Agah et al. 2002). This delay was attributed to a deficit in a chemotactic chemokine in the null wound, but the absence in the null of pro-angiogenic activity from cleaved N-domain of TSP1 would also be consistent with this phenotype.

Integrins have also been identified as receptors for COMP. The integrin α7β1 was identified as a receptor that mediates effects of COMP on contractility of VSMC (Wang et al. 2010). This involves inhibition by COMP of PDGF signaling. The domain of COMP involved in binding this integrin was not identified. In another study using ligament cells and chondrocytes, αvβ3 integrin was identified as a COMP receptor (Rock et al. 2010). This interaction was inhibited by both RGD peptides and a VVM-containing peptide derived from COMP. The CD47 blocking antibody B6H12 also inhibited adhesion, suggesting a role for CD47. However, it is unclear whether CD47 served as a direct receptor for

COMP or simply as the integrin-associated activating protein for αvβ3 integrin. Integrin αvβ3 binding is consistent with the presence of an RGD sequence in type 3 repeat 5N and other exposed Asp residues in the crystallized signature domain of COMP (Tan et al. 2009). Chondrocyte adhesion on COMP was also inhibited by α5 and β1 blocking antibodies, which further implicates a functional RGD site (Chen et al. 2005). Thus, TSP1 and COMP share putative αvβ3 and α5β1 integrin-binding sites, but this shared receptor interaction may be the result of convergent evolution rather than conservation of a fixed integrin-binding site.

Less is known about the role of integrins as receptors for other B group TSPs. αMβ2 was identified as a receptor that mediates migration of polymorphonuclear leukocytes on immobilized TSP4 (Pluskota et al. 2005). β2 and β3 integrins were similarly implicated as receptors for TSP4-mediated macrophage migration (Frolova et al. 2010).

8.10 CD36

CD36 is the major plasma membrane fatty acid translocase and also serves a promiscuous scavenger receptor for proteins including TSP1 (Silverstein and Febbraio 2009). CD36 is an evolutionarily ancient receptor, which presumably reflects its important roles in lipid metabolism and sensing and in host defense independent of being a TSP1 receptor (Ozbek et al. 2010; Means et al. 2009; Martin et al. 2011). CD36 is not an essential gene in humans, mice, or rats. CD36 null humans (type I CD36 deficiency) have pathologies related to defects in fatty acid transport including decreased insulin sensitivity and postprandial hypertriglyceridemia as adults and hypoglycemia as children (Hirano et al. 2003), but phenotypes related to its proposed function as a platelet TSP1 receptor have not been found (Greenwalt et al. 1992). The interactions of TSP1 with CD36 have been extensively investigated, and CD36 mediates some of the anti-angiogenic activities of TSP1 (Dawson et al. 1997). This activity has been localized to the TSR repeats of TSP1, and CD36-binding peptides derived from these repeats have been utilized as templates to develop anti-angiogenic TSP1 mimetic drugs (Haviv et al. 2005).

Interaction of TSP1 or TSP1 mimetics with CD36 on endothelial cells results in inhibition of cell migration, inhibition of fatty acid uptake, and induction of apoptosis. Various signaling pathways include Src kinases, Akt, p38, JNK, cGMP signaling, VEGFR2, Syk, and caspase activation (Jimenez et al. 2000, 2001; Isenberg et al. 2005; Kazerounian et al. 2011; Sun et al. 2009). Similar to CD47, CD36 may function as part of a supramolecular complex involving its lateral association in the plasma membrane with specific integrins and CD9 (Miao et al. 2001). Structure-function studies of TSP1 mimetics and TSP1 peptides indicate that different modes of binding to CD36 or engagement of distinct signaling pathways mediate its regulation of chemotaxis vs. apoptosis (Haviv et al. 2005; Isenberg et al. 2008e).

CD36 is also reported to be a receptor for TSP2. The anti-angiogenic activity of TSP2 was partially blocked by a CD36 antibody (Simantov et al. 2005). CD36 was necessary for TSP2 to inhibit the angiogenic activity of FGF2 in a corneal pocket assay. A trimeric N-terminal construct of TSP2 extending through its TSR repeats inhibited angiogenic responses and induced endothelial cell apoptosis in a CD36-dependent manner, implying that TSP1 and TSP2 share CD36-binding sites in their TSR repeats (Koch et al. 2011). Because TSP3, TSP4, and COMP lack TSRs, they are presumed to not engage CD36.

8.11 CD47

CD47 orthologs are not found in fish, and likewise eNOS is not present in fish. Instead, vascular NO regulation in fish derives from nerve endings that express nNOS and release NO at sites of function (Toda and Ayajiki 2006). Divergence of eNOS from nNOS appeared in amphibians and higher land dwelling vertebrates and could reflect their need for more finely tuned regional control of vascular tone to adapt to the vertical pressure gradient in the circulatory system that appears when an animal moves into the low density atmosphere from the uniform hydrostatic pressure on a body in an aqueous environment.

Because CD47 is the only receptor known to interact with TSP1 via an immunoglobulin-like domain, it is unclear how this interaction arose late in evolution. Interactions between TSP1 and CD47 were first discovered based on a peptide screen of cell interacting sequences in the G domain. Two sequences that share the motif VVM were identified (Kosfeld and Frazier 1992), and one of these was used as an affinity label to identify a membrane protein as a receptor candidate for TSP1 (Gao and Frazier 1994). That 50 KDa membrane protein was then identified as CD47 using a panel of anti-CD47 mAbs generated by Brown and coworkers (Gao et al. 1996a, 1996b). Because one of the two VVM motif peptides in the G1 domain was well conserved in the other TSPs, it was first predicted that all TSPs would bind to CD47. However, subsequent studies using recombinant signature domains from TSP1, TSP2, and TSP4 showed that only TSP1 bound with high affinity as assessed by competitive displacement of the CD47 ligand SIRPα (Isenberg et al. 2009a). TSP4 and TSP2 showed minimal binding in this assay and in displacing radiolabeled TSP1 signature domain from cells expressing CD47. In a cGMP signaling assay, TSP1 correspondingly showed the most activity. TSP4 showed modest activity, and TSP2 was the weakest inhibitor of NO-stimulated cGMP synthesis. Therefore, inhibition of NO/cGMP signaling is a selective activity of TSP1. In agreement with these in vitro data, TSP2 null mice showed no advantage over WT in maintaining cutaneous perfusion of ischemic skin flaps or hindlimbs within the first hour following injury. This does not exclude some effects of TSP2 on NO signaling, but in these ischemic injury models the expression level and activity of TSP2 does not result in a significant contribution relative to TSP1 to limit ischemic tissue survival. Given the importance of NO/cGMP signaling in cardiovascular homeostasis, it should not be surprising that activation of a major negative regulator of this pathway is limited primarily to a single member of the TSP family.

The selective expression and storage of TSP1 in platelets is also consistent with its prime role as an inhibitor of NO/cGMP signaling. NO is a potent inhibitor of platelet activation, and we found that activation of platelets by thrombin is defective in TSP1 and CD47 null platelets (Isenberg et al. 2008c, d). During activation, TSP1 is rapidly released, and its binding to CD47 on the platelet surface rapidly decreases cGMP levels leading to downstream inactivation of cGMP-dependent protein kinase and loss of phosphorylation of its targets Rap1GAP2 and VASP. This leads to activation of the platelet integrin $\alpha IIb\beta 3$ and drives platelet aggregation. Coincident with accelerating thrombosis of platelets, released TSP1 at a site of injury presumably causes local vasoconstriction that limits bleeding until a stable clot can be formed.

These combined activities provide a rationale for the evolutionary conservation of the G domain of TSP1 in vertebrates and its high expression in platelets. However, the price for this short-term survival benefit is that chronic conditions where TSP1 is expressed can result in NO insufficiency. Such conditions are increasingly being recognized and include aging, diabetes, atherosclerosis, pulmonary hypertension, and sickle cell disease (Novelli et al. 2012; Bauer et al. 2012; Choi et al. 2012; Hugo and Daniel 2009). Unfortunately, these diseases have relatively little impact on evolution, so a dark side of the evolutionary pressure to maintain potent inhibition of NO signaling via CD47 is that the same inhibitory pathway may contribute to the pathogenesis of many common diseases of aging.

Another manifestation of CD47's regulation of NO signaling was found in studies of melanoma bone metastasis. In a model in which labeled B16 melanoma cells are injected directly into the heart (to avoid trapping in the lung which occurs with venous injection), CD47 KO mice were found to be markedly resistant to tumor development in bone marrow (Uluckan et al. 2009). They were also resistant to tumor growth when the melanoma cells were injected directly in the bone marrow cavity, but subcutaneous injections resulted in similar tumor growth in WT and CD47 KO mice. This suggested that some property of the bone environment was altered in the CD47 null animals. Ex vivo studies of osteoclast (OC) development from bone marrow macrophages (BMM) revealed that CD47 null BMMs could not form functional OCs in culture under conditions that result in robust WT OC differentiation. However, CD47 KO BMMs could differentiate normally when treated with L-NAME to inhibit NOS activity. As with many cellular functions, an optimal level of NO signaling is required for OC differentiation and the increased level of NO signaling in CD47 null BMMs inhibits differentiation. Melanomas and other osteotropic tumors secrete factors that stimulate OC development and function which in turn releases growth factors from the bone matrix that stimulate and support tumor growth, the so-called vicious cycle of bone destruction. The blockade of OC differentiation imposed by increased NO signaling in the absence of CD47 inhibition thus abrogated this vicious cycle and inhibited tumor growth in bone. That CD47 normally regulates bone turnover through influencing OC differentiation was evidenced by the increased bone volume and decreased level of bone degradation products in the CD47 KO mice (Uluckan et al. 2009).

8.12 Is CD47 Coming of Age?

While regulation of NO-cGMP signaling is clearly of far-reaching physiological importance, it is useful to consider that CD47 also impacts cAMP signaling. This ability to coordinately suppress both of these wide-ranging signaling systems is perhaps a unique feature of CD47 biology. Assays of cAMP and cGMP levels in skeletal muscle and heart of the CD47 KO mice revealed that steady-state levels of both cyclic nucleotides were significantly elevated compared to those in WT animals. A physiological role in which both cAMP and cGMP signaling have been shown to be important is the regulation of mitochondrial biogenesis. Both cAMP and cGMP have been shown to induce biogenesis of new mitochondria (Nisoli and Carruba 2006); therefore it might be predicted that CD47 KO mice would have a greater volume density of mitochondria, at least in some tissues. Simple visual inspection of EM images of skeletal muscle from TSP1 KO, CD47 KO, and WT mice revealed a dramatic increase in the density of mitochondria in muscle of both knockout strains compared to WTs. Morphometric analysis showed that both knockouts had nearly twice the volume density of mitochondria as WT muscle. Further, quantitative PCR analysis showed dramatic upregulation of mRNAs encoding mitochondrial proteins (both nuclear and mitochondrial encoded) primarily in skeletal muscle (Frazier et al. 2011).

Concordant with the increased load of mitochondria, young CD47 KO mice have twice the treadmill endurance of WTs and their fast twitch muscles (e.g., gastrocnemius) have gene expression profiles more characteristic of slow twitch muscle (e.g., soleus). The KO mice appear to be more metabolically efficient in that they burn less oxygen than WTs and produce less ROS in spite of the increased complement of mitochondria. CD47 nulls are also leaner than WTs and remain so into old age. In fact, many of the phenotypes discovered for the CD47 null mice are concordant with those seen in calorie-restricted mice (Lopez-Lluch et al. 2006) or mice treated with compounds such as resveratrol (Guarente 2008). There has been a massive research effort to discover the mechanism by which resveratrol and other polyphenolic compounds affect the aging process. Resveratrol and chemical derivatives thereof are often referred to as antioxidants, but have also been thought to act as sirtuin activators, a controversial claim. Recently, it was reported that resveratrol reverses metabolic aging phenotypes because it is a good inhibitor of cAMP phosphodiesterases including PDEs 1, 3, and 4 with little effect on PDEs 2 and 5 (Park et al. 2012). Furthermore, treatment of myotubes with the PDE 4 inhibitor rolipram induced the same metabolic alterations as resveratrol and chronic treatment of mice with resveratrol induced improvements in metabolism and performance similar to those achieved with calorie restriction. Thus it seems that chronically elevating cAMP, whether genetically by knocking out CD47 or by administering rolipram or resveratrol, leads to a beneficial shift in metabolism which ameliorates the effects of aging on many critical biological functions. Notably, some of these PDEs mediate cross-talk between cAMP and cGMP levels, and consequently the ability of CD47 signaling to differentially modulate levels of these two cyclic nucleotides depends in part on which PDEs are expressed in each cell type (Yao et al. 2011).

Experimental therapeutics targeting CD47 or TSP1 have now demonstrated efficacy in several rodent and non-rodent injury models. Based on the functions of TSP1 and CD47 described here, such therapeutics could be useful for treating common diseases associated with aging. However, in developing such therapeutic applications we must remain aware of the potential impact of such therapeutics on the fundamental physiological roles of CD47 and TSP1 that have been preserved throughout evolution of higher animals. In other words, "first, do no harm."

Acknowledgments This work was supported by the Intramural Research Program of the NIH/NCI (D.D.R.) and by grants from the NIH (W.A.F.).

References

Adams JC, Lawler J (1993) Diverse mechanisms for cell attachment to platelet thrombospondin. J Cell Sci 104(Pt 4):1061–1071

Adams JC, Monk R et al (2003) Characterisation of Drosophila thrombospondin defines an early origin of pentameric thrombospondins. J Mol Biol 328(2):479–494

Agah A, Kyriakides TR et al (2002) The lack of thrombospondin-1 (TSP1) dictates the course of wound healing in double-TSP1/TSP2-null mice. Am J Pathol 161(3):831–839

Anilkumar N, Annis DS et al (2002) Trimeric assembly of the C-terminal region of thrombospondin-1 or thrombospondin-2 is necessary for cell spreading and fascin spike organisation. J Cell Sci 115(Pt 11):2357–2366

Annes JP, Munger JS et al (2003) Making sense of latent TGFbeta activation. J Cell Sci 116(Pt 2): 217–224

Baenziger NL, Brodie GN et al (1971) A thrombin-sensitive protein of human platelet membranes. Proc Natl Acad Sci U S A 68(1):240–243

Baenziger NL, Brodie GN et al (1972) Isolation and properties of a thrombin-sensitive protein of human platelets. J Biol Chem 247(9):2723–2731

Bauer EM, Qin Y et al (2010) Thrombospondin-1 supports blood pressure by limiting eNOS activation and endothelial-dependent vasorelaxation. Cardiovasc Res 88(3):471–481

Bauer PM, Bauer EM et al (2012) Activated CD47 promotes pulmonary arterial hypertension through targeting caveolin-1. Cardiovasc Res 93(4):682–693

Calzada MJ, Sipes JM et al (2003) Recognition of the N-terminal modules of thrombospondin-1 and thrombospondin-2 by alpha6beta1 integrin. J Biol Chem 278(42):40679–40687

Calzada MJ, Annis DS, Zeng B, Marcinkiewicz C, Banas B, Lawler J, Mosher DF, Roberts DD (2004a) Identification of novel b1 integrin binding sites in the type 1 and type 2 repeats of thrombospondin-1. J Biol Chem 279:41734–41743

Calzada MJ, Zhou L, Sipes JM, Zhang J, Krutzsch HC, Iruela-Arispe ML, Annis DS, Mosher DF, Roberts DD (2004b) a4b1 integrin mediates selective endothelial cell responses to thrombospondins 1 and 2 and modulates angiogenesis in vivo. Circ Res 94:462–470

Calzada MJ, Kuznetsova SA et al (2008) Calcium indirectly regulates immunochemical reactivity and functional activities of the N-domain of thrombospondin-1. Matrix Biol 27(4):339–351

Carlson CB, Bernstein DA, Annis DS, Misenheimer TM, Hanna BA, Mosher DF, Keck JL (2005) Structure of the calcium-rich signature domain of human thrombospondin-2. Nat Struct Biol Mol Biol 12:910–914

Chandrasekaran L, He CZ et al (2000) Cell contact-dependent activation of alpha3beta1 integrin modulates endothelial cell responses to thrombospondin-1. Mol Biol Cell 11(9):2885–2900

Chen FH, Thomas AO et al (2005) Cartilage oligomeric matrix protein/thrombospondin 5 supports chondrocyte attachment through interaction with integrins. J Biol Chem 280(38):32655–32661

Chen FH, Herndon ME et al (2007) Interaction of cartilage oligomeric matrix protein/thrombospondin 5 with aggrecan. J Biol Chem 282(34):24591–24598

Choi KY, Kim DB et al (2012) Higher plasma thrombospondin-1 levels in patients with coronary artery disease and diabetes mellitus. Korean Circ J 42(2):100–106

Chung J, Gao A-G, Frazier WA (1997) Thrombospondin acts via integrin-associated protein to activate the platelet integrin $a_{IIb}b_3$. J Biol Chem 272:14740–14746

Chung J, Wang X-Q, Lindberg FP, Frazier WA (1999) Thrombospondin-1 acts via CD47/CD47 to synergize with collagen in a2b1 -mediated platelet activation. Blood 94:642–648

Crawford SE, Stellmach V, Murphy-Ullrich JE, Ribeiro SM, Lawler J, Haynes RO, Boivin GP, Bouck N (1998) Thrombospondin-1 is a major activator of TGFb-1 in vivo. Cell 93:1159–1170

Daniel C, Wiede J et al (2004) Thrombospondin-1 is a major activator of TGF-beta in fibrotic renal disease in the rat in vivo. Kidney Int 65(2):459–468

Daniel C, Schaub K et al (2007) Thrombospondin-1 is an endogenous activator of TGF-beta in experimental diabetic nephropathy in vivo. Diabetes 56(12):2982–2989

Dawson DW, Pearce FA, Zhong R, Silverstein RL, Frazier WA, Bouck NP (1997) CD36 mediates the inhibitory effects of thrombospondin on endothelial cells. J Cell Biol 138:707–717

Derrick JM, Shattil SJ, Poncz M, Gruppo RA, Gartner TK (2001) Distinct domains of $a_{IIb}b3$ support different aspects of outside-in signal transduction and platelet activation induced by LSARLAF, an $a_{IIb}b3$ interacting peptide. Thromb Haemost 86:894–901

Dixit VM, Grant GA et al (1984) Isolation and characterization of a heparin-binding domain from the amino terminus of platelet thrombospondin. J Biol Chem 259(16):10100–10105

Dixit VM, Haverstick DM et al (1985) Inhibition of platelet aggregation by a monoclonal antibody against human fibronectin. Proc Natl Acad Sci U S A 82(11):3844–3848

Dixit VM, Galvin NJ et al (1986) Monoclonal antibodies that recognize calcium-dependent structures of human thrombospondin. Characterization and mapping of their epitopes. J Biol Chem 261(4):1962–1968

Frazier WA, Gao A-G, Dimitry J, Chung J, Lindberg FP, Brown EJ, Linder ME (1999) The thrombospondin receptor integrin-associated protein (CD47) functionally couples to heterotrimeric Gi. J Biol Chem 274:8554–8560

Frazier EP, Isenberg JS et al (2011) Age-dependent regulation of skeletal muscle mitochondria by the thrombospondin-1 receptor CD47. Matrix Biol 30(2):154–161

Frolova EG, Pluskota E et al (2010) Thrombospondin-4 regulates vascular inflammation and atherogenesis. Circ Res 107(11):1313–1325

Galvin NJ, Dixit VM et al (1985) Mapping of epitopes for monoclonal antibodies against human platelet thrombospondin with electron microscopy and high sensitivity amino acid sequencing. J Cell Biol 101(4):1434–1441

Galvin NJ, Vance PM et al (1987) Interaction of human thrombospondin with types I-V collagen: direct binding and electron microscopy. J Cell Biol 104(5):1413–1422

Gao A-G, Frazier WA (1994) Identification of a receptor candidate for the carboxyl-terminal cell binding domain of thrombospondins. J Biol Chem 269:29650–29657

Gao A-G, Lindberg FP, Finn MB, Blystone SD, Brown EJ, Frazier WA (1996a) Integrin-associated protein is a receptor for the C-terminal domain of thrombospondin. J Biol Chem 271(1):21–24

Gao A-G, Lindberg FP, Dimitry JM, Brown EJ, Frazier WA (1996b) Thrombospondin modulates avb3 function through integrin associated protein. J Cell Biol 135:533–544

Gardai SJ, Xiao YQ, Dickinson M, Nick JA, Voelker DR, Greene KE, Henson PM (2003) By binding SIRP-alpha or calreticulin/CD91, lung collectins act as dual function surveillance molecules to suppress or enhance inflammation. Cell 115:13–23

Ginsberg MH, Partridge A, Shattil SJ (2005) Integrin regulation. Curr Opin Cell Biol 17:509–516

Greenwalt DE, Lipsky RH, Ockenhouse CF, Ikeda H, Tandon NN, Jamieson GA (1992) Membrane glycoprotein CD36: a review of its role sin adherence, signal transduction and transfusion medicine. Blood 80:1105–1115

Guarente L (2008) Mitochondria—a nexus for aging, calorie restriction, and sirtuins? Cell 132(2): 171–176

Haverstick DM, Dixit VM et al (1984) Localization of the hemagglutinating activity of platelet thrombospondin to a 140 000-dalton thermolytic fragment. Biochemistry 23(23):5597–5603

Haverstick DM, Dixit VM et al (1985) Characterization of the platelet agglutinating activity of thrombospondin. Biochemistry 24(13):3128–3134

Haviv F, Bradley MF et al (2005) Thrombospondin-1 mimetic peptide inhibitors of angiogenesis and tumor growth: design, synthesis, and optimization of pharmacokinetics and biological activities. J Med Chem 48(8):2838–2846

Hennessy SW, Frazier BA et al (1989) Complete thrombospondin mRNA sequence includes potential regulatory sites in the 3′ untranslated region. J Cell Biol 108(2):729–736

Hirano K-I, Kuwasako T, Nakagawa-Toyama Y, Janabi M, Yamashita S, Matsuzawa Y (2003) Pathophysiology of human genetic CD36 deficiency. TCM 13:136–414

Hogg PJ, Owensby DA et al (1993) Thrombospondin 1 is a tight-binding competitive inhibitor of neutrophil cathepsin G. Determination of the kinetic mechanism of inhibition and localization of cathepsin G binding to the thrombospondin 1 type 3 repeats. J Biol Chem 268(29): 21811–21818

Hohenstein B, Daniel C et al (2008) PDE-5 inhibition impedes TSP-1 expression, TGF-beta activation and matrix accumulation in experimental glomerulonephritis. Nephrol Dial Transplant 23(11):3427–3436

Hotchkiss KA, Matthias LJ et al (1998) Exposure of the cryptic Arg-Gly-Asp sequence in thrombospondin-1 by protein disulfide isomerase. Biochim Biophys Acta 1388(2):478–488

Hugo C, Daniel C (2009) Thrombospondin in renal disease. Nephron Exp Nephrol 111(3):e61–e66

Isenberg JS, Ridnour LA, Perruccio EM, Espey MG, Wink DA, Roberts DD (2005) Thrombospondin-1 inhibits endothelial cell responses to nitric oxide in a cGMP-dependent manner. Proc Natl Acad Sci U S A 102:13141–13146

Isenberg JS, Ridnour LA et al (2006a) CD47 is necessary for inhibition of nitric oxide-stimulated vascular cell responses by thrombospondin-1. J Biol Chem 281(36):26069–26080

Isenberg JS, Wink DA et al (2006b) Thrombospondin-1 antagonizes nitric oxide-stimulated vascular smooth muscle cell responses. Cardiovasc Res 71(4):785–793

Isenberg JS, Hyodo F et al (2007a) Blocking thrombospondin-1/CD47 signaling alleviates deleterious effects of aging on tissue responses to ischemia. Arterioscler Thromb Vasc Biol 27(12): 2582–2588

Isenberg JS, Jia Y et al (2007b) Thrombospondin-1 inhibits nitric oxide signaling via CD36 by inhibiting myristic acid uptake. J Biol Chem 282(21):15404–15415

Isenberg JS, Romeo MJ et al (2007c) Increasing survival of ischemic tissue by targeting CD47. Circ Res 100(5):712–720

Isenberg JS, Maxhimer JB et al (2008a) Thrombospondin-1 and CD47 limit cell and tissue survival of radiation injury. Am J Pathol 173(4):1100–1112

Isenberg JS, Maxhimer JB et al (2008b) Treatment of liver ischemia-reperfusion injury by limiting thrombospondin-1/CD47 signaling. Surgery 144(5):752–761

Isenberg JS, Romeo MJ et al (2008c) Gene silencing of CD47 and antibody ligation of thrombospondin-1 enhance ischemic tissue survival in a porcine model: implications for human disease. Ann Surg 247(5):860–868

Isenberg JS, Romeo MJ et al (2008d) Thrombospondin-1 stimulates platelet aggregation by blocking the antithrombotic activity of nitric oxide/cGMP signaling. Blood 111(2):613–623

Isenberg JS, Yu C et al (2008e) Differential effects of ABT-510 and a CD36-binding peptide derived from the type 1 repeats of thrombospondin-1 on fatty acid uptake, nitric oxide signaling, and caspase activation in vascular cells. Biochem Pharmacol 75(4):875–882

Isenberg JS, Annis DS et al (2009a) Differential interactions of thrombospondin-1, -2, and -4 with CD47 and effects on cGMP signaling and ischemic injury responses. J Biol Chem 284 (2):1116–1125

Isenberg JS, Qin Y et al (2009b) Thrombospondin-1 and CD47 regulate blood pressure and cardiac responses to vasoactive stress. Matrix Biol 28(2):110–119

Jimenez B, Volpert OV, Crawford SE, Febbraio M, Silverstein RL, Bouck N (2000) Signals leading to apoptosis-dependent inhibition of neovascularization by thrombospondin-1. Nat Med 6:41–48

Jimenez B, Volpert OV, Reiher F, Chang L, Munoz A, Karin M, Bouck N (2001) C-Jun N-terminal kinase activation is required for the inhibition of neovascularization by thrombospondin-1. Oncogene 20:3443–3448

Kaur S, Martin-Manso G et al (2010) Thrombospondin-1 inhibits VEGF receptor-2 signaling by disrupting its association with CD47. J Biol Chem 285(50):38923–38932

Kaur S, Kuznetsova SA et al (2011) Heparan sulfate modification of the transmembrane receptor CD47 is necessary for inhibition of T cell receptor signaling by thrombospondin-1. J Biol Chem 286(17):14991–15002

Kazerounian S, Duquette M et al (2011) Priming of the vascular endothelial growth factor signaling pathway by thrombospondin-1, CD36, and spleen tyrosine kinase. Blood 117(17): 4658–4666

Koch M, Hussein F et al (2011) CD36-mediated activation of endothelial cell apoptosis by an N-terminal recombinant fragment of thrombospondin-2 inhibits breast cancer growth and metastasis in vivo. Breast Cancer Res Treat 128(2):337–346

Kosfeld MD, Frazier WA (1992) Identification of active peptide sequences in the carboxyl-terminal cell binding domain of human thrombospondin-1. J Biol Chem 267:16230–16236

Krutzsch HC, Choe BJ, Sipes JM, Guo N, Roberts DD (1999) Identification of an a3b1 integrin recognition sequence in thrombospondin-1. J Biol Chem 274:24080–24086

Kuznetsova SA, Day AJ et al (2005) The N-terminal module of thrombospondin-1 interacts with the link domain of TSG-6 and enhances its covalent association with the heavy chains of inter-alpha-trypsin inhibitor. J Biol Chem 280(35):30899–30908

Kuznetsova SA, Issa P et al (2006) Versican-thrombospondin-1 binding in vitro and colocalization in microfibrils induced by inflammation on vascular smooth muscle cells. J Cell Sci 119(Pt 21): 4499–4509

Kvansakul M, Adams JC, Hohenester E (2004) Structure of a thrombospondin C-terminal fragment reveals a novel calcium core in the type 3 repeats. EMBO J 23:1223–1233

Lawler JW, Slayter HS et al (1978) Isolation and characterization of a high molecular weight glycoprotein from human blood platelets. J Biol Chem 253(23):8609–8616

Lawler J, McHenry K et al (1995) Characterization of human thrombospondin-4. J Biol Chem 270 (6):2809–2814

Lawler J, Sunday M, Thibert V, Duquette M, George EL, Rayburn H, Hynes RO (1998) Thrombospondin-1 is required for normal murine pulmonary homeostatis and its absence causes pneumonia. J Clin Invest 101:982–992

Lefkowitz RJ, Caron MG (1988) Adrenergic receptors. Models for the study of receptors coupled to guanine nucleotide regulatory proteins. J Biol Chem 263(11):4993–4996

Lopez-Lluch G, Hunt N et al (2006) Calorie restriction induces mitochondrial biogenesis and bioenergetic efficiency. Proc Natl Acad Sci U S A 103(6):1768–1773

Manna PP, Frazier WA (2003) The mechanism of CD47-dependent killing of T cells: heterotrimeric Gi-dependent inhibition of protein kinase A. J Immunol 170:3544–3553

Margosio B, Rusnati M et al (2008) Fibroblast growth factor-2 binding to the thrombospondin-1 type III repeats, a novel antiangiogenic domain. Int J Biochem Cell Biol 40(4):700–709

Margossian SS, Lawler JW et al (1981) Physical characterization of platelet thrombospondin. J Biol Chem 256(14):7495–7500
Martin C, Chevrot M et al (2011) CD36 as a lipid sensor. Physiol Behav 105(1):36–42
Martin-Manso G, Calzada MJ et al (2011) sFRP-1 binds via its netrin-related motif to the N-module of thrombospondin-1 and blocks thrombospondin-1 stimulation of MDA-MB-231 breast carcinoma cell adhesion and migration. Arch Biochem Biophys 509(2):147–156
Maxhimer JB, Shih HB et al (2009a) Thrombospondin-1/CD47 blockade following ischemia-reperfusion injury is tissue protective. Plast Reconstr Surg 124(6):1880–1889
Maxhimer JB, Soto-Pantoja DR et al (2009b) Radioprotection in normal tissue and delayed tumor growth by blockade of CD47 signaling. Sci Transl Med 1(3):3ra7
McDonald JF, Zheleznyak A, Frazier WA (2004) Cholesterol-independent interactions with CD47 enhance a_vb_3 avidity. J Biol Chem 279:17301–17311
McKenzie P, Chadalavada SC et al (2006) Phylogenomic analysis of vertebrate thrombospondins reveals fish-specific paralogues, ancestral gene relationships and a tetrapod innovation. BMC Evol Biol 6:33
Means TK, Mylonakis E et al (2009) Evolutionarily conserved recognition and innate immunity to fungal pathogens by the scavenger receptors SCARF1 and CD36. J Exp Med 206(3):637–653
Miao W-M, Vasile E, Lane WS, Lawler J (2001) CD36 associates with CD9 and integrins on human blood platelets. Blood 97:1689–1696
Mii Y, Taira M (2011) Secreted Wnt "inhibitors" are not just inhibitors: regulation of extracellular Wnt by secreted Frizzled-related proteins. Dev Growth Differ 53(8):911–923
Miller TW, Isenberg JS et al (2011) Thrombospondin-1 is an inhibitor of pharmacological activation of soluble guanylate cyclase. Br J Pharmacol 159(7):1542–1547
Nisoli E, Carruba MO (2006) Nitric oxide and mitochondrial biogenesis. J Cell Sci 119(Pt 14): 2855–2862
Novelli EM, Kato GJ et al (2012) Plasma thrombospondin-1 is increased during acute sickle cell vaso-occlusive events and associated with acute chest syndrome, hydroxyurea therapy, and lower hemolytic rates. Am J Hematol 87(3):326–330
Ozbek S, Balasubramanian PG et al (2010) The evolution of extracellular matrix. Mol Biol Cell 21 (24):4300–4305
Panetti TS, Kudryk BJ et al (1999) Interaction of recombinant procollagen and properdin modules of thrombospondin-1 with heparin and fibrinogen/fibrin. J Biol Chem 274(1):430–437
Park SJ, Ahmad F et al (2012) Resveratrol ameliorates aging-related metabolic phenotypes by inhibiting cAMP phosphodiesterases. Cell 148(3):421–433
Pearce SFA, Wu J, Silverstein RL (1995) Recombinant GST/CD36 fusion proteins define a thrombospondin binding domain. J Biol Chem 270:2981–2986
Pluskota E, Stenina OI et al (2005) Mechanism and effect of thrombospondin-4 polymorphisms on neutrophil function. Blood 106(12):3970–3978
Qabar AN, Lin Z et al (1994) Thrombospondin 3 is a developmentally regulated heparin binding protein. J Biol Chem 269(2):1262–1269
Ramanathan S, Mazzalupo S et al (2011) Thrombospondin-1 and angiotensin II inhibit soluble guanylyl cyclase through an increase in intracellular calcium concentration. Biochemistry 50 (36):7787–7799
Reissen R, Kearney M, Lawler J, Isner JM (1998) Immunolocalization of thrombospondin-1 in human atherosclerotic and restenotic arteries. Am Heart J 135:357–364
Ren Y, Savill J (1995) Proinflammatory cytokines potentiate thrombospondin-mediated phagocytosis of neutrophils undergoing apoptosis. J Immunol 154(5):2366–2374
Ridnour LA, Isenberg JS, Espey MG, Thomas DD, Roberts DD, Wink DA (2005) Nitric oxide regulates angiogenesis through a functional switch involving thrombospondin-1. Proc Natl Acad Sci U S A 102:13147–13152
Roberts DD, Haverstick DM et al (1985) The platelet glycoprotein thrombospondin binds specifically to sulfated glycolipids. J Biol Chem 260(16):9405–9411

Roberts DD, Miller TW et al (2012) The matricellular protein thrombospondin-1 globally regulates cardiovascular function and responses to stress via CD47. Matrix Biol 31(3):162–169
Rock MJ, Holden P et al (2010) Cartilage oligomeric matrix protein promotes cell attachment via two independent mechanisms involving CD47 and alphaVbeta3 integrin. Mol Cell Biochem 338(1–2):215–224
Silverstein RL, Febbraio M (2009) CD36, a scavenger receptor involved in immunity, metabolism, angiogenesis, and behavior. Sci Signal 2(72):re3
Simantov R, Febbraio M, Crombie R, Asch AS, Nachman RL, Silverstein RL (2001) Histidine-rich glycoprotein inhibits the antiangiogenic effect of thrombospondin-1. J Clin Invest 107:45–52
Simantov R, Febbraio M et al (2005) The antiangiogenic effect of thrombospondin-2 is mediated by CD36 and modulated by histidine-rich glycoprotein. Matrix Biol 24(1):27–34
Staniszewska I, Zaveri S et al (2007) Interaction of alpha9beta1 integrin with thrombospondin-1 promotes angiogenesis. Circ Res 100(9):1308–1316
Subramanian A, Wayburn B et al (2007) Thrombospondin-mediated adhesion is essential for the formation of the myotendinous junction in Drosophila. Development 134(7):1269–1278
Sun J, Hopkins BD et al (2009) Thrombospondin-1 modulates VEGF-A-mediated Akt signaling and capillary survival in the developing retina. Am J Physiol Heart Circ Physiol 296(5): H1344–H1351
Sweetwyne MT, Murphy-Ullrich JE (2012) Thrombospondin1 in tissue repair and fibrosis: TGF-beta-dependent and independent mechanisms. Matrix Biol 31(3):178–186
Tan K, Duquette M, Liu J-h, Dong Y, Zhang R, Joachimiak A, Lawler J, Wang J-h (2002) Crystal structure of the TSP-1 type 1 repeats: a novel layered fold and its biological implication. J Cell Biol 159(2):373–382
Tan K, Duquette M, Liu J-h, Zhang R, Joachimiak A, Wang J-h, Lawler J (2006) The structures of the thrombospondin-1N-terminal domain and its complex with a synthetic pentameric heparin. Structure 14:33–42
Tan K, Duquette M et al (2009) The crystal structure of the signature domain of cartilage oligomeric matrix protein: implications for collagen, glycosaminoglycan and integrin binding. FASEB J 23(8):2490–2501
Thakar CV, Zahedi K et al (2005) Identification of thrombospondin 1 (TSP-1) as a novel mediator of cell injury in kidney ischemia. J Clin Invest 115(12):3451–3459
Toda N, Ayajiki K (2006) Phylogenesis of constitutively formed nitric oxide in non-mammals. Rev Physiol Biochem Pharmacol 157:31–80
Uluckan O, Becker SN et al (2009) CD47 regulates bone mass and tumor metastasis to bone. Cancer Res 69(7):3196–3204
van Almen GC, Verhesen W et al (2011) MicroRNA-18 and microRNA-19 regulate CTGF and TSP-1 expression in age-related heart failure. Aging Cell 10(5):769–779
Wang L, Zheng J et al (2010) Cartilage oligomeric matrix protein maintains the contractile phenotype of vascular smooth muscle cells by interacting with alpha(7)beta(1) integrin. Circ Res 106(3):514–525
Yao M, Roberts DD et al (2011) Thrombospondin-1 inhibition of vascular smooth muscle cell responses occurs via modulation of both cAMP and cGMP. Pharmacol Res 63(1):13–22
Yurchenco PD (2011) Basement membranes: cell scaffoldings and signaling platforms. Cold Spring Harb Perspect Biol Feb 1;3(2)

Chapter 9
Evolution of Cell Adhesion to Extracellular Matrix

Mark S. Johnson, Jarmo Käpylä, Konstantin Denessiouk, Tomi Airenne, Bhanupratap Singh Chouhan, and Jyrki Heino

Abstract The extracellular matrix and cell adhesion receptors, especially the integrins, have played a major role in the emergence of multicellular animals. The members of the integrin family can be found in all present-day metazoans, and they actually predate the origin of the animal kingdom. Chordate integrins show structural and functional diversity, and they gather around themselves a large number of adaptor and signaling proteins, an adhesome. This chapter reviews the early evolution of integrin-type protein domains, the origin of integrin-dependent adhesion mechanisms, and the later developments in chordate-specific integrins.

9.1 Extant Integrins and Adhesomes

9.1.1 Mammalian Integrins

In mammals, 24 integrin-type transmembrane receptors bind to the extracellular matrix (ECM), plasma proteins, or the immunoglobulin-like counter receptors (Hynes 2002). In addition to anchoring cells to their surrounding tissues, these receptors participate in cell migration and mediate information about the structure and the organization of the ECM (Legate et al. 2009) as well as physical forces, such as mechanical tension (Schwartz and DeSimone 2008). The large extracellular

M.S. Johnson • T. Airenne • B.S. Chouhan
Department of Biosciences, Åbo Akademi University, Turku, Finland

J. Käpylä • J. Heino (✉)
Department of Biochemistry and Food Chemistry, University of Turku, Turku, Finland
e-mail: jyrki.heino@utu.fi

K. Denessiouk
Centre for Biotechnology, University of Turku and Åbo Akademi University, Turku, Finland

F.W. Keeley and R.P. Mecham (eds.), *Evolution of Extracellular Matrix*, Biology of Extracellular Matrix, DOI 10.1007/978-3-642-36002-2_9,

"ectodomains" of the integrins have evolved to recognize major EMC components, including collagens, laminins, tenascins, thrombospondins, and fibronectin. The short intracellular tails of integrins interact with tens of different proteins (Legate and Fässler 2009) and form stable complexes, such as focal adhesions and hemidesmosomes. Adaptor proteins, including tensin, filamin, talin, and α-actinin, can link the receptors to actin microfilaments. Plectin can also interact with intermediate filament proteins, and integrin-linked kinase (ILK, pseudokinase) may connect integrins to microtubules (Wickström et al. 2010). Talin and kindlins also regulate the conformation and the activity of the integrins (Moser et al. 2009). Protein tyrosine kinases, e.g., focal adhesion kinase (FAK), p130Cas, and Src, as well as GTPases, e.g., Rho, Rac, and Cdc42, activate cellular signaling pathways, which regulate cell migration, survival, differentiation, and proliferation (Legate et al. 2009). The recent high-resolution microscopy observations (Kanchanawong et al. 2010) have allowed the creation of even more detailed models concerning, how integrin-related adhesion sites have been organized. Integrins are connected to cellular mechanisms that control endocytosis, and they can orchestrate the recycling of many other receptor systems, including the growth factors receptors (Ivaska and Heino 2011). Integrin endocytosis and recycling may also play an important role in cell migration. The fact that integrins are in constant interplay with many growth factors and their receptors even increases the complexity of their action (Ivaska and Heino 2011). Therefore, it is not surprising that the integrin-related protein complex, or the adhesome, has been reported to include more than 150 different proteins and 700 interactions (Zaidel-Bar et al. 2007).

The 24 mammalian integrin-type receptors are composed of 1 out of 18 α-subunits and 1 out of 8 β-subunits (Hynes 2002). The careful analysis of knockout mouse phenotypes (Bouvard et al. 2001) has shed light on the role of each individual receptor in the development and the physiology of the vertebrates. The β1-subunit is the most "promiscuous" of all integrin subunits, since it can form a heterodimeric receptor with 12 different α-subunits, and mouse embryos lacking β1 undergo peri-implantation lethality before day 5.5 of gestation (E5.5; Stephens et al. 1995). The α5-subunit is one β1 partner, and the α5β1 heterodimer is a fibronectin receptor that recognizes the RGD (arginine–glycine–asparagine; throughout we will use the single-letter and the three-letter amino acid codes) motif of its ligand. Fibronectin-deficient mouse embryos have severe neural tube deformations and defects in mesodermally derived tissues and die at E8.5. Given the critical role of fibronectin during the early development of the vertebrates (George et al. 1993), it is not surprising that α5 knockouts also show embryonic lethality at E9.5–10, a little later than fibronectin-null embryos (Yang et al. 1993). This is also the case with the RGD-independent fibronectin receptors, α4β1 and α4β7 integrins (Yang et al. 1995), where α4 null embryos die at E10–12. They are also receptors for VCAM-1 (vascular cell adhesion molecule), and the most prominent developmental defects may be due to the lack of the VCAM-1 interaction with integrins having the α4-subunit. RGD-independent α4 integrins recognize LDV motifs and variants thereof (Newham and Humphries 1996; Newham et al. 1997).

The integrins containing the αV-subunit are RGD-dependent fibronectin and vitronectin receptors. Integrin αV can form a functional receptor with five β-subunits, and the simultaneous deficiency of all these heterodimers caused by αV deletion leads to embryonic (80%, around E10.5) or perinatal (20%) lethality, and mice that survive until birth suffer from several defects, e.g., intracerebral hemorrhage (Bader et al. 1998).

The knockout phenotypes of other mouse integrins are less severe (Bouvard et al. 2001). Integrin α9β1 is an RGD-independent fibronectin receptor. Deficiency of the α9-subunit is associated with defective lymphatic valve morphogenesis (Bazigou et al. 2009). Laminin-receptor integrins are clearly involved in the development of specific organs, and their deficiency often leads to neonatal lethality. For example, α3 knockouts have kidney and lung defects (Kreidberg et al. 1996), and α7 knockouts suffer from muscle dystrophy (Mayer et al. 1997). Integrin α6 can form a heterodimer with the β1- and β4-subunits, and integrin α6β4 is a critical structural component in hemidesmosomes (Stepp et al. 1990), and the deletion of either one of the subunits leads to a skin blistering disorder.

The deficiency of the αV-subunit's β partners causes mild developmental defects only. For example, β5 null mice generate age-related blindness (Nandrot et al. 2004), and β3 nulls show impaired osteoclast (McHugh et al. 2000) and platelet function (Hodivala-Dilke et al. 1999). Mice lacking integrins with β3-subunits, or both β3- and β5-subunits, are reported to develop enhanced pathological angiogenesis (Reynolds et al. 2002). Integrins αVβ6 and αVβ8 play an important role in the activation of latent transforming growth factor β and are therefore involved in the generation of fibrotic conditions (Margadant and Sonnenberg 2010). In addition to the αV-subunit, β3 also has another α partner, namely, platelet-specific αIIb. This heterodimer acts as an RGD-dependent fibrinogen receptor and is critical for proper platelet clot formation.

Nine out of 18 mammalian α-subunits fall into a structurally unique subgroup of integrins due to the fact that they contain an extra, "inserted," domain (αI domain). Four of these α-subunits (α1, α2, α10, and α11) partner with the β1-subunit and act as collagen receptors (Heino et al. 2009). The β2-subunit can associate with four other αI domain subunits (αD, αL, αM, and αX), while the β7-subunit is a partner for αE. Integrins with β2 and β7 are usually found on inflammatory cells and often called leukocyte αI domain integrins (Evans et al. 2009). Mice lacking any of the nine αI domain integrins are still born alive and without major developmental defects (Bouvard et al. 2001). They are also fertile. The deficiency of the collagen receptors may lead to impaired function of platelets, mesenchymal stem cells, acquired and native immunity, and tissue repair after injury (Heino et al. 2009). The leukocyte integrins act as receptors for, e.g., intercellular adhesion molecule-1 (ICAM-1) and C3b component of the complement system and participate in the related immune functions (Evans et al. 2009).

The knockout phenotypes reflect the emergence of vertebrate-specific organs and physiological functions, but they may also tell something about the evolutionary history of integrins in general. The important role of the β1-subunit (Stephens et al. 1995) is mainly due to its multiple α partners, whereas none of the vertebrate

α/β heterodimers alone seem to have a critical function during early embryogenesis. Despite the fact that the fibronectin gene is found in vertebrates only, fibronectin and its receptors are essential for the development of many fetal organs (George et al. 1993; Yang et al. 1993). Thus, the important function of $\alpha5\beta1$ may reflect the essential role of RGD-related adhesion mechanisms. In addition to fibronectin, many other adhesion glycoproteins harbor the RGD motif too. For example, fruit fly integrins recognize tigrin and thrombospondin in an RGD-dependent manner (Johnson et al. 2009). The mild knockout phenotypes of αI domain integrins are in accordance with the fact that these integrins have evolved quite late and that they are found only in chordates.

9.1.2 Non-vertebrate Integrins

All extant multicellular animals express integrins. Sponges are generally considered to be among the simplest and maybe the oldest lineage of metazoans, despite the fact that their exact phylogenic position has remained unresolved (Adamska et al. 2011). Both integrin α- and β-subunits have been identified in sponges (Brower et al. 1997; Pancer et al. 1997; Müller 1997). Furthermore, sponge integrins have been reported to interact with the RGD motif (Wimmer et al. 1999). The accumulation of new genomic information about the morphologically simplest metazoan species, e.g., *Trichoplax adhaerens* (Srivastava et al. 2008; Schierwater et al. 2009), indicates that the diversity of the integrin family is larger than previously expected. This has been obvious, for example, in studies focused on anthozoan cnidarians, such as the coral *Acropora millepora* and the sea anemone *Nematostella vectensis*. The Acropora integrins are expressed in the presumptive endoderm during gastrulation, but their exact ligands and functions are not known (Knack et al. 2008).

Integrin expression in non-vertebrate model organisms has also pointed out the important role of adhesion mechanisms during organ development. The genome of the fruit fly, *Drosophila melanogaster*, codes for two integrin β-subunits and five α-subunits (Bökel and Brown 2002). Fly βPS integrin is a mammalian β1-like integrin that forms a heterodimer with, at least, two different α-subunits, namely, laminin-receptor-like αPS1 and RGD-binding αPS2. The deficiency of βPS leads to muscle detachment and embryonic lethality. These flies also have wing blisters. Three other fruit fly integrin α-subunits form a unique subgroup within metazoan integrins. The expression of the β-neu-subunit is restricted to the mid gut, and its function is not known. Recent observations have also stressed the role of the fruit fly integrin–laminin interaction in the attachment of the dendrites to the ECM and in the control of dendrite positioning (Han et al. 2012; Kim et al. 2012).

The nematode *Caenorhabditis elegans* has two integrin α-subunits (ina-1 and pat-2) and one β (pat-3) generating two heterodimeric integrins. The heterodimer composed of αINA-1 and βPAT-3 is predicted to bind laminins, and it participates in commissural axon navigation. Integrin αPAT-2 may recognize the RGD motif, promote the formation of dense bodies (focal adhesions), and participate in cell migration (Cox et al. 2004).

9.1.3 Adhesomes

Another approach to integrin evolution has been taken, when the origin of the adhesome has been considered (Zaidel-Bar 2009). These studies have revealed that the core adhesome components can be found in the present-day primitive metazoans. In sponge, RGD peptides can activate GTPases indicating that their integrins carry the signaling capability (Wimmer et al. 1999). Indeed, many signaling proteins such as Rho GTPases, serine/threonine kinases, and phosphatases have existed already before the integrins and then adopted new functional roles at the integrin-related adhesion sites (Zaidel-Bar 2009). Later, gene duplications have increased the diversity and complexity of integrin signaling, and especially protein tyrosine phosphorylation has evolved concomitantly with the integrins (Zaidel-Bar 2009).

9.2 Integrin Structure and Conformational Regulation

9.2.1 Domain Structure of Integrins

Integrins are dynamic, multidomain cell-surface receptors that undergo a range of conformational changes as part of their inside-out and outside-in signaling (Hynes 2002). A key to their function is the dynamic interplay between several domains of the noncovalently linked α- and β-subunits. The α-subunits of all integrins have the β-propeller, Thigh, Genu (or knee), Calf-1, Calf-2, transmembrane (TM), and a small cytosolic tail domain, whereas the β-subunits have the PSI (Plexin-Semaphorin-Integrin), Hybrid, βI, integrin beta epidermal growth factor-like domains (I-EGF-1, I-EGF-2, I-EGF-3, and I-EGF-4), tail, TM, and a small cytosolic tail domain. Moreover, the α1-, α2-, α10-, α11-, αD-, αE-, αL-, αM-, and αX-subunits are equipped with the additional αI domain budding out of the β-propeller domain, which is crucial for ligand recognition and binding in integrins containing these subunits. The basic domain structure of integrins is depicted in Fig. 9.1.

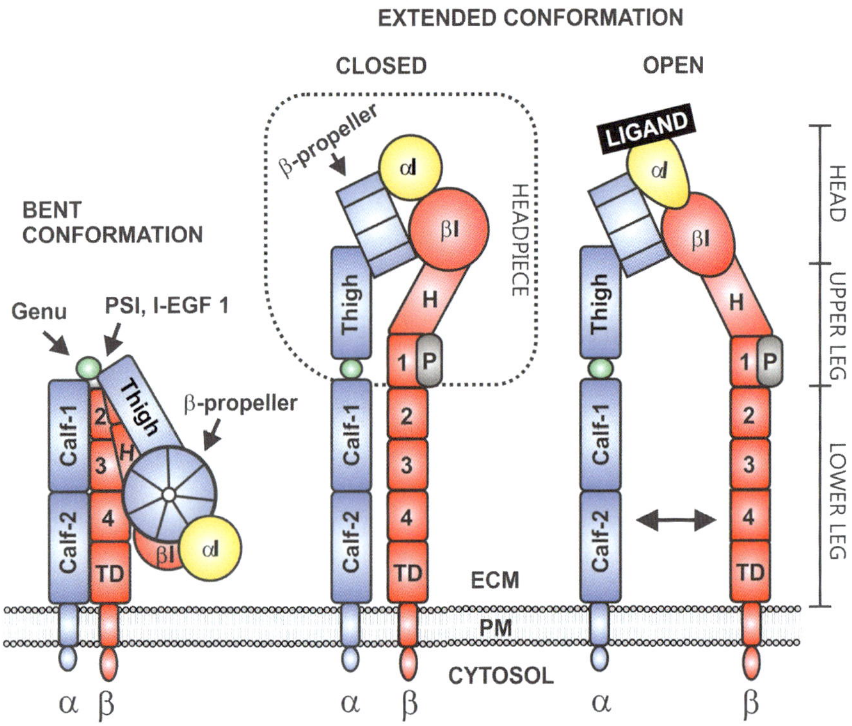

Fig. 9.1 Integrin conformational states and domain organization. Simplified, schematic view of the domain architecture of the α- (*blue*) and β (*red*)-subunits is presented. The αI domain (*yellow*) is the major ligand-binding site in αI-containing integrins, whereas in the αI-lacking integrins, the major ligand-binding site is located at the βI domain (*red*). Genu (or knee) is drawn as a green sphere, and the EGF-like domains 1–4 of the β-subunit are numbered. H, Hybrid domain; P, PSI domain (*gray*); TD, beta-tail domain; PM, plasma membrane; and ECM, extracellular matrix. This figure is based on Xiong et al. (2001) and Luo et al. (2007). An extended, intermediate conformation suggested recently by Yu et al. (2012), is not shown (see text for details)

9.2.2 αI Domain

Structural studies have centered on the human integrins, the earliest structure determinations focusing on the αI domain, known also as the inserted (I) domain or A domain after the von Willebrand factor A domain (for a list of X-ray and NMR structures of integrins deposited to the Protein Data Bank (PDB; Berman et al. 2000), see Tables 9.1 and 9.2). The first structure solved was the X-ray structure of the human αI domain of integrin αMβ2 (Lee et al. 1995a, b). This structure and subsequent crystal structures of other αI domains have revealed that they are built up of a six-stranded β-sheet in the center of the domain surrounded by seven (non-collagen receptors) or eight (collagen receptors) α-helices together folding into an open, classical α/β sheet fold known also as

Table 9.1 X-ray structures of human integrins

Fragment	α/β	PDB	Å	References	Comments
αI-domain:	α1	1QC5	2.00	Rich et al. (1999) J Biol Chem 274: 24906–24913	Mg^{2+}; T137R, Q138S, P139S, T338G
		1CK4[a]	2.20	Nolte et al. (1999) FEBS Lett 452: 379–385	No cation
		1MHP[a]	2.80	Karpusas et al. (2003) J Mol Biol. 327: 1031–1041	G217V, R218Q, Q219R, L222R. Fab complex
		1PT6	1.87	Nymalm et al. (2004) J Biol Chem 279: 7962–7970	Mg^{2+}
		1QCY	2.30	Nymalm et al. (2004) J Biol Chem 279: 7962–7970	Mg^{2+}; K170E
		2B2X[a]	2.20	Clark et al (2006) Protein Sci 15: 949–960	Mg^{2+}; G217V, R218Q, Q219R, L222R. Fab complex
		4A0Q	1.90	Lahti et al. (2011) J Biol Chem 286: 43343–43351	Mg^{2+}; E317A
		1DZI	2.10	Emsley et al. (2000) Cell 101: 47–56	Co^{2+}, in complex with a collagen peptide
		1V7P	1.90	Horii et al (2004) J Mol Biol 341: 519–527	Mn^{2+}; EMS16 complex; Q138G, P139S, C140S
	αL	1LFA	1.80	Qu and Leahy (1995) Proc Natl Acad Sci USA 92: 10277–10281	Mn^{2+}; W189R
		1ZON	2.00	Qu and Leahy (1996) Structure 4: 931–942	No cation, W189R
		1ZOO	3.00	Qu and Leahy (1996) Structure 4: 931–942	Mg^{2+}; W189R
		1ZOP	2.00	Qu and Leahy (1996) Structure 4: 931–942	Mn^{2+}; W189R
		1CQP	2.60	Kallen et al. (1999) J Mol Biol 292: 1–9	Mg^{2+}, in complex with lovastatin
		1MQ8	3.30	Shimaoka et al. (2003) Cell 112: 99–111	Mg^{2+}; in complex with ICAM-1; L186C, F324C
		1MQ9	2.00	Shimaoka et al. (2003) Cell 112: 99–111	Mn^{2+}; K287C, K294C
		1MQA	2.50	Shimaoka et al. (2003) Cell 112: 99–111	K287C, K294C
		1MJN	1.30	Shimaoka et al. (2003) Cell 112: 99–111	Mg^{2+}; L161C, F299C
		1RD4	2.40	Crump et al (2004) Biochemistry 43: 2394–2404	Inhibitor complex
		1XDD	2.20	Weitz-Schmidt et al (2004) J Biol Chem 279: 46764–71	Mg^{2+}, in complex with LFA703
		1XDG	2.10	Weitz-Schmidt et al (2004) J Biol Chem 279: 46764–71	Mg^{2+}, in complex with LFA878
		1XUO	1.80	Wattanasin et al (2005) Bioorg Med Chem Lett. 15: 1217–1220	Mg^{2+}, inhibitor complex
		1T0P	1.66	Song et al. (2005) Proc Natl Acad Sci USA 102: 3366–3371	Mg^{2+}; ICAM-3 D1 complex
		2ICA	1.56	Potin et al. (2006) J Med Chem 49: 6946–6949	Antagonist complex
		2O7N	1.75	Dodd et al. (2007) Bioorg Med Chem Lett 17: 1908–1911	Antagonist complex
		2BN3	2.10	Zhang et al. (2008) Mol Cell 31: 432–437	Mg^{2+}; ICAM-5 complex; F265S, F292G

(continued)

Table 9.1 (continued)

Fragment	α/β	PDB	Å	References	Comments
		3BQM	1.95	Guckian et al. (2008) Bioorg Med Chem Lett 18: 5249–5251	Inhibitor complex
		3BQN	1.80	Lin et al. (2008) Bioorg Med Chem Lett 18: 5245–5248	Inhibitor complex
		3E2M	2.00	Lin et al. (2008) Bioorg Med Chem Lett 18: 5245–5248	Inhibitor complex
		3EOA	2.80	Li et al (2009) Proc Natl Acad Sci USA 106: 4349–4354	In complex with the Efalizumab Fab
		3EOB	3.60	Li et al (2009) Proc Natl Acad Sci USA 106: 4349–4354	Zn^{2+}, in complex with the Efalizumab Fab
		3F74	1.70	Zhang et al. (2009a) Faseb J 23: 2735–2740	Mg^{2+}
		3F78	1.60	Zhang et al. (2009a) FASEB J 23: 2735–2740	Mg^{2+}, isoflurane complex
		3HI6	2.30	Zhang et al. (2009b) Proc Natl Acad Sci USA 106: 18345–18350	Mn^{2+}, in complex with Fab fragment of AL-57
		3M6F	1.85	Watterson et al. (2010) J Med Chem 53: 3814–3830	Antagonist complex
		3TCX	3.60	In PDB: Kang et al. (2011). To be published	Mg^{2+}; engineered ICAM-1 D1, high affinity to αL
	αM	1IDO	1.70	Lee et al. (1995b) Cell 80: 631–638	Mg^{2+}, the first X-ray structure of integrins
		1JLM	2.00	Lee et al. (1995a) Structure 3: 1333–1340	Mn^{2+}
		1BHO	2.70	Baldwin et al. (1998) Structure 6: 923–935	Mn^{2+}
		1BHQ	2.70	Baldwin et al. (1998) Structure 6: 923–935	Cd^{2+}
		1IDN	2.70	Baldwin et al. (1998) Structure 6: 923–935	No cation
		1M1U	2.30	Xiong et al. (2000) J Biol Chem 275: 38762–38767	C128S
		1MF7	1.25	McCleverty and Liddington (2003) Biochem J 372:121–127	A318C
		1N9Z	2.50	McCleverty and Liddington (2003) Biochem J 372:121–127	C128A, D132C, K315C
		1NA5	1.50	McCleverty and Liddington (2003) Biochem J 372:121–127	
	αX	3Q3G	2.70	Mahalingam et al. (2011) J Immunol 187: 6393–6401	Antibody complex
		3QA3	3.00	Mahalingam et al. (2011) J Immunol 187: 6393–6401	Antibody complex
		1N3Y	1.65	Vorup-Jensen et al. (2003) Proc Natl Acad Sci USA 100: 1873–78	T192S, T234A
Headpiece	αIIbβ3	2VC2	3.10	Springer et al. (2008) J Cell Biol 182: 791–800	Mg^{2+}, antagonist L-739758, antibody 10E5
		2VDK	2.80	Springer et al. (2008) J Cell Biol 182: 791–800	Mg^{2+}, antibody 10E5
		2VDL	2.75	Springer et al. (2008) J Cell Biol 182: 791–800	Mg^{2+}, antibody 10E5
		2VDM	2.90	Springer et al. (2008) J Cell Biol 182: 791–800	Mg^{2+}, antagonist tirofiban, antibody 10E5
		2VDN	2.90	Springer et al. (2008) J Cell Biol 182: 791–800	Mg^{2+}, antagonist eptifibatide, antibody 10E5

		2VDO	2.51	Springer et al. (2008) J Cell Biol 182: 791–800	Mg^{2+}, fibrinogen γ chain pep: HHLGGAKQAGDV, 10E5
		2VDP	2.80	Springer et al. (2008) J Cell Biol 182: 791–800	Mg^{2+}, fibrinogen γ chain pep: LGGAKQAGDV, 10E5
		2VDQ	2.59	Springer et al. (2008) J Cell Biol 182: 791–800	Mg^{2+}, chimeric γ chain pep: HHLGGAKQRGDV, 10E5
		2VDR	2.40	Springer et al. (2008) J Cell Biol 182: 791–800	Mg^{2+}, chimeric γ chain peptide: LGGAKQRGDV, 10E5
		3FCU	2.90	Zhu et al. (2009) Mol Cell 32: 849–861	Mg^{2+}
		3NID	2.3	Zhu et al. (2010) Blood 116: 5050–5059	Mg^{2+}; 10E5 antibody complex
		3NIF	2.4	Zhu et al. (2010) Blood 116: 5050–5059	Mg^{2+}, antagonist and 10E5 antibody complex
	α4β7	3V4P	3.15	Yu et al. (2012) J Cell Biol 196: 131–146	Mg^{2+}, antibody complex
		3V4V	3.10	Yu et al. (2012) J Cell Biol 196: 131–146	Mg^{2+}, a small molecule and antibody complex
	α5β1	3VI3	2.9	Nagae et al. (2012) J Cell Biol 197:131–140	Mg^{2+}; SG/19 Fab bound to the β1 chain
Ectodomain	αIIbβ3	1TYE	2.90	Xiao et al. (2004) Nature 432: 59–67	Mg^{2+}
		3FCS	2.55	Zhu et al. (2008) Mol Cell 32: 849–861	Mg^{2+}, complete ectodomain
	αVβ3	1JV2	3.10	Xiong et al. (2001) Science 294: 339–345	The first ectodomain structure
		1M1X	3.30	Xiong et al. (2002) Science 296: 151–155	Mn^{2+}
		1L5G	3.20	Xiong et al. (2002) Science 296: 151–155	Mn^{2+}, in complex with cyclic RGD
		1U8C	3.10	Xiong et al. (2004) J Biol Chem 279: 40252–40254	
		3IJE	2.9	Xiong et al. (2009) J Cell Biol 186: 589–600	Complete ectodomain with TM stretches
	αXβ2	3K6S	3.5	Xie et al. (2010) EMBO J 29: 666–679	Mg^{2+}, the first ectodomain structure with I-domain
		3K71	3.95	Xie et al. (2010) EMBO J 29: 666–679	Mg^{2+}, I-domain not visible
		3K72	3.7	Xie et al. (2010) EMBO J 29: 666–679	Mg^{2+}, I-domain not visible
Cytosolic/ TM	β1D	3G9W	2.16	Anthis et al. (2009) EMBO J 28:3623–32	In complex with talin2 domains F2–F3
	β2	2V7D	2.50	Takala et al. (2008) Blood 112: 1853–62	In complex with 14-3-3
		2JF1	2.20	Takala et al. (2008) Blood 112: 1853–62	In complex with filamin A repeat 21
	β3	1MIZ	1.90	Garcia-Alvarez et al. (2003) Mol Cell 11:49–58	A chimera of fragments of β3 tail and talin
		1MK7	2.20	Garcia-Alvarez et al. (2003) Mol Cell 11:49–58	A chimera of fragments of β3 tail and talin
		1MK9	2.80	García-Alvarez et al. (2003) Mol Cell 11:49–58	A chimera of fragments of β3 tail and talin
		2Q6W	2.25	Parry et al. (2007) J Mol Biol 371: 435–446	In complex with leukocyte antigen DR52a

(continued)

Table 9.1 (continued)

Fragment	α/β	PDB	Å	References	Comments
	β4	1QG3	2.15	de Pereda et al. (1999) EMBO J 18: 4087–4095	Tandem pair of fibronectin type III domains
		3F7P	2.75	de Pereda et al. (2009) EMBO J. 28: 1180–1190	Plectin complex
		3F7Q	1.75	de Pereda et al. (2009) EMBO J. 28: 1180–1190	In complex with a fibronectin fragment
		3F7R	2.04	de Pereda et al. (2009) EMBO J. 28: 1180–1190	In complex with a fibronectin fragment
		3FQ4	1.49	Alonso-Garcia et el. (2009) Acta C D65: 858–871	The Calx-beta motif of β4
		3FSO	1.40	Alonso-Garcia et el. (2009) Acta C D65: 858–871	The Calx-beta motif of β4 (Ca^{2+} soak)
		3H6A	1.61	Alonso-Garcia et el. (2009) Acta C D65: 858–871	The Calx-beta motif of β4 (co-crystallized with Ca^{2+})
	β7	2BRQ	2.10	Kiema et al. (2006) Mol Cell 21: 337–47	Filamin A repeat 21 complex
Legs	β2	2P26	1.75	Shi et al. (2007) J Biol Chem 282: 30198–30206	PSI, Hybrid, I-EGF-1, and I-EGF-2 domain
		2P28	2.20	Shi et al. (2007) J Biol Chem 282: 30198–30206	PSI, Hybrid, I-EGF-1, I-EGF-2, I-EGF-3 domain
	β4	1YUK	1.80	Shi et al. (2005) J Biol Chem 280: 30586–30593	PSI, Hybrid, and I-EGF1 domain

The list is based on text search of the Protein Data Bank (http://www.rcsb.org; 02/07/2012)
[a]Rat protein

Table 9.2 NMR structures of human integrins

Fragment	α/β	PDB	Reference	Comments
αI-domain	αL	1DGQ	Legge et al. (2000) J Mol Biol 295: 1251–1264	Magnesium complex
Cytoplasmic tail	αIIb	1DPK	Vinogradova et al. (2000) Proc Natl Acad Sci USA 97: 1450–1455	
		1DPQ	Vinogradova et al. (2000) Proc Natl Acad Sci USA 97: 1450–1455	Constitutively active P1029A/P1030A mutant
		1S4W	Vinogradova et al. (2004) Proc Natl Acad Sci USA 101: 4094–99	In DPC micelles
		2LM5	Huang and Vogel (2012) J Am Chem Soc 134:3864–72	In complex with Ca^{2+}-CIB1
	αIIbβ3	1M8O	Vinogradova et al. (2002) Cell 110: 587–597	Membrane-proximal helices of α and β cytoplasmic tails
		1KUP	Weljie et al. (2002) Proc Natl Acad Sci USA 99: 5878–5883	Complex of membrane-proximal regions
		1KUZ	Weljie et al. (2002) Proc Natl Acad Sci USA 99: 5878–5883	Complex of membrane-proximal regions
	αL	2K8O	Bhunia et al. (2009) J Biol Chem 284: 3873–3884	Mutation M14I
	αM	2LKE	Chua et al. (2011) J Biol Chem 286:43842–54	In DPC micelles, myristoylated
		2LKJ	Chua et al. (2011) J Biol Chem 286:43842–54	In DPC micelles, serine phosphorylated
	β3	1S4X	Vinogradova et al. (2004) Proc Natl Acad Sci USA 101: 4094–99	In DPC micelles
		2H7D	Wegener et al. (2007) Cell 128: 171–82	Talin F3 and chimeric β3-PIP kinase peptide
		2H7E	Wegener et al. (2007) Cell 128: 171–82	Talin F3 and chimer, β3-PIP kinase pep. (min. averaged structure)
		2KV9	Metcalf et al. (2010) Proc Natl Acad Sci USA 107: 22481–22486	In phospholipid bilayers
		2L1C	Deshmukh et al. (2010) J Biol Chem 285:34875–84	Biphosphorylated β3, Shc PTB domain complex
		2LJD	Deshmukh et al. (2011) J Biol Chem 286:40943–53	Monophosphorylated (747pY), membrane mimetic conditions
		2LJE	Deshmukh et al. (2011) J Biol Chem 286:40943–53	Diphosphorylated (747pY and 759pY)
		2LJF	Deshmukh et al. (2011) J Biol Chem 286:40943–53	Monophosphorylated (747pY), under aqueous conditions
	β4	2YRZ	In PDB: Abe et al. (2008). To be published	Fibronectin type III domain
TM	αIIb	2K1A	Lau et al. (2008a) J Biol Chem 283: 16162–16168	Bicelle-embedded

(continued)

Table 9.2 (continued)

Fragment	α/β	PDB	Reference	Comments
	αIIbβ3	2K9J	Lau et al. (2009) EMBO J 28: 1351–1361	The first structure of a heterodimeric TM receptor complex
	β3	2RMZ	Lau et al. (2008b) Biochemistry 47: 4008–4016	Bicelle-embedded
		2RN0	Lau et al. (2008b) Biochemistry 47: 4008–4016	Micelle-embedded
		2L91	Kim et al. (2012) Nature 481: 209–213	C687S, A711P, K716A
Cytopl. and TM	α1	2L8S	In PDB: Lai et al. (2012). To be published	TM and cytosolic regions of α1 in detergent micelles
	αIIbβ3	2KNC	Yang et al. (2009) Proc Natl Acad Sci USA 106: 17729–17734	TM-cytoplasmic heterocomplex
Legs	β2	1L3Y	Beglova et al. (2002) Nat Struct Biol 9: 282–287	I-EGF-3

The list is based on text search of the Protein Data Bank (http://www.rcsb.org; 02/07/2012)

the Rossmann fold or the dinucleotide-binding domain. The metal ion-dependent adhesion site (MIDAS) is a key, conserved structural feature of αI domains, forming the ligand-binding site in αI-containing integrins; the extra αC helix in the collagen receptor integrins (α1β1, α2β1, α10β1, and α11β1) is located close to MIDAS.

Numerous αI domain structures with different conformations have now been obtained in both the presence and absence of divalent metal ions (Table 9.1). One significant milestone in the structural studies was the αI domain structure of α2β1 integrin in complex with a triple helical GFOGER peptide that mimics a bound collagen ligand (Emsley et al. 2000). This peptide-bound structure is thought to represent an open/activated or high-affinity conformation, and comparison with the human α2I ligand-free structure (PDB code: 1AOX; Emsley et al. 1997) revealed significant conformational changes due to ligand binding: a water molecule is displaced and the ligand directly coordinates to the metal ion at MIDAS, the α1 helix shifts its position, the αC helix partly unfolds and the α6 helix correspondingly extends itself, and a large movement is seen for the α7 helix (Emsley et al. 2000). Despite being solved separately from the ectodomain context, these conformational changes in these structures suggested a mechanism for communicating inside-out and outside-in signaling throughout the intact integrin.

While "intended" ligands favor the open/activated form of αI and bind to MIDAS, echovirus-1 (EV-1) exploits human α2β1 to infect cells, but structural reconstructions from cryoelectron microscopy and crystal structures together with binding studies are consistent with a preference for the closed conformation of α2I and no direct involvement of MIDAS (Xing et al. 2004; Jokinen et al. 2010). A snake venom peptide from the metalloproteinase of *Bothrops jararaca* venom binds at MIDAS on α2I but not to the metal; a negatively charged residue is not involved in binding, but binding further increases the strength of the α2I–EV1 complex (Ivaska et al. 1999; Pentikäinen et al. 1999).

The structure of αIL was solved in a high-affinity conformation in complex with ICAM-1, in an intermediate-affinity conformation (ligand-free and with a bound ICAM-1), and in a low-affinity conformation (ligand-free), reflecting the conformational range and regulation of the αI domain (Qu and Leahy 1996; Shimaoka et al. 2003). Recently, Lahti et al. (2011) have published an α1I domain structure resembling the intermediate conformation observed in the αLI structure. However, unlike in the αLI intermediate-affinity structure, neither the α7 helix moved nor structural changes took place at the metal-binding site of the α1I structure. Moreover, based on computational analyses (modeling and energy calculations), previous papers have considered it unlikely that α1I or α2I could have a similar intermediate conformation as the αLI or αMI domains (Jin et al. 2004). Whether the intermediate conformations observed in the isolated αLI domain are physiologically relevant remains to be seen.

9.2.3 Ectodomain Structure

Because of the large number of domains, compact interactions among some domains in the heterodimer, and conformational dynamics, different sets of domains have been given descriptive names. Together with the β-propeller (with or without the αI domain) and the Thigh domains of the α-subunit, the βI, Hybrid, PSI, and the I-EGF1 domains of the β-subunit constitute the integrin "headpiece." The extracellular ectodomain portion of the integrins (Fig. 9.1) is also divided to regions called the "head" (β-propeller with or without αI, and βI), "upper leg" (Thigh, Hybrid, PSI, I-EGF1), and "lower leg" (Calf-1 and 2, I-EGF2-4, β-tail). The leg regions of the β-subunit seem to be more flexible than the corresponding regions of the α-subunit, which is thought to be important for conformational regulation (Xie et al. 2010).

The first ectodomain crystal structure revealed details of the entire extracellular region of the αVβ3 integrin heterodimer (Xiong et al. 2001). The α- and β-subunits were found to exist in a bent conformation, whereby the headpiece would point towards the cell membrane, and in a manner thought to indicate an inactive conformation. A similar conformation was observed for the αVβ3 structure in complex with an RGD-containing peptide (Xiong et al. 2002) as well as for the subsequent ectodomain structures of αIIbβ3 (Xiao et al. 2004; Zhu et al. 2008) and αXβ2 (Xie et al. 2010). At present, the αXβ2 structure is the only ectodomain X-ray structure having an αI domain, and it revealed that the N- and C-terminal links between the β-propeller domain and the αI domain appear highly flexible, thus enabling easy access for ligands. The ectodomain structures (Table 9.1) with or without αI domain are generally quite similar, but some local differences can be found, for example, in the relative orientation of the I-EGF4 and β-tail domains.

Headpiece/ectodomain structures have recently been solved for α5β1 and α4β7. With α5β1, only minor changes were observed between the RGD-ligand bound and unbound structures, but the structures are constrained by the presence of a bound antibody fragment known to force the structure into the inactive conformation (Nagae et al. 2012). X-ray and EM analyses of the α4β7 integrin headpiece/ectodomain structures combined with functional studies revealed unique features as compared to the other known integrin structures (Yu et al. 2012). For example, structural differences in the ligand-binding environment, especially in the β-propeller domain, and a wide groove well suited for lengthwise binding of protein domains were found. The domains of α4β7 were also less densely packed in the bent conformation. Interestingly, a novel extended intermediate conformation supporting rolling adhesion was reported, separating the α4β7 integrin from other integrins that are activated via the open extended conformation as a result of ligand binding.

The α-subunit consists of four or five extracellular domains. Starting from the amino-terminal end, these are the β-propeller domain having seven blades and 3–4 Ca^{2+}-binding sites (in human integrin αV-subunit, residues 1–438; PDB code: 1JV2; Xiong et al. 2001), and, where present, the αI domain is inserted between the second and third blades of the β-propeller; next are the Thigh domain (residues

439–592) and the Calf-1 (residues 602–738) and Calf-2 (residues 739–956) domains, which all have the immunoglobulin-like β-sandwich fold. Between the Thigh and Calf-1 domains resides what is known as Genu or knee, which is a calcium-binding loop and an important "hinge" region involved in conformational regulation. Generally, the β-propeller, the Thigh, and the two calf domains are structurally rigid, whereas the αI domain is a more dynamic structure due to its role in ligand binding and affinity regulation.

The β-subunit has a more complex domain structure than the α-subunit, with a total of seven domains. At the amino terminus is located the PSI (residues 1–54 of human β3, PDB code 1JV2, Xiong et al. 2001), the Hybrid domain (residues 55–108 and 353–432), and then the βI domain (residues 109–352) being inserted within the Hybrid domain. Thus, based on the sequence, the amino-terminal domain is the PSI domain, but in the known structures (Table 9.1, Fig. 9.1), the PSI and other Hybrid regions pack in such a way that the βI domain is the most terminally exposed domain, whether in the extended conformation or facing the membrane in the bent conformation. The disulfide-bonded (3 each) I-EGF domains 1–4 (residues 453–605) follow the three first domains; the final β-tail domain (residues 606–690), located just before the transmembrane region and cytoplasmic tail, has a unique alpha–beta–loop–beta fold so far not observed elsewhere.

The βI domain is structurally similar to the αI domain and has the MIDAS Mg^{2+}-binding site, too; βI also contains two Ca^{2+}-binding sites, one referred to as the synergistic metal ion-dependent site (SyMBS) and the other as the adjacent to metal ion-dependent adhesion site (ADMIDAS). The functional role of the βI domain depends on whether the αI domain is present in the integrin heterodimer: with αI domains, the βI domain appears to bind a glutamate residue (e.g., E336 in α2) from αI as an "intrinsic ligand" as a part of the conformational regulatory mechanism (Alonso et al. 2002; Yang et al. 2004; Xie et al. 2010; Jokinen et al. 2010), whereas in integrins lacking the αI domain, the βI domain serves as part of the ligand-recognition site, e.g., for the RGD motif (Xiao et al. 2004; Xiong et al. 2002), in addition to transferring conformational changes towards the β-tail domain.

9.2.4 *Transmembrane Domains*

The transmembrane domain (TMD) of both integrin subunits consists of a stretch of 20 hydrophobic amino acids consistent with the formation of one transmembrane α-helix for each subunit. Earlier studies on the TMD regions were inconclusive regarding the local structure and interactions between the TMDs (Hughes et al. 1996; Li et al. 2001, 2005; Vinogradova et al. 2002; Luo et al. 2004a; Gottschalk 2005; Partridge et al. 2005), but notable progress was achieved when NMR structures in bicelles were determined for TMDs of αIIb- and β3-subunits separately (Lau et al. 2008a, b) and as a complex (Lau et al. 2009). These results suggest that the helix in αIIb is oriented perpendicular to the cell membrane, whereas the helix in β3 is somewhat tilted, contacting the αIIb helix near the extracellular face of the membrane. Stabilizing salt bridges are also proposed to be located near the cytosol where the cytoplasmic tails protrude (Lau et al. 2009). Similar results were also obtained by combining disulfide scanning analysis with the so-called Rosetta computational modeling (Zhu et al. 2009).

9.2.5 Cytoplasmic Tails

The integrin cytoplasmic domains are very short with the exception of the β4-subunit, which has a large extended Calx β cytoplasmic domain, with a possible role in protein–protein interactions, whose structure has been solved (Alonso-García et al. 2009).

As with TMD structures, structural studies on the cytoplasmic tails has not led to a consistent view, which may reflect their highly dynamic role related to recognizing multiple cytoplasmic proteins and in responding to inside-out and outside-in signaling events. However, the latest NMR and X-ray structures, as well as structural modeling, predict that interactions between the cytoplasmic α- and β-subunits do take place (Lau et al. 2008a, b, 2009; Zhu et al. 2009), but it is still unclear as to the range of accessible structures and how stabile or transient they are within the cellular environment.

9.2.6 Integrin Dynamics and Conformational Regulation

In Fig. 9.1 we present a scenario for the dynamic behavior of integrins. Integrins exist in at least three conformational states according to the current view, referred to as the bent (closed), extended closed, and extended open states. Ligands can bind to each of the states but the affinities are clearly different. The first two states are considered to be low-affinity states in terms of ligand binding. The third state is a high-affinity state. Extension from the bent conformation to the extended conformation happens at the α-knee and β-knee, breaking interactions between the headpiece and lower legs.

In the αI domain-containing integrins, ligand binding to the extended closed conformation causes a conformational change in the αI domain that is conferred to β-subunit via the βI domain; the conformational change in the βI domain in turn generates a pistonlike movement of the PSI/Hybrid domain separating the β-subunit from the α-subunit, subunit separation starting at the stalk, then the transmembrane region, and finally progressing to the cytoplasmic tails.

In this scenario of outside-in signaling, ligand binding thus triggers spreading of the cytoplasmic tails, exposes binding sites for cytoplasmic proteins, and induces cytoplasmic signaling events. Also in this scenario, the β-subunit changes position most in the process. Inside-out signaling is thought to be essentially the same process but where the order of similar conformational changes begins from the cytoplasmic tails and proceeds towards the amino-terminal domains. Thus, both inside-out and outside-in signaling processes are thought to be similar structurally, but they are triggered by ligand interactions and conformational changes initiated from the opposite sides of the membrane. Thus, we will only describe, below, the process of outside-in signaling with regard to an integrin structure having an αI domain.

9.2.7 Regulation of the αI Domain

In αI domain-containing integrins, ligand binding triggers the outside-in signaling process. Structural studies of αI domains have defined the different conformations along the route of activation. In addition to structural studies, numerous constructs have been designed mainly based on mutagenesis, for example, generating a disulfide bridge in a suitable place to lock the conformation in a particular way or designing mutants that break key interactions and thus allow the conformation to change. Gain-of-function mutations have been used to open the αI domain structure and experimentally mimic the open conformation, and activated I domains have been reported for αXI (I314G) (Vorup-Jensen et al. 2003), αMI (I316R and I316G) (Xiong et al. 2000), αMI (F302W) (Li et al. 1998), as well as α1I (E317A) (Lahti et al. 2011) and α2I (E318W) (Aquilina et al. 2002; Tulla et al. 2008). Cysteine substitutions stabilizing open conformations via disulfide bonds have also been used to activate integrin αI domains: α2I G172C-L322C (Siljander et al. 2004), αLI K287C/K294C, αLI E284C/E301C (Shimaoka et al. 2003), and αMI C128A/D132C/K315C (McCleverty and Liddington 2003).

Based on the available data, when a ligand binds to MIDAS of the αI domain, typically using a negatively charged residue to displace water and coordinate Mg^{2+}, it triggers movement of the metal ion, a shift of the α1 helix, and finally a large downward movement of the α7 helix that activates the αI domain. Some data suggest that collagen receptor integrin αI domains may have a somewhat different mechanism or a more complicated mechanism. For instance, collagen receptor I domains differ by having the additional αC helix near MIDAS, which changes the structure on activation (in α2I, local unfolding at one end and extending the helix at the other end within the α6 helix, with a large movement of Y285; Emsley et al. 2000). Despite this difference, the end result is the same, i.e., α7 helix movement (Lahti et al. 2011), and it is this movement resulting from αI domain activation that is communicated to the βI domain and subsequently to the interior of the cell. This communication is mediated by a specific glutamate residue located on the α7 helix that serves as an intrinsic ligand for the βI domain: E310 in αLI (Alonso et al. 2002; Yang et al. 2004), E318 in αXI (Xie et al. 2010), and E336 in α2I (Jokinen et al. 2010).

The αXβ2 structure provided more detailed information regarding the role of the αI domain in activation (Xie et al. 2010). The αI domain has relatively few contacts with the β-propeller domain from which it buds out, and it does form a few contacts with the specificity-determining loop (SDL) of the βI domain. The N-terminal linker of the αI domain is also attached near where it exits the β-propeller by a disulfide bridge to blade 2 of the β-propeller; this feature may be common to the αI domain-containing integrins since these cysteines are well conserved. The likely flexibility at the linker region has led to an updated model, adding more options for activation of αI domain-containing integrins, i.e., in which three αI domain conformational states can couple with two βI domain conformational states (Xie et al. 2010)

9.2.8 Regulation of the βI Domain

Integrins lacking αI domains bind their ligand on the interface between α- and β-subunits. For example, with integrins binding ligands with the RGD motif, arginine binds to the β-propeller of the α-subunit, and aspartate binds to Mg^{2+} at MIDAS of the βI domain (Xiong et al. 2002). The resulting conformational changes in the βI domain cause a downward movement of the α7 helix of that domain. Evidence for this comes from numerous studies, including mutations introduced into the α7 helix in α5β1 (Mould et al. 2003a), disulfide-locked high- and low-affinity conformations in αIIbβ3 (Luo et al. 2004b), β-subunit mutations in Drosophila βPS (Jannuzi et al. 2004), and activating and inactivating mutations in the βI domain in α4β1 and α5β1 (Barton et al. 2004). Additionally, the structure of the headpiece of αIIbβ3 with a synthesized small-molecule ligand confirmed that a conformational change occurred in the βI domain (Xiao et al. 2004). Despite differences, the αI and βI domains share similar structures, and conformational changes have already been seen in the αI domains (see above). However, in the βI domain, the Mg^{2+} at MIDAS does not move as is seen in the αI domain, but, instead, Ca^{2+} at ADMIDAS moves closer to MIDAS. All in all, these conformational changes in the βI domain produce the same end result as in αI domain, a significant downward movement of the α7 helix, which enables signal transduction in outside-in signaling. With inside-out signaling, cytoplasmic protein interactions act to spread the legs converting the bent conformation into the extended and then activated extended conformation, with similar local changes occurring as in outside-in signaling, but in a reversed order. The formation of the bent inactive integrin conformation with and without I domains is thought to be the reverse of the inside-out and outside-in activation process.

9.2.9 Hybrid Domain Swing-Out

Because of the unique structural arrangements of the PSI, Hybrid, and βI domains, small local rearrangements, starting from MIDAS of the βI domain, generate a large downward motion of the α7 helix. In automotive parlance, this is analogous to a movement of the connecting rod between the piston and the *crank*shaft. This movement leads to a large shift in the orientation between the βI and Hybrid domains, and the latter swings out about 60° in the open headpiece relative to the closed headpiece; the knees of α- and β-subunits separate by 70 Å (Xiao et al. 2004), but the latter distance may vary widely as part of the dynamic movements of the leg regions as part of the bidirectional signaling processes. Numerous studies support this observation of Hybrid domain "swing-out" (Takagi et al. 2001, 2003; Luo et al. 2004b; Mould et al. 2003a, b; Xiong et al. 2004). It is also important that the PSI/Hybrid domain interface does not change during the activation process and remains as a rigid unit together with the I-EGFI domains during activation (Xiao et al. 2004; Shi et al. 2005).

9.2.10 Transmembrane and Cytoplasmic Domains

The ectodomains, from which the X-ray structures have been obtained, lack the TMDs and the cytoplasmic tails. This may be one reason why all of the structures are in the bent conformation. The structures solved so far, including results from EM studies in the presence and absence of metal ions like Mn^{2+} known to be very effective activators of integrins, mainly represent inactive states according to the dynamic model presented herein. Because of this, it may not be enough to study isolated ectodomains, but new methods may be needed to monitor integrin conformational changes (Kim et al. 2011, 2012) that, e.g., include the missing domains and the membrane context.

Both transmembrane spanning segments of the integrin subunits consist of one α-helix having some 20 hydrophobic amino acids. It is a well-known fact that interaction between these helices produces an inactive integrin conformational state. Many studies have also shown the importance of the membrane-proximal region with the GFFKR motif, which is found in most integrin α-subunits and is highly conserved. These studies include truncating the cytoplasmic domain of αIIb (O'Toole et al. 1991), mutation, and deletion of the Gly–Phe–Phe–Lys–Arg (GFFKR) motif of αIIbβ3 and in αLβ2 (O'Toole et al. 1994; Hughes et al. 1996; Lu and Springer 1997; Lu et al. 2001). FRET studies have also shown that a significant separation between α- and β-subunit cytoplasmic domains occurs during activation (Kim et al. 2003). Integrin activation as part of an inside-out signaling event is supported by talin head domain (Tadokoro et al. 2003) and RAPL (Katagiri et al. 2003) binding to the cytoplasmic tails, and the TMDs separate after activation in inside-out signaling (Luo et al. 2004a).

Although a complete structure of an integrin has not been solved as yet, based on the available crystal data, it has been possible to model the structure of entire integrin heterodimers in both the extended and bent conformations, active and inactive conformations, and with and without inserted I domains. For example, models based on different crystal structures and the cryo-EM data of EV1 have been used to illustrate the possible interactions between α2β1 and EV1 that could result in integrin clustering and virus internalization (Jokinen et al. 2010).

9.3 Early Evolution of Integrin Domains

As described above, the α- and β-subunits of integrins are multidomain structures (Luo et al. 2007; Arnaout et al. 2007). Each integrin domain *sans* the TMDs and cytoplasmic tails is a globular protein with its own characteristic fold, often found in many other unrelated proteins or multidomain complexes (Murzin et al. 1995). As also seen in many other proteins, the integrin domains often contain a set of characteristic structural or sequence signature motifs, which are either specific for integrins or common for several proteins that share the same domain (Xiong et al. 2001; Xiao et al. 2004). Analysis of such motifs allows us to classify the integrin domains within the groups of similar domains in unrelated proteins and trace their earliest origins. We find examples of most domains reported for prokaryotic species, although some are rare, and several have not been observed.

9.3.1 The von Willebrand Factor Type A and βA Domains

The integrin αI domain, also known as the von Willebrand factor type A (vWA) domain (Larson et al. 1989), belongs to a larger group of von Willebrand factor type A protein ECM (vWA_ECM), according to *sequence* data, which are well studied and classified within the NCBI Conserved Domain Database (CDD, Marchler-Bauer et al. 2011). The integrin αI domains belong to the cd01469 (vWA_integrins_α_subunit) sequence cluster, one of nine subfamilies within the "parent" group (CDD ID: cd01450, vWFA_subfamily_ECM) that currently includes more than 110 different eukaryotic domains (Fig. 9.2). Ligand binding is mediated via MIDAS, with the characteristic DxSxS-x(60)T-x(24)-DG MIDAS motif (Lee et al. 1995a, b). Across the vW*F* protein sequences, the MIDAS motif may only be partially conserved, such as in the Sec 24 protein (CDD ID: cd01479), or even absent as in the Ku70/Ku80 N-terminal domain (CDD ID: cd01458).

The evolution of I domains and vWFA domains have been previously reviewed (Colombatti and Bonaldo 1991; Colombatti et al. 1993; Tuckwell 1999; Whittaker and Hynes 2002; Johnson and Tuckwell 2003), and proteins containing the vWA domain are widely distributed across all domains of life (Ponting et al. 1999). They are found in viruses, such as the EsV-1-176 protein in the *Ectocarpus siliculosus* virus 1, NCBI Ref.: NP_077661.1; in Archaea, such as the Mg^{2+}-chelatase subunit ChlI/Chld in *M. kandleri* AV19, NCBI Ref.: NP_613479.1; in numerous Bacteria; and virtually in all branches of Eukaryota (Pfam ID: pfam00092; InterPro ID: IPR002035; http://pfam.sanger.ac.uk; http://www.ebi.ac.uk/interpro).

In terms of the structural *fold*, the αI domain of the integrin α-subunit has the von Willebrand factor type A-like fold (vWA-like, the SCOP database, Murzin et al. 1995), formed from three α/β/α layers and a mixed β-sheet of six strands, which is not equivalent to the vWA protein *sequence* family described above. The vWA-like fold is common to five vWA and vWF protein fold subfamilies: (1) the integrin A

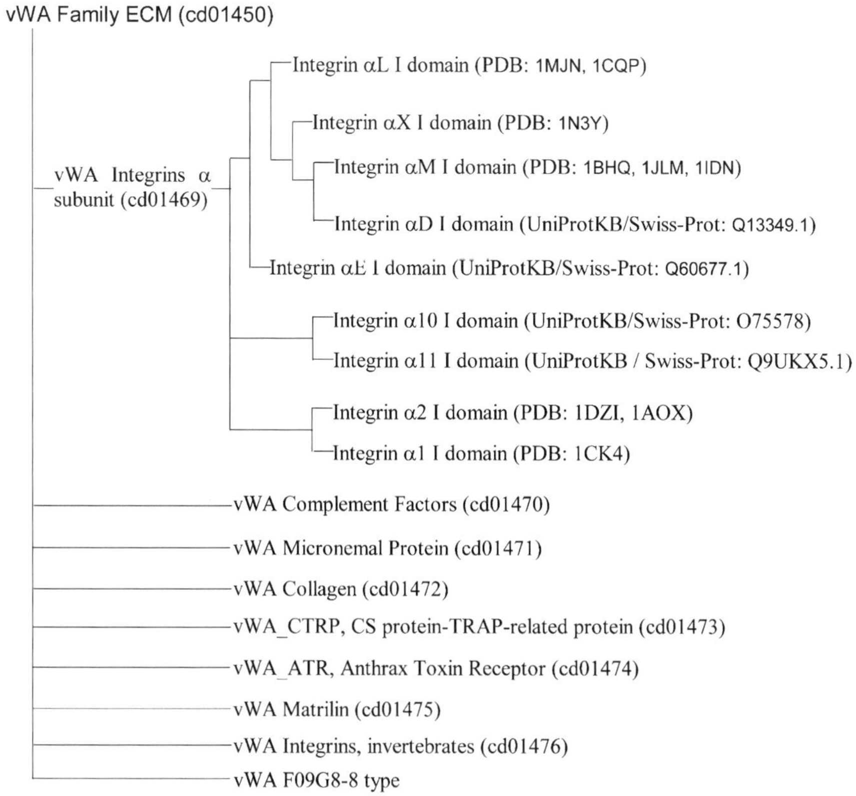

Fig. 9.2 The integrin αI domain (von Willebrand factor type A domain, Larson et al. 1989) belongs to a larger group of von Willebrand factor type A proteins that have been classified within the NCBI Conserved Domain Database (CDD; Marchler-Bauer et al. 2011)

(or I) domain fold, (2) the trunk domain of Sec23/24 fold, (3) the Ku70 subunit N-terminal domain fold, (4) the Ku80 subunit N-terminal domain fold, and (5) the RoRNP C-terminal domain-like fold. The integrin A (or I) domain fold includes all known vWA integrin αI-subunits as well as the von Willebrand factors A1–A3 domains, vWA integrin βA-subunits, and several other vWA proteins (see SCOP: Family: Integrin A (or I) domain). There are more than 80 three-dimensional structures from 17 different proteins, which contain a domain with the vWA-like fold and deposited within the PDB.

The βA domain of the integrin β-subunit also shares the von Willebrand factor type A-like fold (vWA-like, the SCOP database, Murzin et al. 1995), but there are significant differences: in addition to MIDAS, the βA domain contains two additional characteristic metal ion-binding sites, ADMIDAS and SyMBS (formerly known as LIMBS) (Arnaout et al. 2007; Zhu et al. 2008), which are specific to integrin βA domains. On the basis of sequence, the βA domain has been combined

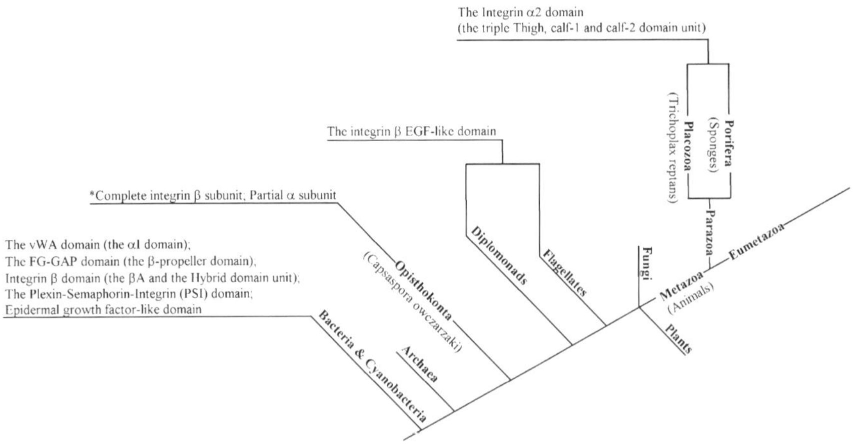

Fig. 9.3 The integrin α- and β-subunits are composed of multiple domains, and complete β-subunits and a partial α-subunit have been seen in the protozoa *C. owczarzaki* (Sebé-Pedrós et al. 2010). Here, we have indicated the presence of domains (in the earliest diverging species) having similar sequences and folds to those found in integrin subunits as compiled by the CDD. The β-tail domains have only been observed in integrins

with the Hybrid domain to provide an integrin-specific signature: Integrin_β (Pfam ID: pfam00362; InterPro ID: IPR002369), and the two domains could be identified together in many sequences, in non-integrin eukaryotes, as well as very early in evolution (Fig. 9.3). The InterPro and Pfam databases list at least 15 bacterial sequences, two cyanobacteria, and one archaea. It is not clear, however, how conserved the MIDAS, ADMIDAS, and SyMBS motifs are in these sequences or how similar these sequences are to the integrin domains.

9.3.2 The β-Propeller Domain of the α-Subunit

The integrin β-propeller domain belongs to the "Integrin α N-terminal domain" superfamily of the 7-bladed β-propeller fold, which is a meander of seven sequential 4-stranded β-sheet motifs called blades (see SCOP: Fold: 7-bladed β-propeller). Similarly to the vWA containing proteins, different β-propellers are found in a wide variety of organisms from bacteria to human (Adindla et al. 2007; Quistgaard and Thirup 2009; Redruello et al. 2010). The evolution of the sequences and structures of the β-propeller proteins has been recently summarized in Chaudhuri et al. (2008). However, unlike the vWA domains described above, which are defined on the level of the entire domain, the β-propeller domains are usually characterized on the level of separate blades, where all the motifs, key features, and determinants are then more or less repeated to form the β-propeller as a whole. There are several key features that can separate integrin β-propeller domains from all the other β-propeller types. These are structural existence of the seven blades plus the unique arrangement of integrin-specific motifs (Chouhan et al. 2011) that include the FG-GAP repeat (Corbi et al. 1987; Springer 1997) and the "Cage" motif (Xiong et al. 2001). The FG-GAP motif has a Phe–Gly pair in the first strand and a Gly–Ala–Pro tripeptide in the second strand (Springer 1997; Xiong et al. 2001). The Cage motif is defined by the ϕϕGϕx13-20Px2-15Gx5-8 (ϕ, aromatic residue; G, glycine; x, any residue; P, proline) consensus sequence (Xiong et al. 2001). A combined computer-based generalized FG-GAP motif spans over the entire length of a blade (FG-GAP motif, pfam01839; IPR013517). The Ca^{2+}-binding motif, found in 3–4 of the repeats in integrins, is a combination of the well-known "DxDxDG-like Ca^{2+}-binding loop", present in many other unrelated Ca^{2+}-binding proteins and a separate glycine residue (in the 3,4-loop of the blade), which coordinates a water molecule bound to Ca^{2+} (Rigden and Galperin 2004; Chouhan et al. 2011).

Integrin-like β-propeller sequences were identified in prokaryotes (Johnson et al. 2009; Chouhan et al. 2011), but not as part of an intact integrin subunit. While the evolution of several non-integrin β-propeller domains could also be identified in bacterial species (Adindla et al. 2007; Quistgaard and Thirup 2009; Redruello et al. 2010), the integrin-type 7-bladed β-propellers were much more difficult to study and relied on inferred structural and sequence features. Up to 538 instances of one or more contiguous blades could be identified throughout prokaryote species, but only a few examples of complete 7-bladed or 2 × 7 bladed β-propellers were identified. In the presumptive integrin of the unicellular eukaryote *Capsaspora owczarzaki*, 3 of 7 blades are clearly defined along with a transmembrane region and distinctive KXGFFXR motif, thus having several hallmarks of integrin α-subunits (Sebé-Pedrós et al. 2010).

The prokaryotic integrin-like 7-bladed β-propeller sequences are more conserved than the human sequences in that they have seven conserved Ca^{2+}-binding motifs, instead of 3–4 in human, and the loops were shorter and more consistent in length. This conservation and consistency might argue for a prokaryotic origin of the metazoan integrin-specific motifs within the seven repeats; alternatively, the paucity of 7-bladed examples might argue in favor of lateral gene transfer (Fig. 9.3; Chouhan et al. 2011).

9.3.3 Immunoglobulin-Like Domains: Thigh, Calf-1, Calf-2, and Hybrid Domains

The Thigh, Calf-1, and Calf-2 domains of the α-subunit and the Hybrid domain of the β-subunit are immunoglobulin-like β-sandwich domains. The Pfam integrin-specific motif for the α-subunit, Integrin_α2 (pfam08441), spans all three domains and contains conserved cysteines and the other conserved nonpolar residues (see the HMM logo presentation in Schuster-Böckler et al. 2004 for pfam08441). The Integrin_α2 three-domain unit is not reported among prokaryotes although immunoglobulin-type folds are represented throughout the domains of life

(Bork et al. 1994; Halaby et al. 1999; and see below) and in numerous cellular receptors. The Thigh and calf domain types are not identical, but here are several key features that can discriminate among them:

1. The topology of the Thigh domain resembles the topology of the constant-2 set (C2-set; InterPro ID: IPR008424) of immunoglobulins with three β-strands packed against four β-strands in one bundle (Xiong et al. 2001), while the Calf-1 and Calf-2 domains have four β-strands packed against five β-strands in one bundle, which is somewhat similar to the immunoglobulin variable V-set (InterPro ID: IPR013106), but the connectivity of strands differs.
2. The Thigh, Calf-1, and Calf-2 domains differ in the number of disulfide bonds and their location, creating a unique pattern of disulfide bridges.
3. The Genu loop interface between the Thigh and Calf-1 domains has a disulfide bond, which locks its ends, and it binds and coordinates Ca^{2+} (Xie et al. 2004).

The Hybrid domain of the β-subunit of integrins resembles the immunoglobulin I-set family (pfam07679; IPR013098), rather than the immunoglobulin C2-set or V-set domains (Xiong et al. 2001). The immunoglobulin I-set domains are widely observed throughout evolution with more than 15,000 sequences known among eukaryotes, over 110 sequences in bacteria, 13 sequences in viruses, and one sequence from Archaea (see IPR013098 Immunoglobulin I-set). Integrin specificity in searches can be enforced by using a sequence pattern (Integrin_β: pfam00362) that includes both the Hybrid and the βA domain.

9.3.4 The Plexin–Semaphorin–Integrin Domain

The Plexin-Semaphorin-Integrin (PSI) domain of the integrin β-subunit is a small, ~50 amino acid domain belonging to the PSI family (pfam01437) with the Trefoil/Plexin domain-like fold, a disulfide-rich fold with the common α + β core, and several conserved disulfide bridges (see, e.g., PDB files 1TYE, chain B, residues 1–57 and 1U8C, chain B, residues 1,001–1,057). In integrins, the PSI domain forms a two-stranded antiparallel β-sheet flanked by two short α-helices and contains four cysteine pairs connected in a 1–4, 2–8, 3–6, 5–7 pattern (Xiong et al. 2004). One or several copies of the PSI domain were found in more than 3,200 eukaryotic sequences, most of which (2,800) are metazoan, protist (200) or plant (40) (see InterPro: IPR003659). In addition, there are two bacterial sequences, the Actinobacteria *Actinomyces* sp. oral taxon 180 str. F0310 and the Proteobacteria *Plesiocystis pacifica* SIR-1, that were predicted to contain one copy of the PSI domain (UniProtKB/TrEMBL ID: E6KTM4_9ACTO and A6GE46_9DELT, respectively).

9.3.5 The EGF Domains

There are four cysteine-rich EGF domains in the β-subunit, which are located adjacent to the PSI domain. The integrin EGF domains have six cysteines forming three disulfide bonds (PDB: 1JV2_B:532–562; Pfam family EGF, pfam00008; IPR006209), which represent a separate "Integrin β EGF-like domain" family with the "Disulfide-bound" fold (see SCOP, Fold: Knottins, small inhibitors, toxins, lectins). The cysteine-rich EGF domains are usually found within extracellular parts of larger membrane-bound complexes and in secreted proteins. Bacterial EGF domains have not been reported for this domain family, but they have been identified in a number of unicellular and colonial flagellate eukaryotes including reported for the "Epidermal growth factor-like domain" family (IPR000742).

The integrin β EGF-like domain fold is well known and belongs to the EGF/Laminin superfamily. There are many functionally unrelated protein domains with the same fold and whose structures have been solved: e.g., epidermal growth factor (EGF, PDB code: 1IVO) and fibrillin-1 (PDB code: 1UZJ), the EGF-like domain of nidogen-1 (PDB code: 1GL4), the laminin-type module (PDB code: 1KLO), the follistatin module N-terminal domain (PDB code: 1NUB), the merozoite surface protein 1 (PDB code: 1B9W), and the mouse Cripto EGF-like domain (PDB code: 2J5H).

9.3.6 Integrin β-Subunit: The β-Tail Domain

The β-tail domain (βTD) (e.g., PDB code: 1JV2, residues 606–690), pfam07965; IPR012896, with an α–β–loop–β fold, so far appears to be unique to the integrins since it has not been seen in other proteins (see SCOP: Fold: Integrin β-tail domain). The interactions between the α-helix and the β–loop–β structure are mainly hydrophobic; there is also a single disulfide bond (Xiong et al. 2001). Functionally, the β-tail domain has been shown to have a role in ligand binding and signaling (Gupta et al. 2007).

9.4 Evolution of Metazoan Integrins

Although homologues of the individual domains from which the integrin subunits are composed can be identified in prokaryotes (Fig. 9.3) (e.g., May and Ponting 1999; Jenkins et al. 2002, Whittaker and Hynes 2002; Johnson and Tuckwell 2003; Johnson et al. 2009; Chouhan et al. 2011), integrins appear to be a hallmark of multicellular metazoan organisms in that they have been reported across the wide range of sequenced genomes of metazoans (Fig. 9.4). Furthermore, the number and diversity of subunits has expanded with increases in the complexity of the metazoans. Johnson et al. (2009) proposed that integrin functions would predate

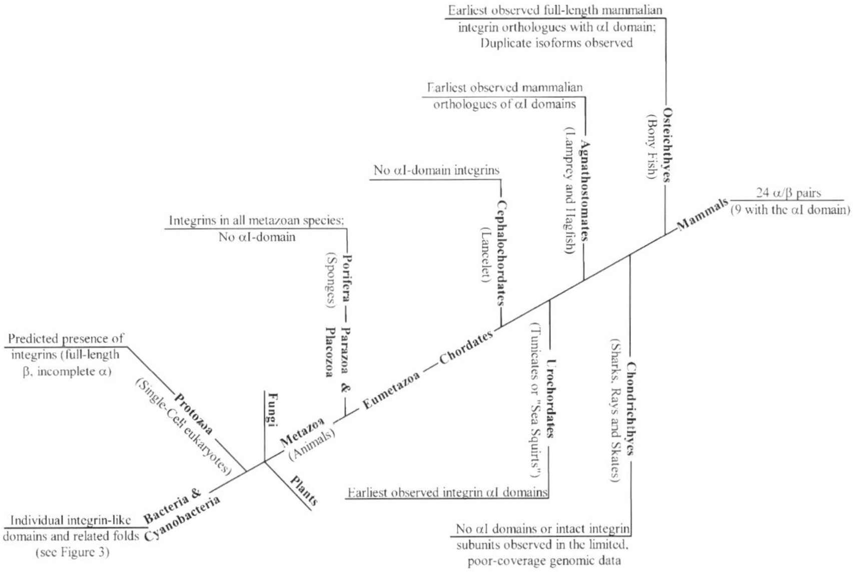

Fig. 9.4 Timeline for integrin evolution reflecting our current knowledge

the metazoans and, indeed, integrin subunits and fragments thereof have been reported in a single-cell eukaryote (i.e., *Capsaspora owczarzaki*; Sebé-Pedrós et al. 2010).

In mammals, as typified by the human sequence data, there are 18 α-subunits and 8 β-subunits and the functional and phylogenetic relationships extend throughout the vertebrates, i.e., even fish have representatives of subunits present in mammals, but bony fish also typically have isoforms that represent the additional round of whole-genome duplication proposed for fish (Fig. 9.4). Early phylogenetic analyses on the integrin subunit sequences (Hughes 1992) have been refined over the past 20 years as a result of the genomic sequence studies, although there are still outstanding areas of uncertainty (e.g., α I domains in sharks/rays).

9.4.1 *The α-Subunit*

Based on sequence alignments and phylogenetic reconstructions, the α- and β-subunits can be divided into multiple groups (Fig. 9.5), and these groups can differ depending on personal preferences although the phylogenetic trees are in close agreement with each other (Hughes 1992, 2001; Burke 1999; Johnson and Tuckwell 2003; Huhtala et al. 2005; Johnson et al. 2009). Hughes (2001), for example, segregates the α chains into five groups that include three groups named with respect to the drosophila α-subunits: PS1 (also includes, e.g., bony fish/mammalian α3/6/7), PS2 (e.g., bony fish/mammalian αV/IIb/5/8), and PS3; a

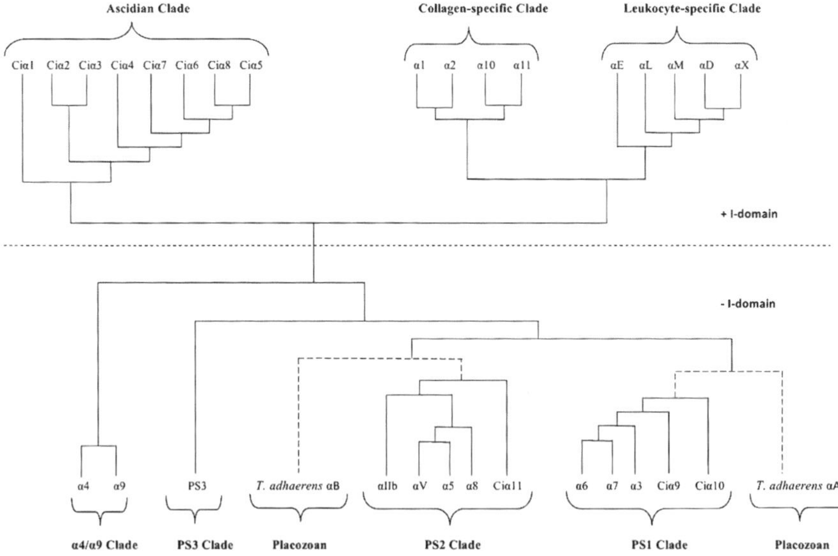

Fig. 9.5 A phylogenetic representation of the relationships among integrin α-subunits from human and the tunicate *C. intestinalis* ("Ci") based on a larger tree derived from the comparison of 42 vertebrate and invertebrate sequences. This summary tree reflects the clusterings already reported from a number of sources (see text). The earliest diverging sequences of integrin α-subunits containing an additional I-domain appear with the tunicates, but these sequences are not orthologous to the human collagen and the leukocyte-specific subunits. The PS3 cluster only contains invertebrate sequences. Three other *C. intestinalis* sequences do not have αI domains. The position where the two sequences of the placozoan *T. adhaerens* branch off is indicated. The branch lengths in this tree are arbitrary

group containing the bony fish/mammalian α4/9 sequences; and large I-domain-containing group. The nine mammalian integrins with I domains segregate into two separate groups, the collagen receptors (α1/2/10/11) and the leukocyte receptors (αL/M/D/X/E), and a third cluster is composed of only tunciate sequences from *Ciona intestinalis*. The two sequences of *T. adhaerens* branch from the tree in the vicinity of the PS1 and PS2 clades.

9.4.2 The β-Subunit

Comparison of the integrin β-subunits show that they segregate into several broad groups (Fig. 9.6). The vertebrate A clade contain, e.g., the human β1/2/7-subunits, all of which have interactions with α-subunits containing I domains, although β1 also has eight non-I domain α-subunit partners and β7 also pairs with the α7-subunit. The vertebrate B group contains the remainder of the mammalian chains, β3-8, with Ciβ5 from Ciona clustering with human β4. (Note that we have

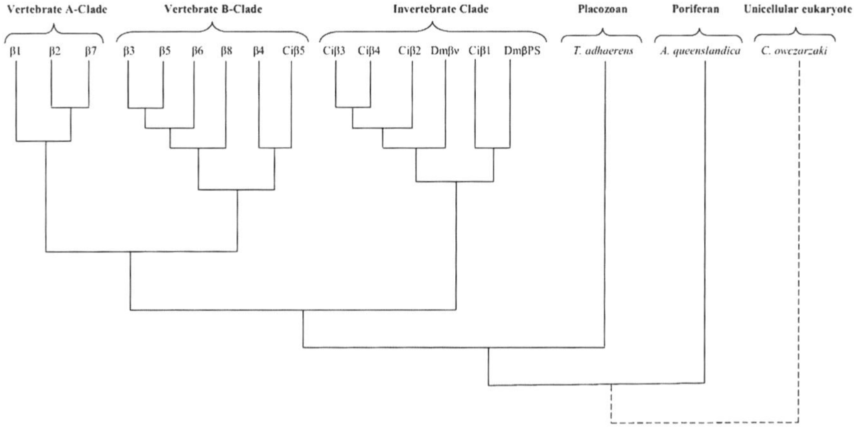

Fig. 9.6 Phylogenetic relationships among integrin β-subunits, including sequences from human, *C. intestinalis*, *D. melanogaster* (Dm), *T. adhaerens*, and the poriferan *A. queenslandica*. The tree is derived from a comparison of 42 sequences. Sequences of the integrin β-subunits from the unicellular eukaryote *C. owczarzaki* have been utilized to root the tree. This summary tree reflects the clusterings already reported from a number of sources (see text). The branch lengths in this tree are arbitrary

used the naming of Ciona sequences from Ewan et al. 2005 and these differ from those used by Huhtala et al. 2005.) The remaining Ciona β-subunits cluster within the invertebrate group. The full-length sequence of the integrin β-subunit from *Capsaspora owczarzaki* can be used to root the tree (11–21 % identity with the metazoan β-subunits in the comparison). The sequences of early metazoans, e.g., *T. adhaerens* and *A. queenslandica*, branch after the root but before the invertebrate and vertebrate clusters (Fig. 9.6) and in agreement with the relationships based on the genomic sequencing study of Srivastava et al. (2008) on *T. adhaerens*.

9.4.3 The αI Domain

The αI domain, inserted between the second and third blade of the β-propeller domain in the α-subunit, requires special mention since it represents a major paradigm shift to the integrin structure with increased specialization and functional diversification. In human, nine α-subunits contain the αI domain. This major structural change occurred over 500 million years ago as the chordate line took root from the branch leading to the invertebrate deuterostomes (i.e., echinoderms). The relatively "new" and independent αI domain functions to shift the ligand-binding site away from the interface between the β-propeller of the α-subunit and the βA domain of the β-subunit. Until then, integrins had been limited in their recognition to motifs such as RGD and LVD, and the αI domain added the capability to recognize new types of ligands, expanding the repertoire of interactions facilitated by the integrin family. The αI domain is also a dynamic unit in itself (see above), consistent with the nature of the α- and β-subunits to function mechanically to transduce signaling and ligand binding bidirectionally via additional conformational changes within the αI domain (Emsley et al. 2000; Lahti et al. 2011) and as part of the dynamic heterodimer structure.

The origin of the αI domain and the origin of the mammalian-type orthologues of αI-domain-containing subunits have been outstanding questions. The emergence of the αI domain in integrin α-subunits took place among the earliest diverging members of the chordate line (Ewan et al. 2005; Huhtala et al. 2005). The origin seems to be close to the divergence of the tunicates, where αI domains are present. The genome (11.5× redundant) of the lancelet *Branchiostoma floridae* (amphioxus, Cephalochordata), once considered a species diverging *after* the tunicates, has been sequenced (Putnam et al. 2008), and αI domains have not been observed. A phylogeny of the deuterostomes (echinoderms through to humans), based on an alignment of 1,029 concatenated sequences, places the divergence of the lancelet after the echinoderms but prior to the tunicates (Putnam et al. 2008).

The absence of an αI domain in an α-subunit from the lancelet and presence in tunicates provide a clear demarcation for the earliest origin of the αI domains, since αI domains have not been observed in echinoderms (non-chordate deuterostomes) or in any earlier diverging species, but they are present in the urochordates—the tunicates—first reported in *Halocynthia roretzi* (Miyazawa et al. 2001).

We can bracket the appearance of *orthologues* of the mammalian α-subunits with αI domains by their absence in the tunicates and their presence observed in bony fish. Osteichthyes (bony fish) contain orthologues of mammalian integrins with αI domains, and, like with many other protein families in bony fish, isoforms exist due to the additional round of whole-genome duplication that has likely occurred within the bony fish branch. There are not many candidates in between the tunicates and bony fish: the jawless cartilaginous vertebrates (Agnatha), hagfish and lamprey, and the cartilaginous fish (Chondrichthyes)—the sharks and rays.

The genomes of *Petromyzon marinus* (sea lamprey, 5.9× coverage) and inshore hagfish (*Eptatretus burgeri*, ESTs available) were found to contain identifiable fragments of integrin subunits and recognizable I domain sequences (Chouhan et al. 2012), including three full-length sequences from the sea lamprey (Pma_f1, Pma_f2, and Pma_f3), and a fragment from the inshore hagfish (Ebu_f). Unlike the Ciona sequences, the three lamprey sequences contain sequence matching MIDAS and the αC helix, the later being diagnostic of the mammalian collagen-binding αI domains such as those present in integrins α2β1 and α1β1. The hagfish sequence contains the MIDAS motif but ends just prior to the αC helix region.

Despite many extant cartilaginous fish species, genomic data from sharks/rays is sparse, with some data available: the genome from *Callorhinchus milii* (ghost shark/elephant shark, 1.4× coverage) and expressed sequence tags (ESTs) from *Raja erinacea* (little skate) and *Squalus acanthias* (dogfish shark). However, searches (Chouhan et al. 2011) of the data unexpectedly did not reveal the unambiguous presence of αI domains, but we attribute this to the quality and assembly state of the data, fully expecting mammalian-type orthologues to be present throughout the cartilaginous fish.

Thus, the current data support the notion that integrin subunits have a very early history, preceding the origin of metazoan multicellularity and likely a necessary component to manage an increasing number of tasks among different cell types and with the ECM. Invertebrates have a smaller set of integrin subunits, but these rapidly expand within the chordate line along with increasing complexity of the organism and specialization of tissues, organs, and systems. Thus, the nematode genome was found to contain two α-subunits and one β-subunit, and the drosophila genome contains five α-subunits and two β-subunits. With the chordates, additional complexities such as the tissue remodeling take place in embryogenesis, a complex immune system arises, and with the jawless vertebrates, a circulatory system and fibrinogen-based blood clotting. The appearance of I domains in a common ancestor of the tunicates and more complex vertebrates provides a novel ligand-binding site and additional functional variability that can be harnessed within the organism. Thus, in human there are 18 α-subunits and 8 β-subunits forming 24 observed α/β pairings, but a significant expansion already occurs in some tunicates where, e.g., *C. intestinalis* contains up to ten α-subunits—eight having αI domains—and five β-subunits. The largest number of integrin subunits will likely be in bony fish where subunit isoforms are observed and could at least double the number of both subunit types in comparison to those present in, e.g., human.

9.5 Conclusions

The genome of the unicellular choanoflagellate *Monosiga brevicollis* has been analyzed, and it was reported to code for at least 17 integrin α-subunit-containing proteins, but no integrin β domains (King et al. 2008). Choanoflagellates are the closest known relatives of metazoans. However, another study failed to detect a bona fide integrin α chains in either *M. brevicollis* or *Proterospongia* sp. (another choanoflagellate) (Sebé-Pedrós et al. 2010). Instead the authors found four integrin β and four integrin α genes in *C. owczarzaki* (an amoeba), one β and one α in *Amastigomonas* sp. (an apusozoan), and one of the integrin β-subunits (the extracellular domain) in the cyanobacterium *Trichodesmium erythraeum* (Sebé-Pedrós et al. 2010). In metazoa all integrins are present as αβ heterodimers, and it is still unclear, whether in unicellular organisms the integrin-like protein can form similar functional receptors. However, the published data clearly show that the evolution of integrins has started before multicellularity. Indeed, the integrin-like I domains (A domains) as well as the β-propeller domains can be found in many bacteria.

All metazoans express integrins, suggesting their critical role in the origin of multicellularity. The sponges express both integrin α- and β-subunits which form receptors for proteins containing RGD motifs (Wimmer et al. 1999). Thus, the vertebrate β1-like integrins may represent the original form of an ECM receptor. The new genomic information about "primitive" animal species indicates that the metazoan integrin family has rapidly diverged (Knack et al. 2008), and it is likely that future research will reveal new subgroups of non-vertebrate integrins with specialized functions.

The intracellular tails of integrins bind to multiple cytoskeletal proteins as well as the components of various signaling cascades. The evolution of the adherome has taken place together with the integrin family members and has led to increasing complexity in the integrin functions.

While RGD-binding and laminin-receptor integrins are present in both non-vertebrates and vertebrates, the large subgroup of αI domain integrins is found in chordates only. Their evolution has taken place concomitantly with the development of the skeleton, circulation, the nervous system, and immunity. The knockout mouse phenotypes of mammalian integrins reflect the evolutionary history of integrin subgroups. Interesting questions are only partly resolved, for example, related to the origin of the αC helix -containing collagen receptors.

The rapidly accumulating information about the structural regulation of integrin function will also open new avenues to study the evolution of cell adhesion. In the near future, it may be possible to address questions related to the origin and conservation of structural elements that are known to be critical for the changes between different activation stages of the integrins. This kind of new data would make it possible to estimate the evolution of the cellular signals generated by the adhesion receptors.

References

Adamska M, Degnan BM, Green K, Zwafink C (2011) What sponges can tell us about the evolution of developmental processes. Zoology (Jena) 114:1–10

Adindla S, Inampudi KK, Guruprasad L (2007) Cell surface proteins in archaeal and bacterial genomes comprising "LVIVD", "RIVW" and "LGxL" tandem sequence repeats are predicted to fold as β-propeller. Int J Biol Macromol 41:454–468

Alonso JL, Essafi M, Xiong JP, Stehle T, Arnaout MA (2002) Does the integrin alphaA domain act as a ligand for its betaA domain? Curr Biol 12:R340–R342

Alonso-García N, Inglés-Prieto A, Sonnenberg A, de Pereda JM (2009) Structure of the Calx-beta domain of the integrin beta4 subunit: insights into function and cation-independent stability. Acta Crystallogr D Biol Crystallogr 65:858–871

Anthis NJ, Wegener KL, Ye F, Kim C, Goult BT, Lowe ED, Vakonakis I, Bate N, Critchley DR, Ginsberg MH, Campbell ID (2009) The structure of an integrin/talin complex reveals the basis of inside-out signal transduction. EMBO J 28:3623–3632

Aquilina A, Korda M, Bergelson JM, Humphries MJ, Farndale RW, Tuckwell D (2002) A novel gain-of-function mutation of the integrin alpha2 VWFA domain. Eur J Biochem 269:1136–1144

Arnaout MA, Goodman SL, Xiong JP (2007) Structure and mechanics of integrin-based cell adhesion. Curr Opin Cell Biol 19:495–507

Bader BL, Rayburn H, Crowley D, Hynes RO (1998) Extensive vasculogenesis, angiogenesis, and organogenesis precede lethality in mice lacking all αv integrins. Cell 95:507–519

Baldwin ET, Sarver RW, Bryant GL Jr, Curry KA, Fairbanks MB, Finzel BC, Garlick RL, Heinrikson RL, Horton NC, Kelley LL, Mildner AM, Moon JB, Mott JE, Mutchler VT,

Tomich CS, Watenpaugh KD, Wiley VH (1998) Cation binding to the integrin CD11b I domain and activation model assessment. Structure 6:923–935

Barton SJ, Travis MA, Askari JA, Buckley PA, Craig SE, Humphries MJ, Mould AP (2004) Novel activating and inactivating mutations in the integrin beta1 subunit A domain. Biochem J 380:401–407

Bazigou E, Xie S, Chen C, Weston A, Miura N, Sorokin L, Adams R, Muro AF, Sheppard D, Makinen T (2009) Integrin-α9 is required for fibronectin matrix assembly during lymphatic valve morphogenesis. Dev Cell 17:175–186

Beglova N, Blacklow SC, Takagi J, Springer TA (2002) Cysteine-rich module structure reveals a fulcrum for integrin rearrangement upon activation. Nat Struct Biol 9:282–287

Berman HM, Westbrook J, Feng Z, Gilliland G, Bhat TN, Weissig H, Shindyalov IN, Bourne PE (2000) The Protein Data Bank. Nucleic Acids Res 28:235–242

Bhunia A, Tang XY, Mohanram H, Tan SM, Bhattacharjya S (2009) NMR solution conformations and interactions of integrin alphaLbeta2 cytoplasmic tails. J Biol Chem 284:3873–3884

Bökel C, Brown NH (2002) Integrins in development: moving on, responding to, and sticking to the extracellular matrix. Dev Cell 3:311–321

Bork P, Holm L, Sander C (1994) The immunoglobulin fold. Structural classification, sequence patterns and common core. J Mol Biol 242:309–320

Bouvard D, Brakebusch C, Gustafsson E, Aszódi A, Bengtsson T, Berna A, Fässler R (2001) Functional consequences of integrin gene mutations in mice. Circ Res 89:211–223

Brower DL, Brower SM, Hayward DC, Ball EE (1997) Molecular evolution of integrins: genes encoding integrin β subunits from a coral and a sponge. Proc Natl Acad Sci USA 94:9182–9187

Burke RD (1999) Invertebrate integrins: structure, function and evolution. Int Rev Cytol 191:257–284

Chaudhuri I, Söding J, Lupas AN (2008) Evolution of the β-propeller fold. Proteins 71:795–803

Chouhan B, Denesyuk A, Heino J, Johnson MS, Denessiouk K (2011) Conservation of the human integrin-type β-propeller domain in bacteria. PLoS One 6:e25069

Chouhan B, Denesyuk A, Heino J, Johnson M, Denessiouk K (2012) Evolutionary origin of the alpha C helix in integrins. WASET 65:546–549

Chouhan B, Denesyuk A, Heino J, Johnson M, Denessiouk K (2012) Evolutionary origin of the alpha C helix in integrins. WASET 65:546–549

Chua GL, Tang XY, Amalraj M, Tan SM, Bhattacharjya S (2011) Structures and interaction analyses of integrin αMβ2 cytoplasmic tails. J Biol Chem 286:43842–43854

Clark LA, Boriack-Sjodin PA, Eldredge J, Fitch C, Friedman B, Hanf KJ, Jarpe M, Liparoto SF, Li Y, Lugovskoy A, Miller S, Rushe M, Sherman W, Simon K, Van Vlijmen H (2006) Affinity enhancement of an in vivo matured therapeutic antibody using structure-based computational design. Protein Sci 15:949–960

Colombatti A, Bonaldo P (1991) The superfamily of proteins with von Willebrand factor type A-like domains: one theme common to components of extracellular matrix, hemostasis, cellular adhesion, and defense mechanisms. Blood 77:2305–2315

Colombatti A, Bonaldo P, Doliana R (1993) Type A modules: interacting domains found in several non-fibrillar collagens and in other extracellular matrix proteins. Matrix 13:297–306

Corbi AL, Miller LJ, O'Connor K, Larson RS, Springer TA (1987) cDNA cloning and complete primary structure of the α subunit of a leukocyte adhesion glycoprotein, p150,95. EMBO J 6:4023–4028

Cox EA, Tuskey C, Hardin J (2004) Cell adhesion receptors in *C. elegans*. J Cell Sci 117:1867–1870

Crump MP, Ceska TA, Spyracopoulos L, Henry A, Archibald SC, Alexander R, Taylor RJ, Findlow SC, O'Connell J, Robinson MK, Shock A (2004) Structure of an allosteric inhibitor of LFA-1 bound to the I-domain studied by crystallography, NMR, and calorimetry. Biochemistry 43:2394–2404

Deshmukh L, Gorbatyuk V, Vinogradova O (2010) Integrin {beta}3 phosphorylation dictates its complex with the Shc phosphotyrosine-binding (PTB) domain. J Biol Chem 285:34875–34884

Deshmukh L, Meller N, Alder N, Byzova T, Vinogradova O (2011) Tyrosine phosphorylation as a conformational switch: a case study of integrin β3 cytoplasmic tail. J Biol Chem 286:40943–40953

Dodd DS, Sheriff S, Chang CJ, Stetsko DK, Phillips LM, Zhang Y, Launay M, Potin D, Vaccaro W, Poss MA, McKinnon M, Barrish JC, Suchard SJ, Murali Dhar TG (2007) Design of LFA-1 antagonists based on a 2,3-dihydro-1H-pyrrolizin-5(7aH)-one scaffold. Bioorg Med Chem Lett 17:1908–1911

de Pereda JM, Wiche G, Liddington RC (1999) Crystal structure of a tandem pair of fibronectin type III domains from the cytoplasmic tail of integrin alpha6beta4. EMBO J 18:4087–4095

de Pereda JM, Lillo MP, Sonnenberg A (2009) Structural basis of the interaction between integrin alpha6beta4 and plectin at the hemidesmosomes. EMBO J 28:1180–1190

Emsley J, King SL, Bergelson JM, Liddington RC (1997) Crystal structure of the I domain from integrin alpha2beta1. J Biol Chem 272:28512–28517

Emsley J, Knight CG, Farndale RW, Barnes MJ, Liddington RC (2000) Structural basis of collagen recognition by integrin alpha2beta1. Cell 101:47–56

Evans R, Patzak I, Svensson L, De Filippo K, Jones K, McDowall A, Hogg N (2009) Integrins in immunity. J Cell Sci 122:215–225

Ewan R, Huxley-Jones J, Mould AP, Humphries MJ, Robertson DL, Boot-Handford RP (2005) The integrins of the urochordate Ciona intestinalis provide novel insights into the molecular evolution of the vertebrate integrin family. BMC Evol Biol 5:31

García-Alvarez B, de Pereda JM, Calderwood DA, Ulmer TS, Critchley D, Campbell ID, Ginsberg MH, Liddington RC (2003) Structural determinants of integrin recognition by talin. Mol Cell 11:49–58

George EL, Georges-Labouesse EN, Patel-King RS, Rayburn H, Hynes RO (1993) Defects in mesoderm, neural tube and vascular development in mouse embryos lacking fibronectin. Development 119:1079–1091

Gottschalk KE (2005) A coiled-coil structure of the alphaIIbbeta3 integrin transmembrane and cytoplasmic domains in its resting state. Structure 13:703–712

Guckian KM, Lin EY, Silvian L, Friedman JE, Chin D, Scott DM (2008) Design and synthesis of a series of meta aniline-based LFA-1 ICAM inhibitors. Bioorg Med Chem Lett 18:5249–5251

Gupta V, Gylling A, Alonso JL, Sugimori T, Ianakiev P, Xiong JP, Arnaout MA (2007) The b-tail domain (βTD) regulates physiologic ligand binding to integrin CD11β/CD18. Blood 109:3513–3520

Halaby DM, Poupon A, Mornon J-P (1999) The immunoglobulin fold family: sequence analysis and 3D structure comparisons. Protein Eng 12:563–571

Han C, Wang D, Soba P, Zhu S, Lin X, Jan LY, Jan YN (2012) Integrins regulate repulsion-mediated dendritic patterning of drosophila sensory neurons by restricting dendrites in a 2D space. Neuron 73:64–78

Heino J, Huhtala M, Käpylä J, Johnson MS (2009) Evolution of collagen-based adhesion systems. Int J Biochem Cell Biol 41:341–348

Hodivala-Dilke KM, McHugh KP, Tsakiris DA, Rayburn H, Crowley D, Ullman-Culleré M, Ross FP, Coller BS, Teitelbaum S, Hynes RO (1999) B3-integrin-deficient mice are a model for Glanzmann thrombasthenia showing placental defects and reduced survival. J Clin Invest 103:229–238

Horii K, Okuda D, Morita T, Mizuno H (2004) Crystal structure of EMS16 in complex with the integrin alpha2-I domain. J Mol Biol 6:519–527

Huang H, Vogel HJ (2012) Structural basis for the activation of platelet integrin αIIbβ3 by calcium- and integrin-binding protein 1. J Am Chem Soc 134:3864–3872

Hughes AL (1992) Coevolution of the vertebrate integrin α and β chain genes. J Mol Evol 9:216–234

Hughes AL (2001) Evolution of the integrin alpha and beta protein families. J Mol Evol 52:63–72

Hughes PE, Diaz-Gonzalez F, Leong L, Wu C, McDonald JA, Shattil SJ, Ginsberg MH (1996) Breaking the integrin hinge. A defined structural constraint regulates integrin signaling. J Biol Chem 271:6571–6574

Huhtala M, Heino J, Casciari D, de Luise A, Johnson MS (2005) Integrin evolution: insights from ascidian and teleost fish genomes. Matrix Biol 24:83–95

Hynes RO (2002) Integrins: bidirectional, allosteric signaling machines. Cell 110:673–687

Ivaska J, Käpylä J, Pentikäinen O, Hoffrén A-M, Hermonen J, Huttunen P, Johnson MS, Heino J (1999) A peptide inhibiting the collagen binding function of integrin α2I domain. J Biol Chem 274:3513–3521

Ivaska J, Heino J (2011) Cooperation between integrins and growth factor receptors in signaling and endocytosis. Annu Rev Cell Dev Biol 27:291–320

Jannuzi AL, Bunch TA, West RF, Brower DL (2004) Identification of integrin beta subunit mutations that alter heterodimer function in situ. Mol Biol Cell 15:3829–3840

Jenkins C, Kedar V, Fuerst JA (2002) Gene discovery within the planctomycete division of the domain Bacteria using sequence tags from genomic DNA libraries. Genome Biol 3: research0031–research0031.11

Jin M, Andricioaei I, Springer TA (2004) Conversion between three conformational states of integrin I domains with a C-terminal pull spring studied with molecular dynamics. Structure 12:2137–2147

Johnson MS, Tuckwell D (2003) Evolution of Integrin I-domains. In: Gullberg D (ed) I domains in integrins. Landes Bioscience, Georgetown, TX, pp 1–26

Johnson MS, Lu N, Denessiouk K, Heino J, Gullberg D (2009) Integrins during evolution: evolutionary trees and model organisms. Biochim Biophys Acta 1788:779–789

Jokinen J, White DJ, Salmela M, Huhtala M, Käpylä J, Sipilä K, Puranen JS, Nissinen L, Kankaanpää P, Marjomäki V, Hyypiä T, Johnson MS, Heino J (2010) Molecular mechanism of alpha2beta1 integrin interaction with human echovirus 1. EMBO J 29:196–208

Kallen J, Welzenbach K, Ramage P, Geyl D, Kriwacki R, Legge G, Cottens S, Weitz-Schmidt G, Hommel U (1999) Structural basis for LFA-1 inhibition upon lovastatin binding to the CD11a I-domain. J Mol Biol 292:1–9

Kanchanawong P, Shtengel G, Pasapera AM, Ramko EB, Davidson MW, Hess HF, Waterman CM (2010) Nanoscale architecture of integrin-based cell adhesions. Nature 468:580–584

Karpusas M, Ferrant J, Weinreb PH, Carmillo A, Taylor FR, Garber EA (2003) Crystal structure of the alpha1beta1 integrin I domain in complex with an antibody Fab fragment. J Mol Biol 327:1031–1041

Katagiri K, Maeda A, Shimonaka M, Kinashi T (2003) RAPL, a Rap1-binding molecule that mediates Rap1-induced adhesion through spatial regulation of LFA-1. Nat Immunol 4:741–748

Kiema T, Lad Y, Jiang P, Oxley CL, Baldassarre M, Wegener KL, Campbell ID, Ylänne J, Calderwood DA (2006) The molecular basis of filamin binding to integrins and competition with talin. Mol Cell 21:337–347

Kim M, Carman CV, Springer TA (2003) Bidirectional transmembrane signaling by cytoplasmic domain separation in integrins. Science 301:1720–1725

Kim C, Ye F, Ginsberg MH (2011) Regulation of integrin activation. Annu Rev Cell Dev Biol 27:321–345

Kim C, Schmidt T, Cho EG, Ye F, Ulmer TS, Ginsberg MH (2012) Basic amino-acid side chains regulate transmembrane integrin signalling. Nature 481(7380):209–213

Kim ME, Shrestha BR, Blazeski R, Mason CA, Grueber WB (2012) Integrins establish dendrite-substrate relationships that promote dendritic self-avoidance and patterning in drosophila sensory neurons. Neuron 73:79–91

King N, Westbrook MJ, Young SL, Kuo A, Abedin M, Chapman J, Fairclough S, Hellsten U, Isogai Y, Letunic I, Marr M, Pincus D, Putnam N, Rokas A, Wright KJ, Zuzow R, Dirks W, Good M, Goodstein D, Lemons D, Li W, Lyons JB, Morris A, Nichols S, Richter DJ, Salamov A, Sequencing JG, Bork P, Lim WA, Manning G, Miller WT, McGinnis W, Shapiro H, Tjian

R, Grigoriev IV, Rokhsar D (2008) The genome of the choanoflagellate *Monosiga brevicollis* and the origin of metazoans. Nature 451:783–788

Knack BA, Iguchi A, Shinzato C, Hayward DC, Ball EE, Miller DJ (2008) Unexpected diversity of cnidarian integrins: expression during coral gastrulation. BMC Evol Biol 8:136

Kreidberg JA, Donovan MJ, Goldstein SL, Rennke H, Shepherd K, Jones RC, Jaenisch R (1996) α3β1 integrin has a crucial role in kidney and lung organogenesis. Development 122:3537–3547

Lahti M, Bligt E, Niskanen H, Parkash V, Brandt AM, Jokinen J, Patrikainen P, Käpylä J, Heino J, Salminen TA (2011) Structure of collagen receptor integrin α1I domain carrying the activating mutation E317A. J Biol Chem 286:43343–43351

Larson RS, Corbi AL, Berman L, Springer T (1989) Primary structure of the leukocyte function-associated molecule-1 α subunit: an integrin with an embedded domain defining a protein superfamily. J Cell Biol 108:703–712

Lau TL, Dua V, Ulmer TS (2008a) Structure of the integrin alphaIIb transmembrane segment. J Biol Chem 283:16162–16168

Lau TL, Partridge AW, Ginsberg MH, Ulmer TS (2008b) Structure of the integrin beta3 transmembrane segment in phospholipid bicelles and detergent micelles. Biochemistry 47:4008–4016

Lau TL, Kim C, Ginsberg MH, Ulmer TS (2009) The structure of the integrin alphaIIbbeta3 transmembrane complex explains integrin transmembrane signalling. EMBO J 28:1351–1361

Lee JO, Bankston LA, Arnaout MA, Liddington RC (1995a) Two conformations of the integrin A-domain (I-domain): a pathway for activation? Structure 3:1333–1340

Lee JO, Rieu P, Arnaout MA, Liddington R (1995b) Crystal structure of the A domain from the α subunit of integrin CR3 (CD11b/CD18). Cell 80:631–638

Legate KR, Fässler R (2009) Mechanisms that regulate adaptor binding to β-integrin cytoplasmic tails. J Cell Sci 122:187–198

Legate KR, Wickström SA, Fässler R (2009) Genetic and cell biological analysis of integrin outside-in signaling. Genes Dev 23:397–418

Legge GB, Kriwacki RW, Chung J, Hommel U, Ramage P, Case DA, Dyson HJ, Wright PE (2000) NMR solution structure of the inserted domain of human leukocyte function associated antigen-1. J Mol Biol 295:1251–1264

Li R, Rieu P, Griffith DL, Scott D, Arnaout MA (1998) Two functional states of the CD11b A-domain: correlations with key features of two Mn^{2+}-complexed crystal structures. J Cell Biol 143:1523–1534

Li R, Babu CR, Lear JD, Wand AJ, Bennett JS, DeGrado WF (2001) Oligomerization of the integrin alphaIIbbeta3: roles of the transmembrane and cytoplasmic domains. Proc Natl Acad Sci USA 98:12462–12467

Li W, Metcalf DG, Gorelik R, Li R, Mitra N, Nanda V, Law PB, Lear JD, Degrado WF, Bennett JS (2005) A push-pull mechanism for regulating integrin function. Proc Natl Acad Sci USA 102:1424–1429

Li S, Wang H, Peng B, Zhang M, Zhang D, Hou S, Guo Y, Ding J (2009) Efalizumab binding to the LFA-1 alphaL I domain blocks ICAM-1 binding via steric hindrance. Proc Natl Acad Sci USA 106:4349–4354

Lin EY, Guckian KM, Silvian L, Chin D, Boriack-Sjodin PA, van Vlijmen H, Friedman JE, Scott DM (2008) Structure-activity relationship of ortho- and meta-phenol based LFA-1 ICAM inhibitors. Bioorg Med Chem Lett 18:5245–5248

Lu CF, Springer TA (1997) The alpha subunit cytoplasmic domain regulates the assembly and adhesiveness of integrin lymphocyte function-associated antigen-1. J Immunol 159:268–278

Lu C, Takagi J, Springer TA (2001) Association of the membrane proximal regions of the alpha and beta subunit cytoplasmic domains constrains an integrin in the inactive state. J Biol Chem 276:14642–14648

Luo BH, Springer TA, Takagi J (2004a) A specific interface between integrin transmembrane helices and affinity for ligand. PLoS Biol 2:e153

Luo BH, Takagi J, Springer TA (2004b) Locking the beta3 integrin I-like domain into high and low affinity conformations with disulfides. J Biol Chem 279:10215–10221
Luo BH, Carman CV, Springer TA (2007) Structural basis of integrin regulation and signaling. Annu Rev Immunol 25:619–647
Mahalingam B, Ajroud K, Alonso JL, Anand S, Adair BD, Horenstein AL, Malavasi F, Xiong JP, Arnaout MA (2011) Stable coordination of the inhibitory Ca2+ ion at the metal ion-dependent adhesion site in integrin CD11b/CD18 by an antibody-derived ligand aspartate: implications for integrin regulation and structure-based drug design. J Immunol 187:6393–6401
Marchler-Bauer A, Lu S, Anderson JB, Chitsaz F, Derbyshire MK, DeWeese-Scott C, Fong JH, Geer LY, Geer RC, Gonzales NR, Gwadz M, Hurwitz DI, Jackson JD, Ke Z, Lanczycki CJ, Lu F, Marchler GH, Mullokandov M, Omelchenko MV, Robertson CL, Song JS, Thanki N, Yamashita RA, Zhang D, Zhang N, Zheng C, Bryant SH (2011) CDD: a Conserved Domain Database for the functional annotation of proteins. Nucleic Acids Res 39:D225–D229
Margadant C, Sonnenberg A (2010) Integrin-TGF-β crosstalk in fibrosis, cancer and wound healing. EMBO Rep 11:97–105
May AP, Ponting CP (1999) Integrin α- and β4-subunit-domain homologues in cyanobacterial proteins. Trends Biochem Sci 24:12–13
Mayer U, Saher G, Fässler R, Bornemann A, Echtermeyer F, von der Mark H, Miosge N, Poschl E, von der Mark K (1997) Absence of integrin α7 causes a novel form of muscular dystrophy. Nat Genet 17:318–323
McCleverty CJ, Liddington RC (2003) Engineered allosteric mutants of the integrin alphaMbeta2 I domain: structural and functional studies. Biochem J 372:121–127
McHugh KP, Hodivala-Dilke K, Zheng MH, Namba N, Lam J, Novack D, Feng X, Ross FP, Hynes RO, Teitelbaum SL (2000) Mice lacking β3 integrins are osteosclerotic because of dysfunctional osteoclasts. J Clin Invest 105:433–440
Metcalf DG, Moore DT, Wu Y, Kielec JM, Molnar K, Valentine KG, Wand AJ, Bennett JS, DeGrado WF (2010) NMR analysis of the alphaIIb beta3 cytoplasmic interaction suggests a mechanism for integrin regulation. Proc Natl Acad Sci USA 107:22481–22486
Miyazawa S, Azumi K, Nonaka M (2001) Cloning and characterization of integrin α subunits from the solitary ascidian *Halocynthia roretzi*. J Immunol 166:1710–1715
Moser M, Legate KR, Zent R, Fässler R (2009) The tail of integrins, talin, and kindlins. Science 324:895–899
Mould AP, Barton SJ, Askari JA, McEwan PA, Buckley PA, Craig SE, Humphries MJ (2003a) Conformational changes in the integrin beta A domain provide a mechanism for signal transduction via hybrid domain movement. J Biol Chem 278:17028–17035
Mould AP, Barton SJ, Askari JA, Craig SE, Humphries MJ (2003b) Role of ADMIDAS cation-binding site in ligand recognition by integrin alpha 5 beta 1. J Biol Chem 278:51622–51629
Müller WE (1997) Origin of metazoan adhesion molecules and adhesion receptors as deduced from cDNA analyses in the marine sponge *Geodia cydonium*: a review. Cell Tissue Res 289:383–395
Murzin AG, Brenner SE, Hubbard T, Chothia C (1995) SCOP: a structural classification of proteins database for the investigation of sequences and structures. J Mol Biol 247:536–540
Nagae M, Re S, Mihara E, Nogi T, Sugita Y, Takagi J (2012) Crystal structure of α5β1 integrin ectodomain: atomic details of the fibronectin receptor. J Cell Biol 197:131–140
Nandrot EF, Kim Y, Brodie SE, Huang X, Sheppard D, Finnemann SC (2004) Loss of synchronized retinal phagocytosis and age-related blindness in mice lacking αvβ5 integrin. J Exp Med 200:1539–1545
Newham P, Humphries MJ (1996) Integrin adhesion receptors: structure, function and implications for biomedicine. Mol Med Today 2:304–313
Newham P, Craig SE, Seddon GN, Schofield NR, Rees A, Edwards RRM, Jones EY, Humphries MJ (1997) α4 integrin binding interfaces on VCAM-1 and MAdCAM-1: integrin binding footprints identify accessory binding sites that play a role in integrin specificity. J Biol Chem 272:19429–19440

Nolte M, Pepinsky RB, Venyaminov SY, Koteliansky V, Gotwals PJ, Karpusas M (1999) Crystal structure of the alpha1beta1 integrin I-domain: insights into integrin I-domain function. FEBS Lett 452:379–385

Nymalm Y, Puranen JS, Nyholm TK, Käpylä J, Kidron H, Pentikäinen OT, Airenne TT, Heino J, Slotte JP, Johnson MS, Salminen TA (2004) Jararhagin-derived RKKH peptides induce structural changes in alpha1I domain of human integrin alpha1beta1. J Biol Chem 279:7962–7970

O'Toole TE, Mandelman D, Forsyth J, Shattil SJ, Plow EF, Ginsberg MH (1991) Modulation of the affinity of integrin alpha IIb beta 3 (GPIIb-IIIa) by the cytoplasmic domain of alpha IIb. Science 254:845–847

O'Toole TE, Katagiri Y, Faull RJ, Peter K, Tamura R, Quaranta V, Loftus JC, Shattil SJ, Ginsberg MH (1994) Integrin cytoplasmic domains mediate inside-out signal transduction. J Cell Biol 124:1047–1059

Pancer Z, Kruse M, Müller I, Müller WE (1997) On the origin of Metazoan adhesion receptors: cloning of integrin α subunit from the sponge *Geodia cydonium*. Mol Biol Evol 14:391–398

Parry CS, Gorski J, Stern LJ (2007) Crystallographic structure of the human leukocyte antigen DRA, DRB3*0101: models of a directional alloimmune response and autoimmunity. J Mol Biol 371:435–446

Partridge AW, Liu S, Kim S, Bowie JU, Ginsberg MH (2005) Transmembrane domain helix packing stabilizes integrin alphaIIbbeta3 in the low affinity state. J Biol Chem 280:7294–7300

Pentikäinen O, Hoffrén M-A, Nyrönen T, Ivaska J, Käpyla J, Heino J, Johnson MS (1999) Molecular modeling of the interaction between the peptide derived from the snake venom metalloproteinase of *Bothrops jararaca* and human integrin α2I domain. J Biol Chem 274:31493–31505

Ponting CP, Aravind L, Schultz J, Bork P, Koonin EV (1999) Eukaryotic signalling domain homologues in archaea and bacteria. Ancient ancestry and horizontal gene transfer. J Mol Biol 289:729–745

Potin D, Launay M, Monatlik F, Malabre P, Fabreguettes M, Fouquet A, Maillet M, Nicolai E, Dorgeret L, Chevallier F, Besse D, Dufort M, Caussade F, Ahmad SZ, Stetsko DK, Skala S, Davis PM, Balimane P, Patel K, Yang Z, Marathe P, Postelneck J, Townsend RM, Goldfarb V, Sheriff S, Einspahr H, Kish K, Malley MF, DiMarco JD, Gougoutas JZ, Kadiyala P, Cheney DL, Tejwani RW, Murphy DK, Mcintyre KW, Yang X, Chao S, Leith L, Xiao Z, Mathur A, Chen BC, Wu DR, Traeger SC, McKinnon M, Barrish JC, Robl JA, Iwanowicz EJ, Suchard SJ, Dhar TG (2006) Discovery and development of 5-[(5S,9R)-9-(4-cyanophenyl)-3-(3,5-dichlorophenyl)-1-methyl-2,4-dioxo-1,3,7-tria zaspiro[4.4]non-7-yl-methyl]-3-thiophenecarboxylic acid (BMS-587101)—a small molecule antagonist of leukocyte function associated antigen-1. J Med Chem 49:6946–6949

Putnam NH, Butts T, Ferrier DE, Furlong RF, Hellsten U, Kawashima T, Robinson-Rechavi M, Shoguchi E, Terry A, Yu JK, Benito-Gutiérrez EL, Dubchak I, Garcia-Fernàndez J, Gibson-Brown JJ, Grigoriev IV, Horton AC, de Jong PJ, Jurka J, Kapitonov VV, Kohara Y, Kuroki Y, Lindquist E, Lucas S, Osoegawa K, Pennacchio LA, Salamov AA, Satou Y, Sauka-Spengler T, Schmutz J, Shin-I T, Toyoda A, Bronner-Fraser M, Fujiyama A, Holland LZ, Holland PW, Satoh N, Rokhsar DS (2008) The amphioxus genome and the evolution of the chordate karyotype. Nature 453:1064–1071

Qu A, Leahy DJ (1995) Crystal structure of the I-domain from the CD11a/CD18 (LFA-1, alpha L beta 2) integrin. Proc Natl Acad Sci USA 92:10277–10281

Qu A, Leahy DJ (1996) The role of the divalent cation in the structure of the I domain from the CD11a/CD18 integrin. Structure 4:931–942

Quistgaard EM, Thirup SS (2009) Sequence and structural analysis of the Asp-box motif and Asp-box β-propellers; a widespread propeller-type characteristic of the Vps10 domain family and several glycoside hydrolase families. BMC Struct Biol 9:46

Redruello B, Louro B, Anjos L, Silva N, Greenwell RS, Canario AV, Power DM (2010) CRTAC1 homolog proteins are conserved from cyanobacteria to man and secreted by the teleost fish pituitary gland. Gene 456:1–14
Reynolds LE, Wyder L, Lively JC, Taverna D, Robinson SD, Huang X, Sheppard D, Hynes RO, Hodivala-Dilke KM (2002) Enhanced pathological angiogenesis in mice lacking β3 integrin or β3 and β5 integrins. Nat Med 8:27–34
Rich RL, Deivanayagam CC, Owens RT, Carson M, Höök A, Moore D, Symersky J, Yang VW, Narayana SV, Höök M (1999) Trench-shaped binding sites promote multiple classes of interactions between collagen and the adherence receptors, alpha(1)beta(1) integrin and Staphylococcus aureus cna MSCRAMM. J Biol Chem 274:24906–24913
Rigden DJ, Galperin MY (2004) The DxDxDG motif for calcium binding: multiple structural contexts and implications for evolution. J Mol Biol 343:971–984
Schierwater B, Eitel M, Jakob W, Osigus HJ, Hadrys H, Dellaporta SL, Kolokotronis SO, Desalle R (2009) Concatenated analysis sheds light on early metazoan evolution and fuels a modern "urmetazoon" hypothesis. PLoS Biol 7:e20
Schuster-Böckler B, Schultz J, Rahmann S (2004) Logos for visualization of protein families. BMC Bioinformatics 5:7
Schwartz MA, DeSimone DW (2008) Cell adhesion receptors in mechanotransduction. Curr Opin Cell Biol 20:551–556
Sebé-Pedrós A, Roger AJ, Lang FB, King N, Ruiz-Trillo I (2010) Ancient origin of the integrin-mediated adhesion and signaling machinery. Proc Natl Acad Sci USA 107:10142–10147
Shi M, Sundramurthy K, Liu B, Tan SM, Law SK, Lescar J (2005) The crystal structure of the plexin-semaphorin-integrin domain/hybrid domain/I-EGF1 segment from the human integrin beta2 subunit at 1.8-A resolution. J Biol Chem 280:30586–30593
Shi M, Foo SY, Tan SM, Mitchell EP, Law SK, Lescar J (2007) A structural hypothesis for the transition between bent and extended conformations of the leukocyte beta2 integrins. J Biol Chem 282:30198–30206
Shimaoka M, Xiao T, Liu JH, Yang Y, Dong Y, Jun CD, McCormack A, Zhang R, Joachimiak A, Takagi J, Wang JH, Springer TA (2003) Structures of the alpha L I domain and its complex with ICAM-1 reveal a shape-shifting pathway for integrin regulation. Cell 112:99–111
Siljander PR, Hamaia S, Peachey AR, Slatter DA, Smethurst PA, Ouwehand WH, Knight CG, Farndale RW (2004) Integrin activation state determines selectivity for novel recognition sites in fibrillar collagens. J Biol Chem 279:47763–47772
Song G, Yang Y, Liu JH, Casasnovas JM, Shimaoka M, Springer TA, Wang JH (2005) An atomic resolution view of ICAM recognition in a complex between the binding domains of ICAM-3 and integrin alphaLbeta2. Proc Natl Acad Sci USA 102:3366–3371
Springer TA (1997) Folding of the N-terminal, ligand-binding region of integrin α-subunits into a β-propeller domain. Proc Natl Acad Sci USA 94:65–72
Springer TA, Zhu J, Xiao T (2008) Structural basis for distinctive recognition of fibrinogen gammaC peptide by the platelet integrin alphaIIbbeta3. J Cell Biol 182:791–800
Srivastava M, Begovic E, Chapman J, Putnam NH, Hellsten U, Kawashima T, Kuo A, Mitros T, Salamov A, Carpenter ML, Signorovitch AY, Moreno MA, Kamm K, Grimwood J, Schmutz J, Shapiro H, Grigoriev IV, Buss LW, Schierwater B, Dellaporta SL, Rokhsar DS (2008) The Trichoplax genome and nature of placozoans. Nature 454:955–960
Stephens LE, Sutherland AE, Klimanskaya IV, Andrieux A, Meneses J, Pedersen RA, Damsky CH (1995) Deletion of β1 integrins in mice results in inner cell mass failure and peri-implantation lethality. Genes Dev 9:1883–1895
Stepp MA, Spurr-Michaud S, Tisdale A, Elwell J, Gipson IK (1990) α6β4 integrin heterodimer is a component of hemidesmosomes. Proc Natl Acad Sci USA 87:8970–8974
Tadokoro S, Shattil SJ, Eto K, Tai V, Liddington RC, de Pereda JM, Ginsberg MH, Calderwood DA (2003) Talin binding to integrin beta tails: a final common step in integrin activation. Science 302:103–106

Takagi J, Erickson HP, Springer TA (2001) C-terminal opening mimics 'inside-out' activation of integrin alpha5beta1. Nat Struct Biol 8:412–416

Takagi J, Strokovich K, Springer TA, Walz T (2003) Structure of integrin alpha5beta1 in complex with fibronectin. EMBO J 22:4607–4615

Takala H, Nurminen E, Nurmi SM, Aatonen M, Strandin T, Takatalo M, Kiema T, Gahmberg CG, Ylänne J, Fagerholm SC (2008) Beta2 integrin phosphorylation on Thr758 acts as a molecular switch to regulate 14-3-3 and filamin binding. Blood 112:1853–1862

Tuckwell D (1999) Evolution of von Willebrand factor A (VWA) domains. Biochem Soc Trans 27:835–840

Tulla M, Lahti M, Puranen JS, Brandt AM, Käpylä J, Domogatskaya A, Salminen TA, Tryggvason K, Johnson MS, Heino J (2008) Effects of conformational activation of integrin alpha 1I and alpha 2I domains on selective recognition of laminin and collagen subtypes. Exp Cell Res 314:1734–1743

Vinogradova O, Haas T, Plow EF, Qin J (2000) A structural basis for integrin activation by the cytoplasmic tail of the alpha IIb-subunit. Proc Natl Acad Sci USA 97:1450–1455

Vinogradova O, Velyvis A, Velyviene A, Hu B, Haas T, Plow E, Qin J (2002) A structural mechanism of integrin alpha(IIb)beta(3) "inside-out" activation as regulated by its cytoplasmic face. Cell 110:587–597

Vinogradova O, Vaynberg J, Kong X, Haas TA, Plow EF, Qin J (2004) Membrane-mediated structural transitions at the cytoplasmic face during integrin activation. Proc Natl Acad Sci USA 101:4094–4099

Vorup-Jensen T, Ostermeier C, Shimaoka M, Hommel U, Springer TA (2003) Structure and allosteric regulation of the alpha X beta 2 integrin I domain. Proc Natl Acad Sci USA 100:1873–1878

Wattanasin S, Kallen J, Myers S, Guo Q, Sabio M, Ehrhardt C, Albert R, Hommel U, Weckbecker G, Welzenbach K, Weitz-Schmidt G (2005) 1,4-Diazepane-2,5-diones as novel inhibitors of LFA-1. Bioorg Med Chem Lett 15:1217–1220

Watterson SH, Xiao Z, Dodd DS, Tortolani DR, Vaccaro W, Potin D, Launay M, Stetsko DK, Skala S, Davis PM, Lee D, Yang X, McIntyre KW, Balimane P, Patel K, Yang Z, Marathe P, Kadiyala P, Tebben AJ, Sheriff S, Chang CY, Ziemba T, Zhang H, Chen BC, DelMonte AJ, Aranibar N, McKinnon M, Barrish JC, Suchard SJ, Murali Dhar TG (2010) Small molecule antagonist of leukocyte function associated antigen-1 (LFA-1): structure-activity relationships leading to the identification of 6-((5S,9R)-9-(4-cyanophenyl)-3-(3,5-dichlorophenyl)-1-methyl-2,4-dioxo-1,3,7-tria zaspiro[4.4]nonan-7-yl)nicotinic acid (BMS-688521). J Med Chem 53:3814–3830

Wegener KL, Partridge AW, Han J, Pickford AR, Liddington RC, Ginsberg MH, Campbell ID (2007) Structural basis of integrin activation by talin. Cell 128:171–182

Weitz-Schmidt G, Welzenbach K, Dawson J, Kallen J (2004) Improved lymphocyte function-associated antigen-1 (LFA-1) inhibition by statin derivatives: molecular basis determined by x-ray analysis and monitoring of LFA-1 conformational changes in vitro and ex vivo. J Biol Chem 279:46764–46771

Weljie AM, Hwang PM, Vogel HJ (2002) Solution structures of the cytoplasmic tail complex from platelet integrin alpha IIb- and beta 3-subunits. Proc Natl Acad Sci USA 99:5878–5883

Whittaker CA, Hynes RO (2002) Distribution and evolution of von Willebrand/integrin A domains: widely dispersed domains with roles in cell adhesion and elsewhere. Mol Biol Cell 13:3369–3387

Wickström SA, Lange A, Hess MW, Polleux J, Spatz JP, Krüger M, Pfaller K, Lambacher A, Bloch W, Mann M, Huber LA, Fässler R (2010) Integrin-linked kinase controls microtubule dynamics required for plasma membrane targeting of caveolae. Dev Cell 19:574–588

Wimmer W, Perovic S, Kruse M, Schröder HC, Krasko A, Batel R, Müller WE (1999) Origin of the integrin-mediated signal transduction. Functional studies with cell cultures from the sponge *Suberites domuncula*. Eur J Biochem 260:156–165

Xiao T, Takagi J, Coller BS, Wang JH, Springer TA (2004) Structural basis for allostery in integrins and binding to fibrinogen-mimetic therapeutics. Nature 432:59–67
Xie C, Shimaoka M, Xiao T, Schwab P, Klickstein LB, Springer TA (2004) The integrin α-subunit leg extends at a Ca2+-dependent epitope in the thigh/genu interface upon activation. Proc Natl Acad Sci USA 101:15422–15427
Xie C, Zhu J, Chen X, Mi L, Nishida N, Springer TA (2010) Structure of an integrin with an alpha I domain, complement receptor type 4. EMBO J 29:666–679
Xing L, Huhtala M, Pietiäinen V, Käpylä J, Vuorinen K, Marjomäki V, Heino J, Johnson MS, Hyypiä T, Cheng RH (2004) Structural and functional analysis of integrin alpha2I domain interaction with echovirus 1. J Biol Chem 279:11632–11638
Xiong JP, Li R, Essafi M, Stehle T, Arnaout MA (2000) An isoleucine-based allosteric switch controls affinity and shape shifting in integrin CD11b A-domain. J Biol Chem 275:38762–38767
Xiong JP, Stehle T, Diefenbach B, Zhang R, Dunker R, Scott DL, Joachimiak A, Goodman SL, Arnaout MA (2001) Crystal structure of the extracellular segment of integrin αVβ3. Science 294:339–345
Xiong JP, Stehle T, Zhang R, Joachimiak A, Frech M, Goodman SL, Arnaout MA (2002) Crystal structure of the extracellular segment of integrin alpha Vbeta3 in complex with an Arg-Gly-Asp ligand. Science 296:151–155
Xiong JP, Stehle T, Goodman SL, Arnaout MA (2004) A novel adaptation of the integrin PSI domain revealed from its crystal structure. J Biol Chem 279:40252–40254
Xiong JP, Mahalingham B, Alonso JL, Borrelli LA, Rui X, Anand S, Hyman BT, Rysiok T, Müller-Pompalla D, Goodman SL, Arnaout MA (2009) Crystal structure of the complete integrin alphaVbeta3 ectodomain plus an alpha/beta transmembrane fragment. J Cell Biol 186:589–600
Yang JT, Rayburn H, Hynes RO (1993) Embryonic mesodermal defects in α5 integrin-deficient mice. Development 119:1093–1105
Yang JT, Rayburn H, Hynes RO (1995) Cell adhesion events mediated by α4 integrins are essential in placental and cardiac development. Development 121:549–560
Yang W, Shimaoka M, Salas A, Takagi J, Springer TA (2004) Intersubunit signal transmission in integrins by a receptor-like interaction with a pull spring. Proc Natl Acad Sci USA 101:2906–2911
Yang J, Ma YQ, Page RC, Misra S, Plow EF, Qin J (2009) Structure of an integrin alphaIIb beta3 transmembrane-cytoplasmic heterocomplex provides insight into integrin activation. Proc Natl Acad Sci USA 106:17729–17734
Yu Y, Zhu J, Mi LZ, Walz T, Sun H, Chen J, Springer TA (2012) Structural specializations of α(4)β(7), an integrin that mediates rolling adhesion. J Cell Biol 196:131–146
Zaidel-Bar R, Itzkovitz S, Ma'ayan A, Iyengar R, Geiger B (2007) Functional atlas of the integrin adhesome. Nat Cell Biol 9:858–867
Zaidel-Bar R (2009) Evolution of complexity in the integrin adhesome. J Cell Biol 186:317–321
Zhang H, Casasnovas JM, Jin M, Liu JH, Gahmberg CG, Springer TA, Wang JH (2008) An unusual allosteric mobility of the C-terminal helix of a high-affinity alphaL integrin I domain variant bound to ICAM-5. Mol Cell 31:432–437
Zhang H, Astrof NS, Liu JH, Wang JH, Shimaoka M (2009a) Crystal structure of isoflurane bound to integrin LFA-1 supports a unified mechanism of volatile anesthetic action in the immune and central nervous systems. FASEB J 23:2735–2740
Zhang H, Liu JH, Yang W, Springer T, Shimaoka M, Wang JH (2009b) Structural basis of activation-dependent binding of ligand-mimetic antibody AL-57 to integrin LFA-1. Proc Natl Acad Sci USA 106:18345–18350
Zhu J, Luo BH, Xiao T, Zhang C, Nishida N, Springer TA (2008) Structure of a complete integrin ectodomain in a physiologic resting state and activation and deactivation by applied forces. Mol Cell 32:849–861

Zhu J, Luo BH, Barth P, Schonbrun J, Baker D, Springer TA (2009) The structure of a receptor with two associating transmembrane domains on the cell surface: integrin alphaIIbbeta3. Mol Cell 34:234–249

Zhu J, Zhu J, Negri A, Provasi D, Filizola M, Coller BS, Springer TA (2010) Closed headpiece of integrin αIIbβ3 and its complex with an αIIbβ3-specific antagonist that does not induce opening. Blood 116:5050–5059

Index

F.W. Keeley and R.P. Mecham (eds.), *Evolution of Extracellular Matrix*,
Biology of Extracellular Matrix, DOI 10.1007/978-3-642-36002-2,

F

G

H

I

Printed by Printforce, the Netherlands